Nanoimaging

Pan Stanford Series on Biomedical Nanotechnology

Series Editors

Vladimir Torchilin and Mansoor Amiji

Titles in the Series

Vol. 1
Handbook of Materials for Nanomedicine
Vladimir Torchilin and Mansoor Amiji, eds.
2010
978-981-4267-55-7 (Hardcover)
978-981-4267-58-8 (eBook)

Vol. 2
Nanoimaging
Beth A. Goins and William T. Phillips, eds.
2011
978-981-4267-09-0 (Hardcover)
978-981-4267-91-5 (eBook)

Vol. 3
Biomedical Nanosensors
Joseph Irudayaraj, ed.
2011
978-981-4303-03-3 (Hardcover)
978-981-4303-04-0 (eBook)

Vol. 4
Inorganic Nanomedicine
Bhupinder Singh Sekhon, ed.
2012

Vol. 5
Nanotechnology for Cancer
Julia Ljubimova, ed.
2012

Vol. 6
Nanotechnology for Delivery of Therapeutic Nucleic Acids
Dan Peer, ed.
2013

Vol. 7
Nanotechnology for Delivery of DNA and Related Materials
Bengt Fadeel, ed.
2013

Vol. 8
Translation Industrial Nanotechnology
Thomas Redelmeier, ed.
2013

Nanoimaging

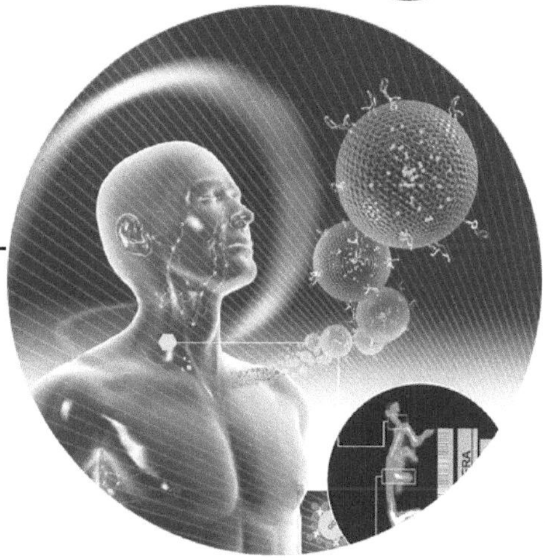

Beth A. Goins ▪ William T. Phillips

Editors

PAN STANFORD PUBLISHING

Published by

Pan Stanford Publishing Pte. Ltd.
Penthouse Level, Suntec Tower 3
8 Temasek Boulevard
Singapore 038988

Email: editorial@panstanford.com
Web: www.panstanford.com

British Library Cataloguing-in-Publication Data
A catalogue record for this book is available from the British Library.

Nanoimaging

ISBN 978-981-4267-09-0 (Hardcover)
ISBN 978-981-4267-91-5 (eBook)

Printed in the USA

Contents

Preface

The last decade has seen significant advances in the fabrication of new nanostructures, including nanoparticles. In parallel, considerable progress in the design of new imaging equipment and development of contrast agents has occurred. The translation of these nanostructures into clinical nanomedicine and diagnostic imaging applications is emerging rapidly with a burgeoning growth of different nanoparticles being tested.

This volume of the Pan Stanford Series on Biomedical Nanotechnology, edited by Vladimir Torchilin and Mansoor Amiji, is devoted especially to reviewing the current status of the use of nanoparticles in *in vivo* imaging applications. The book begins with a review of nanoparticles for oncologic imaging, a major clinical application (Chapter 1). Chapters 2–5 focus on *in vivo* imaging of iron oxide particles and liposomes, some of the first nanoparticles to be used as diagnostic imaging agents clinically. Chapters 6–12 highlight the use of new nanoparticles that are currently at the preclinical stage of testing, such as quantum dots, dendrimers, polymer-based nanoparticles, gold nanoshells, carbon nanotubes, and metal nanorods. Finally, although not technically nanoparticles, no comprehenisive review on imaging contrast agents would be complete without the inclusion of microbubbles for ultrasound contrast (Chapter 13). This volume has attempted to represent the major clinical imaging modalities of magnetic resonance imaging (MRI), computed tomography (CT), positron emission tomography (PET), single photon emission computed tomography (SPECT), and ultrasound as well as optical methods.

The overall thrust of this volume highlights the specific advantages of nanoparticles for drug delivery and imaging applications. Nanoparticles are able to target the delivery of therapeutic and imaging agents and retain these agents at local sites using different approaches not possible with small (free) molecules. In conjunction, noninvasive imaging can monitor the biodistribution of the nanoparticles as drug carriers and contrast agents and more importantly determine how successfully the intended target was reached. Moreover, the versatile nature of nanoparticles allows for their use as multimodality imaging agents as well as theranostics for carrying both diagnostic and therapeutic agents. Thus, readers should find that the principles learned from the *in vivo* imaging of one type of nanoparticle are useful for investigations of other kinds of nanoparticles.

The editors are especially grateful to the numerous authors who agreed to devote their time and effort to share their research on the current use of nanoparticles for imaging applications. We would also like to acknowledge Julie Barker for her invaluable assistance with the organization and formatting of the book chapters and Jonathan Sumner for his help in formatting and labeling many of the images in this book.

Beth A. Goins
William T. Phillips

Chapter 1

COMBINED CONTRAST AND THERAPEUTIC NANOCARRIERS FOR ONCOLOGIC MRI

Yoshinori Kato and Arvind P. Pathak
Johns Hopkins University School of Medicine, Baltimore, MD 21205, USA
ykato@mri.jhu.edu

1.1 INTRODUCTION

Nanocarriers have been used extensively to effectively deliver drug molecules to cancer tissue. While there are several FDA-approved liposomal agents available for clinical use (e.g. Doxil®), neither nanoparticle-based imaging probes (except for superparamagnetic iron oxide [SPIO] nanoparticles) nor actively targeted nanocarriers are currently available for routine use in patients. In addition, technical and biological hurdles preclude the efficacious delivery of imaging agents and drugs into cancer cells. Some of these include our body's innate ability to eliminate foreign substances via the immune system, poor delivery of nanocarriers into the tumor due to their physicochemical properties, and elevated interstitial fluid pressure [1, 2]. However, recent developments in molecular imaging of cancer have increased our understanding of the tumor microenvironment and paved the way for circumventing some of these hurdles [3]. As a result, we foresee the development of nanocarriers loaded with both drugs and imaging agents to enable simultaneous therapy and noninvasive *in vivo* visualization of therapeutic efficacy. This new class of nanocarriers would help us achieve a better understanding of cancer treatment *in vivo*.

Nanoimaging
Edited by Beth Goins and William Phillips

www.panstanford.com

1.2 CHALLENGES IN DRUG/CONTRAST AGENT DELIVERY TO CANCER

The biggest difference between the design of "drug" carriers and "imaging agent" carriers is the release pattern of their respective cargos. Ideally, drug molecules need to be released from passively targeted nanocarriers in the tumor, while imaging agents should always remain conjugated to the nanocarriers in order to track their distribution. Since the pharmacokinetics of cargo molecules and conjugates depends on their "carriers," nanocarriers have to be stable in the blood circulation; otherwise, drugs or imaging agents end up being released prior to their delivery to the tumor, resulting in systemic toxic side effects and high organ-to-tumor accumulation ratios. As is evident, one faces numerous challenges when developing bimodal therapeutic/imaging nanocarriers. However, the benefits of such a nanocarrier to cancer treatment and its potential to improve patients' quality of life make it a goal worth pursuing. This chapter covers the development of such bimodal (therapeutic plus imaging) nanocarriers. Specifically, it describes the design of novel nanocarriers carrying anticancer drugs and contrast agents for magnetic resonance imaging (MRI).

Table 1.1 Preferable release pattern of various nanocarriers

Nanocarriers	*Preferable release pattern*
Passively targeted drug nanocarriers	Drugs should be released in the tumor and not in the systemic circulation.
Actively targeted drug nanocarriers	Drugs should be released only after the nanocarriers reach their intended target cells/receptors.
Nanocarrier-based imaging agents	Imaging agents should not be released and remain stably conjugated to track the distribution of nanocarriers.

1.3 POTENTIAL SOLUTIONS

Table 1.2 Potential of nanocarrier-based MR contrast agents

Advantages	*Mechanism*
Increase in sensitivity	Nanocarriers can carry a large number of contrast agent molecules compared with conventional MRI agents.
Increase in specificity	Nanocarriers tend to accumulate in the solid tumors via the enhanced permeability and retention (EPR) effect [5]. In addition, incorporation of ligands onto the surface of nanocarriers enables specific targeting to particular cells.
Multiple images	One can visualize not only delivery, release, and intratumoral distribution of MR contrast agents, but also therapeutic response noninvasively.

While imaging modalities like positron emission tomography (PET) and single-photon emission computed tomography (SPECT) exhibit greater sensitivity than MRI, the wide variety of MRI contrast mechanisms coupled with its exquisite soft-tissue contrast and high spatial resolution makes it a promising tool for imaging cancer [4]. These strengths enable MRI to be used as an all-in-one imaging modality for visualization of drug delivery, drug release, intratumoral distribution of nanocarriers, and therapeutic response. This has resulted in a new class of agents that combine therapy and imaging in MRI.

1.4 PROPERTIES OF AN "IDEAL" BIMODAL AGENT FOR DRUG DELIVERY

Nanocarriers can incorporate different kinds of agents, such as drugs, imaging probes, and specific ligands, in a single carrier, and deliver them to cancer tissue/cells and tumor vessels [6]. There are several advantages of using nanocarriers, such as (i) predominant delivery of cargo drugs to tumors, which results in minimization of systemic toxicity; (ii) protection from enzymatic or hydrolytic degradation of drug molecules; (iii) conferring of a prodrug property; and (iv) restored anticancer activity against multidrug-resistant cancers. In addition to efficient delivery to the target sites, it is also beneficial for nanocarriers to have prodrug properties and an activatable release characteristic. To achieve this, nanocarriers with an environmentally responsive release property, such as pH-sensitive, thermo-sensitive, and enzyme-cleavable spacers, have been developed and tested *in vivo* [7–9]. Nanocarriers that have been extensively investigated preclinically or used clinically include liposomes, nanoparticles, micelles, and water-soluble polymeric carriers. Ideal nanocarriers vary depending on their intended use as well as characteristics of the drug to be encapsulated. For example, poly(lactic-*co*-glycoric acid) (PLGA) nanoparticles tend to encapsulate more hydrophobic drugs rather than hydrophilic drugs. Therefore, nanocarriers should be selected on the basis of their intended applications.

Table 1.3 Examples of clinically approved micro/nanocarriers

Name	*Active drug*	*Formulation*	*Components*
Doxil	Doxorubicin	Liposomes	HSPC*, cholesterol, DSPE-PEG*
SMANCS	Neocarzinostatin	Polymer conjugates	Poly(stylene-co-maleic acid)
Lupron Depot	Leuprolide acetate	Microspheres	Poly(lactic-co-glycolic acid)
Abraxane	Paclitaxel	Nanoparticles	Albumin

*HSPC: hydrogenated soy phosphatidylcholine; DSPE-PEG: distearoyl phosphatidyl-ethanolamine-poly(ethylene glycol) conjugates

1.5 PHYSICOCHEMICAL PROPERTIES

Nanocarriers can be categorized by their water solubility into water-soluble polymeric carriers of polyethylene glycol, dextran, or *N*-(2-hydroxypropyl) methacrylamide (HPMA) and water-insoluble nanoparticles, such as liposomes, micelles, emulsions, and nanospheres. For passive targeting, a longer circulation time increases the likelihood of the nanocarrier's accumulating in the tumor. This occurs due to leaky tumor vasculature and lack of efficient lymphatic drainage in the solid tumor and is known as the enhanced permeability and retention (EPR) effect [10]. To prolong the circulating half-life of nanocarriers, (i) its molecular size should be greater than 30–45 kDa to limit glomerular filtration [11], (ii) its surface charge should be slightly negative for charge repulsion within Bowman's capsule and the reticuloendothelial system (RES), and (iii) its surface should be modified with polyethylene glycol (PEG) because PEG chains can shield nanocarriers from opsonins that induce phagocytosis by macrophages [12]. Also, since it is best to keep anticancer drugs pharmacologically inactive until they reach cancer tissues, the nanocarrier itself must be inert. Once nanocarriers fulfill their mission, they should eventually be biodegraded and excreted from the body. Altogether, prerequisites for the materials used as nanocarriers include biocompatibility, biodegradability, and inertness.

Table 1.4 Ideal physicochemical properties of nanocarriers for *in vivo* tumor targeting

Molecular size	*Surface charge*	*Surface modification*	*Materials*
>30 kDa <200 nm	Slightly negative	PEG coating	Inert, biodegradable, biocompatible

1.6 IDEAL PROPERTIES FOR IMAGING

Table 1.5 Ideal physicochemical properties of nanocarriers for *in vivo* tumor imaging

Molecular size	*Nanocarrier type*	*Contrast moiety*	*Other requirements*
Can range from several hundred daltons to megadaltons; 8–200 nm	Protein carriers, dendrimers, nanoparticles, micelles, liposomes, nanoemulsions, and carbon nanotubes	T_1 (e.g. Gd-based), T_2 (e.g. Fe-based), or CEST (e.g. polymer-based) contrast enhancing moieties	Contrast moieties must be stable *in vivo*; nonimmunogenic; nontoxic; no metabolic conversion; low accumulation in the RES

As an imaging modality, MRI has a formidable array of available contrast mechanisms to characterize tumor physiology. In conjunction with spectroscopic imaging (MRSI), MRI can provide a wealth of spatio-temporal information about the tumor's microenvironment that is invaluable for treatment planning and tracking drug delivery and therapeutic efficacy. Depending on the method employed, we can obtain information about a variety of tumor physiological parameters, including the tumor vasculature, tumor metabolism, extracellular pH, and metastasis (see Refs. [4] and [13] for an exhaustive review). MRI is, therefore, well suited for characterizing a complex disease like cancer. As mentioned in the introduction, recent advances in the development of targeted contrast agents have significantly increased the versatility of MRI for molecular imaging [14, 15]. In fact, several of these developments can be exploited to design combined contrast and therapeutic nanocarriers for oncologic MRI [6].

Unlike the radiolabeled tracers or chemical agents used in nuclear medicine or X-ray CT that are visualized directly on the image, MRI contrast agents are visualized indirectly due to the changes they induce in water proton relaxation. These relaxation effects in turn modulate the MRI signal. The following section describes the principal relaxation mechanisms and the basis of this image contrast.

1.7 MR RELAXATION MECHANISMS AND THE BASIS OF CONTRAST

Every contrast mechanism in MRI is, in some way, a result of changes in the MR signal intensity brought about by changes in tissue relaxation times (T_1, T_2, or T_2^*). Briefly, T_1, the spin-lattice or longitudinal relaxation time, is the time constant that characterizes the exponential process by which the bulk magnetization returns or "relaxes" to its equilibrium value. It does so by exchanging energy with its surroundings or the lattice, at the Larmor frequency. T_1 relaxation occurs at the molecular level through several pathways, including interactions between protons in tissue water and those on macromolecules or proteins and by interactions with paramagnetic substances, i.e., substances with unpaired electrons in their outermost shells. T_1-based MR contrast results from differences in T_1 dominating the MR signal intensity and tissues with short T_1's appear bright in T_1-weighted MRI, since the transverse magnetization recovers to equilibrium rapidly compared with tissues with long T_1's. Microscopic magnetic field heterogeneities in the main field, as well as local variations in magnetic susceptibility due to the physiological microenvironment, cause protons or "spins" contributing to the transverse magnetization, to lose their phase coherence. The process through which this occurs is known as T_2^* relaxation. However, as protons diffuse through microscopic field inhomogeneities, they also lose phase coherence due to their Brownian random walks through these magnetic field gradients, which results in phase dispersion that cannot be reversed by the application of a refocusing pulse. This process is known as T_2 relaxation. In T_2-weighted MR images, tissues with short T_2's appear dark due to the rapid decay of transverse magnetization compared with those with long T_2's. Similarly, in T_2^*-weighted images, regions with large susceptibility gradients, such as large vessels carrying iron oxide contrast agents, appear hypointense.

While conventional MRI agents affect the relaxation rates of water molecules, recent attention has been directed toward exploiting the frequency spectrum of the NMR phenomenon. One approach has been to develop agents containing hyperpolarized nuclei so that MR images with no background can be acquired [16]. Due to their low abundance in humans, non-hydrogen nuclei exhibit a low signal-to-noise ratio (SNR) in MRI. However, signals from such nuclei can be enhanced by a technique called hyperpolarization [17], in which certain nuclei are polarized, increasing the observed MRI signal by several orders of magnitude [18]. The high cost and fast decay of the hyperpolarization make the setup and use of this method quite challenging. Another method of generating frequency-selective contrast is to exploit the well-characterized magnetization or saturation transfer phenomenon. This has ushered in a new class of chemical exchange saturation transfer (CEST) contrast agents. A CEST agent is a molecule in which exchangeable protons resonate at a chemical shift that is different from the signal of bulk water, which happens when their exchange with bulk water protons is slow relative to the NMR timescale. If one then applies a radio-frequency pulse at the appropriate frequency and power, one can saturate the exchangeable protons of the CEST agent, which in turn "chemically exchange" with the bulk water, reducing its equilibrium magnetization, and decrease the intensity of the MRI signal [16].

1.8 ONCOLOGIC APPLICATION OF MR CONTRAST/THERAPEUTIC NANOCARRIERS

As summarized by Torchilin in an excellent review [3], in addition to combining the requirements listed in Tables 1.4 and 1.5, the next generation of dual contrast/therapeutic nanocarriers must be capable of providing a "real-time" readout about the carrier's and drug's biodistribution and target accumulation as well as the efficacy of therapy by incorporating a contrast moiety within the nanocarrier. Due to their easily modifiable surfaces and excellent pharmacokinetic properties, liposomes, micelles, and dendrimers were the nanocarriers of choice for bimodal oncologic applications [3]. Generally for MRI, the contrast moiety (gadolinium [Gd], dysprosium [Dy], manganese [Mn], and iron [Fe]) is either sequestered within the nanocarrier or anchored to its surface. For MRI, the physicochemical properties of the liposomes and their effect on the ability of MR contrast agent atoms to gain access to bulk water profoundly affect their relaxivities [19–21].

As mentioned earlier, due to the lower inherent sensitivity of MRI relative to that of other imaging modalities such as PET and SPECT, visualization of key receptors and molecular pathways fundamental to cancer was not possible until very recently. Specifically "targeted" imaging was further complicated by factors such as contrast agent delivery to the molecular target and dilution or loss of the contrast label within the blood pool. However, two recent developments have brought the dream of targeted contrast and therapeutic agents for oncologic applications to fruition: The first is the extraordinary advances in the field of nanotechnology. Specifically, the synthesis of novel nanomaterials and drug delivery systems (DDS) has had a profound impact on medicine [6]. The second is the development of new receptor target amplification strategies enabling molecular MR imaging of sparse biomarkers at nanomolar or even picomolar concentrations [4]. Such targeted contrast agents have included the use of modified iron oxide nanoparticles and conjugates of low molecular weight gadolinium (Gd)-diethylene triamine pentaacetic acid (DTPA) with protein carriers, dendrimers, or liposomes [15]. An exciting example of a high-relaxivity targeted contrast agent for molecular MRI of tumor angiogenesis is perfluorocarbon nanoemulsions, which can carry several gadolinium atoms per particle [22]. An additional advantage of these perfluorocarbon constructs is that they are highly flexible and can be used as multimodal contrast agents [23]. In fact, the combination of a diagnostic (imaging) test with therapy or "theranostics" was elegantly demonstrated in a recent study by Schmieder *et al.* in which they characterized angiogenesis in a preclinical breast cancer model using $\alpha_V\beta_1$(RGD)-targeted paramagnetic nanoparticles and used MR molecular imaging to assess the antiangiogenic effectiveness of the same nanoparticles [24]. A large number of studies have focused on targeted MRI of the angiogenic vasculature in tumors because abundant preclinical and indirect clinical data have, for several decades, convincingly supported the notion that antiangiogenesis is an effective strategy for the inhibition of tumor growth [25]. Such studies include imaging inducible E-selectin expression in human endothelial cell culture using cross-linked iron oxide nanoparticles (CLIO) covalently bound to high-affinity antihuman E-selectin F(ab') fragments [26], to the detection of tumor angiogenesis *in vivo* by integrin $\alpha_V\beta_3$-targeted MRI, in which the targeted MR contrast agent is composed of Gd-labeled polymerized liposomes conjugated with biotinylated antibodies targeted against the $\alpha_V\beta_3$ integrin, which is commonly found on tumor endothelial cell surfaces [27]. Although originally developed for applications such as measuring ICAM-1 expression in autoimmune encephalitis [28], or for imaging fibrin-rich clots in an animal model of human thrombus formation [29], these novel molecular imaging probes could be adapted to imaging and treating a wide array of tumors.

1.8.1 Lectin-Based Agents

Due to the significance of the vasculature to the pathophysiology of solid tumors, we and others are developing targeted, drug-payload-bearing contrast agents that bind to all endothelial cells after systemic administration. One class of molecules that fulfills this targeting requirement is lectins [30]. Lectins are sugar-binding proteins that are highly specific for their sugar moieties. They typically play a role in biological recognition phenomena involving cells and proteins. A widely used endothelium-specific lectin isolated from the tomato fruit, with an affinity for

N-acetyl-β-D-glucosamine oligomers, is *Lycopersicon esculentum*, or "tomato" lectin [31]. Paschkunova-Martic *et al.* recently developed a latex nanoparticle-based MRI contrast agent that was functionalized with tomato lectin [32]. They demonstrated that after intravenous administration into mice, a stable, long-lasting, MR-visible label was observable *in vivo*. Additionally, human placental cotyledons and the human vena saphena magna were also successfully labeled and imaged *ex vivo*. Our group has focused on developing targeted agents using the lectin from *Bandeiraea simplicifolia* (GS-1). GS-1 is a plant-derived, carbohydrate-binding protein that is non-immune in origin and has been shown to bind to α-D-galactosyl residues expressed on endothelial cells [33].

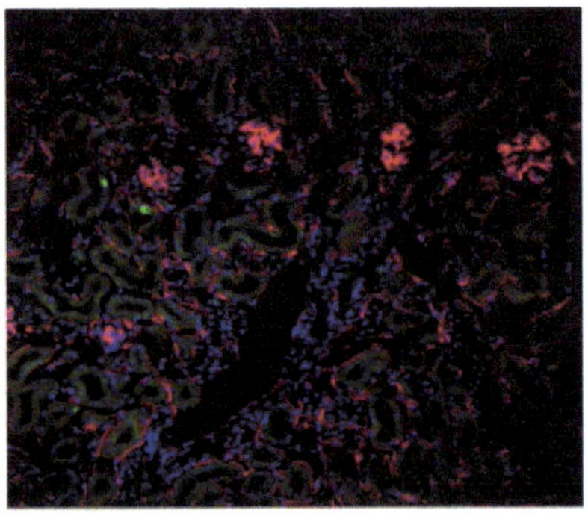

Figure 1.1 Fluorescence microscopy image (20×) of a mouse kidney in which the vasculature has been labeled with GS-1-tetramethyl rhodamine isothiocyanate (TRITC) (red), and the nuclei counterstained with 4'6-diamidino-2-phenylindole (DAPI). It is evident from the labeling of the glomeruli and associated renal vessels that GS-1 is an excellent moiety for targeting the murine endothelium.

In the brain, cardiac muscle, and skeletal muscle, GS-1 lectin conjugates have exhibited prominent binding only to blood vessel endothelia. Similarly, in the liver and kidney cortex, GS-1 conjugates selectively reacted with endothelial cells of the sinusoids and with intertubular and glomerular capillaries (Fig. 1.1). Minamikawa *et al.* conducted "lectin angiography" by injecting the lectin GS-1 into the circulation of ether-anesthetized animals via the left ventricle and found that it bound specifically to the inner surface of endothelial cells [34]. Because of its simplicity and wide applicability, GS-1 has been used as a vascular stain in a wide range of studies, including as a vessel marker in studies of adult neurogenesis [35]. We conducted a series of experiments to determine the feasibility of labeling and imaging murine blood vessels with lectin *in vivo* with optical imaging. For example, to determine if lectin-fluoroscein isothiocyanate (FITC) binds to vessel lumen or leaks out in tumors, we injected an MCF-7 human breast tumor bearing xenograft with 0.2 mL of 1 mg/mL solution of GS-1-FITC (114 kDa) via the tail vein, and an MDA-MB-231 human breast tumor bearing xenograft injected with 0.2 mL of 1 mg/mL solution of BSA-FITC (70 kDa). Three hours later, mice were humanely killed and 10 μm tumor cryosections stained for endothelial cells using CD34. Figure 1.2 shows that bovine serum albumin (BSA-FITC) extravasated from the hyperpermeable tumor vasculature, while the GS-1 labeled the lumen of the tumor vessels without extravasating [36]. Finally, to determine if lectin improved the visibility of brain cortical vessels for MRI, we conducted a two-step labeling protocol with biotinylated-GS-1 lectin and the MRI contrast agent avidin-GdDTPA-FITC.

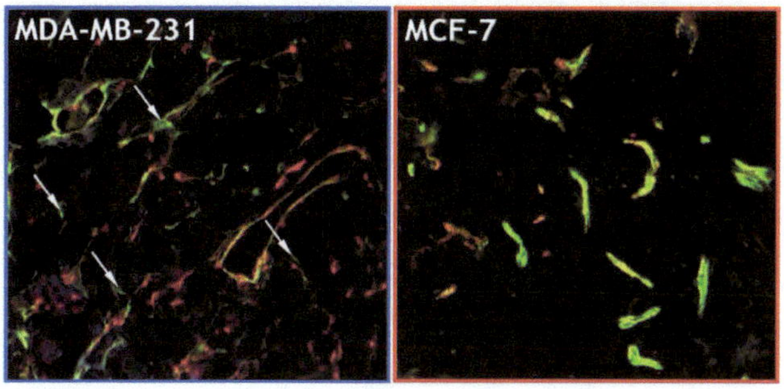

Figure 1.2 Immunofluorescence images of a tissue section from an MDA-MB-231 breast tumor xenograft showing BSA-FITC (green) and blood vessel endothelia stained with CD34 (red). One can clearly see extravasated BSA-FITC (arrows) outside the CD34-stained blood vessel lumina. In contrast, immunofluorescence images of a tissue section from an MCF-7 breast tumor xenograft showing lectin GS1-FITC (green) and blood vessel endothelia stained with CD34 (red) clearly show that lectin-FITC does not extravasate and labels the same endothelia as CD34, as is evident from the excellent colocalization of the two.

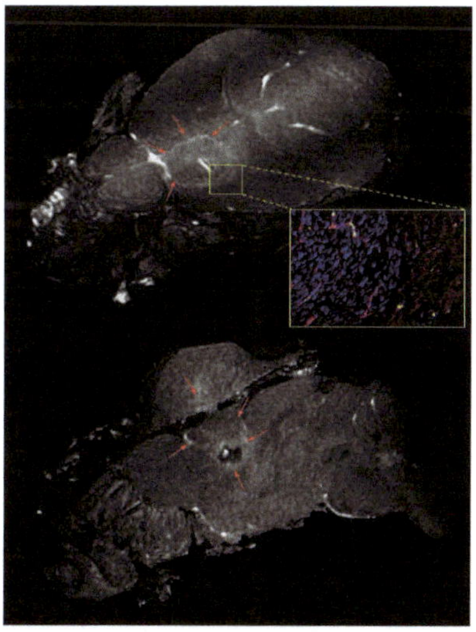

Figure 1.3 MR microscopy images clearly show labeling of the angiogenic rim of the 9L tumor (arrows) in both the upper (3D axial cutaway) and lower (3D sagittal cutaway) views of the mouse brain. This labeling of the tumor vessels is entirely due to the biotin-GS-1-avidin-GdDTPA-FITC label, since any circulating contrast agent will be removed during the perfusion fixation. The inset shows CD34-labeled vessels (red) bearing the FITC-labeled contrast agent complex (green) formed by the two-step label comprising avidin-GdDTPA-FITC attached to the initially administered biotin-GS-1, clearly demonstrating the feasibility of labeling murine blood vessels.

A 9L gliosarcoma bearing female mouse was injected with biotinylated-GS-1 intravenously, and an hour later with avidin-GdDTPA-FITC synthesized in our lab. The animal was humanely killed by perfusion fixation after an hour, and *ex vivo* T_1-weighted 3D spin echo images of the head acquired. Figure 1.3 clearly shows labeling of the angiogenic rim of the 9L tumor. This labeling of the tumor vessels was entirely due to the biotin-GS-1-avidin-GdDTPA-FITC label, since any circulating contrast agent was removed by the perfusion fixation. On the basis of the feasibility of using the GS-1 lectin as a targeting moiety, we are currently in the process of developing liposomal contrast agents that have been functionalized with GS-1 and are capable of carrying a therapeutic cargo (Fig. 1.4).

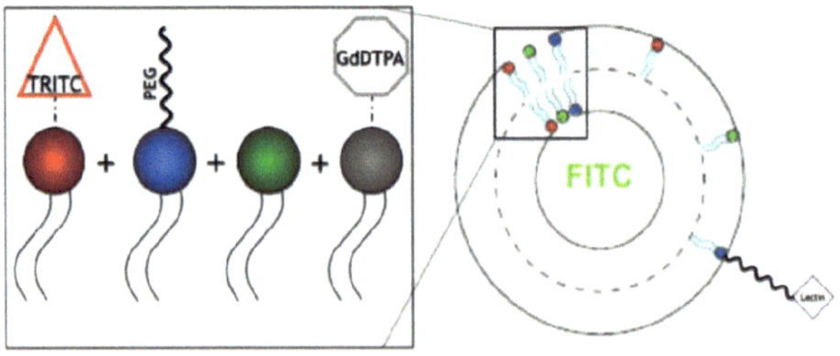

Figure 1.4 Schematic of the dual-purpose liposomes currently under development: (a) Inset shows the four lipids that constitute the liposome contrast agent. The first lipid, 1,2-dihexadecanoyl-*sn*-glycero-phosphoethanolamine (DHPE), incorporates a fluorescent (TRITC) tag; the second, distearyol phosphoethanolamine (DSPE) with polyethylene glycol (PEG) to achieve long circulation *in vivo*; and the fourth, Gd-DTPA for *in vivo* MRI. The liposomes also contain distearyol phosphocholine (DSPC) and cholesterol (Cho). The surface of this liposome is modified so that the lectin GS-1 is attached to the functionalized PEG arm to yield a lectin-targeted liposome nanocarrier that can be used for both imaging and therapy.

1.8.2 Dual Contrast Agents

The dual MR contrast technique was originally introduced for diagnostic imaging of the liver by Suto and Shimatani [37] and Weissleder *et al.* [38]. Recently, Kato and Artemov took this concept further by encapsulating both the positive MR contrast agent Gd-DTPA and the negative contrast agent SPIO into the same nanocarrier. This enabled us to noninvasively visualize delivery, release, and intratumoral distribution of cargo molecules *in vitro* [39]. As specific delivery to tumors and the release of drug molecules from nanocarriers are critical factors in the success of nanocarrier-based chemotherapy, it is important to develop methods that enable the simultaneous noninvasive monitoring of drug delivery and drug release. The concept behind this approach is that intact nanocarriers will greatly attenuate the MR signal due to the dominant T_2/T_2^* effects of SPIO compared with the T_1 effect of GdDTPA. However, upon the release of GdDTPA from the nanocarriers and its diffusion beyond the spatial range of the susceptibility effects of the relatively immobile SPIO (Fig. 1.5a), one will observe positive T_1 contrast. Similarly, the intact dual contrast nanocarrier will exhibit a broad resonance line in the NMR spectrum of an anticancer agent such as 5-fluorouracil (5-FU) due to the T_2^* effect of SPIO. Once the released anticancer agent diffuses beyond the spatial range of the T_2^* effect of SPIO, a narrow resonance line can be detected (Fig. 1.5b). In either case, small cargo molecules such as Gd-based contrast agent and drug molecules need to diffuse away from the influence of SPIO nanoparticles. This would be particularly ideal in tumor tissue, wherein small molecules can easily leak from highly permeable tumor vessels and diffuse through the interstitium, while large particles are mostly restricted to the perivascular space [40]. The average radius of the area with strong T_2 contrast generated by SPIO particles has been estimated at 15 μm, using a typical diffusion coefficient of the fast-diffusing extracellular water in breast tumors *in vivo* [41] and an MRI echo time of 30 ms. From empirical observations, the T_2^* effect has been estimated to extend to a radius of at least 50 times the average size of the SPIO particle size when employing a gradient-echo MR image [42].

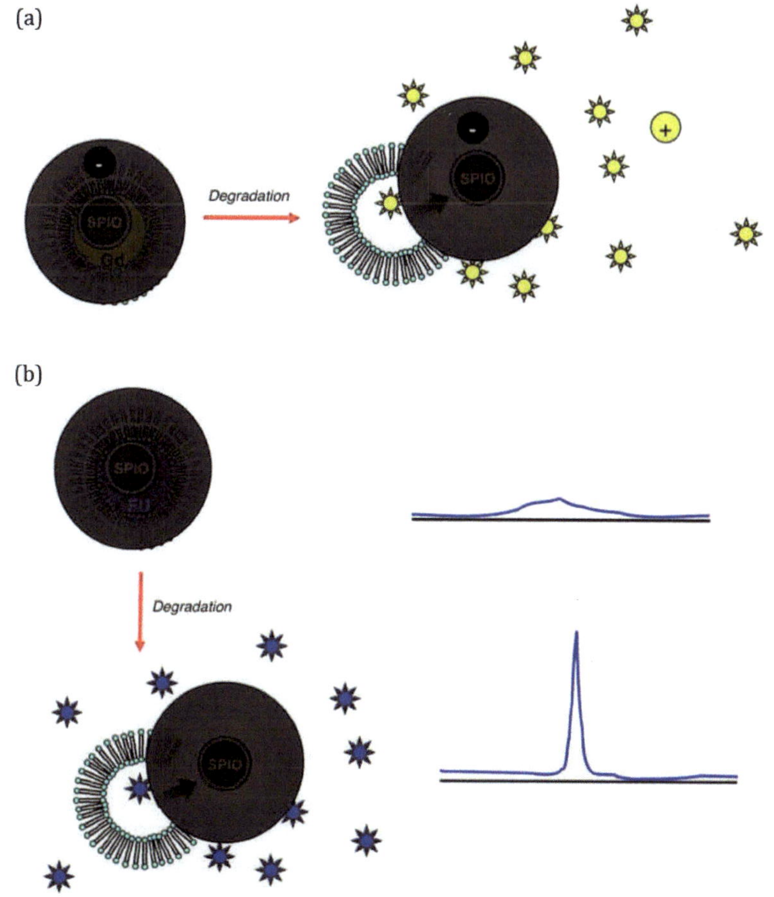

Figure 1.5 Concept of the dual MR contrast technique: (a) MRI and (b) MRS. Gd (GdDTPA) and FU (5-FU) are illustrated as a model for MR contrast agents and anticancer drugs, respectively.

In vitro experimental data validate the concept that the dual contrast approach can differentiate between intact nanocarriers and released cargo molecules by both MRI and MRS (Fig. 1.6) [39]. It is difficult to visualize the distribution of anticancer agents by using MR spectroscopic imaging (MRSI) due to its low sensitivity. Therefore, gadolinium-based positive contrast agents are very useful surrogate tracers for anticancer agents. It is important to note that gadolinium-based contrast agents have to be similar to the drug molecules to be loaded in terms of factors such as molecular weight and hydrophilicity (lipophilicity). Much work remains to be done in the use of MR contrast agents as surrogate tracers for drugs before their widespread clinical use.

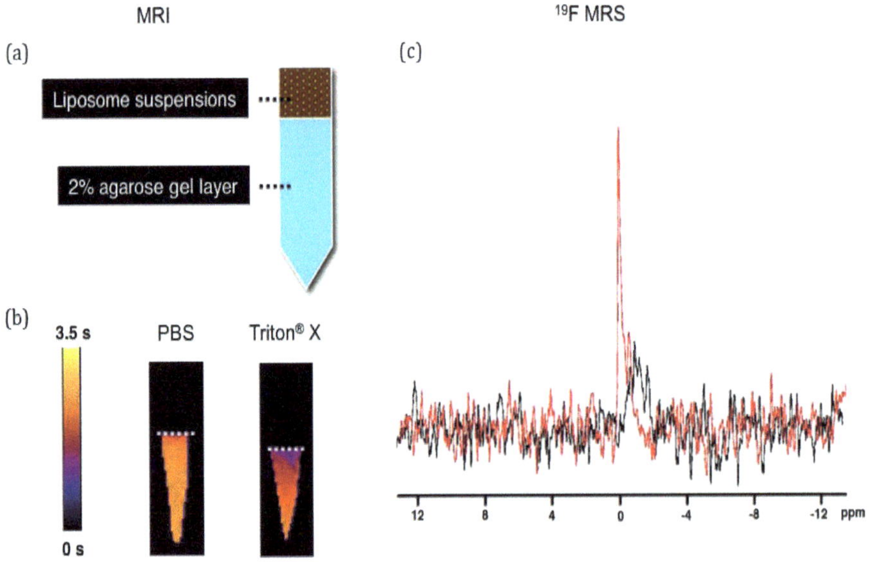

Figure 1.6 Visualization of release of GdDTPA and 5-fluorouracil (5-FU) from liposomes. (a) Schematic of 2% agarose gel phantom. In this experiment, GdDTPA/SPIO or 5-FU/SPIO was encapsulated in liposomes. (b) Quantitative T_1 maps of agarose gel phantoms, 6 hours after the addition of PBS or Triton®X-100 to the liposome suspension. The white dotted lines represent the interface of the suspension/agarose gel layer. Intact liposomes (dispersed in PBS) completely suppressed the MR signal of the suspension in the upper chamber of the phantom due to the significant shortening of T_2/T_2^* relaxation times, while the GdDTPA released from liposomes (dispersed in Triton®X-100) reduced the T_1 value depending on the concentration and the distance from the interface of the liposomes suspension and agarose gel layer. GdDTPA could easily be diffused into the 2% agarose gel, whereas SPIO nanoparticles were not diffused due to their large molecular size. Acquisition parameters are as follows: echo time = 15 ms; repetition time = 250, 500, 1,000, 2,000, 4,000, and 8,000 ms; resolution = 0.3125 × 0.275 mm; slice thickness = 1mm; and number of acquisition = 2. (c) ^{19}F spectra of 5-FU/SPIO-loaded liposome suspension on a 2% agarose gel, pre- and post-Triton®X-100 treatment. Before adding Triton®X-100 to the liposome suspension, the ^{19}F spectrum of 5-FU showed a broad resonance line (black), and a narrow resonance line appeared with the release of 5-FU and subsequent diffusion to agarose gel, 5 hours after the addition of Triton®X-100 (red). Acquisition parameters are as follows: number of acquisition = 4,096; number of points = 2,048; sweep width = 10,000; and repetition time = 1,000 ms.

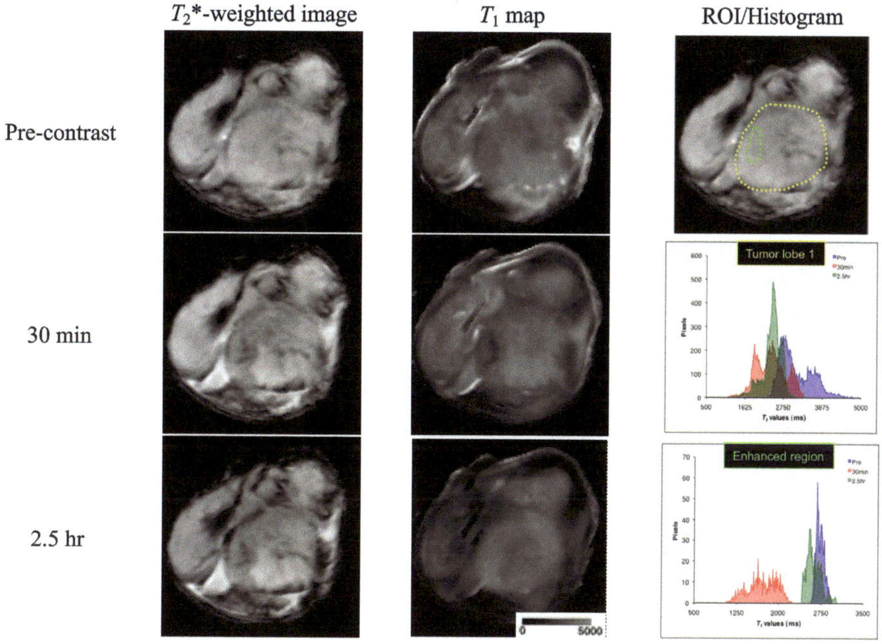

Figure 1.7 *In vivo* release monitoring using the dual contrast technique. In this experiment, GdDTPA/SPIO/5-FU were loaded in nanospheres composed of poly(D,L-lactide-*co*-glycolide) (PLGA). PLGA nanospheres were administered intravenously into female SCID mice that were orthotopically inoculated MCF-7 human breast carcinoma. MRI and MRS were acquired before and after intravenous administration of PLGA nanospheres. SPIO particles appeared as a dark spot in T_2^*-weighted images around the area delineated by green region of interest (ROI) only after administration, which corresponds to the location of nanocarriers, in this case PLGA nanospheres. T_1 reduction was observed in the T_1 map 30 min after administration of nanospheres, while T_1 values were restored 2.5 hours post-administration. This indicates that Gd-DTPA was released from PLGA nanoparticles rapidly, and that most of them were washed out from tumor tissue by 2.5 hours due to their small molecular size, although SPIO particles still remained at the same location where initially distributed. Results obtained by *in vivo* MRI were consistent with *in vitro* release pattern. On the other hand, 5-FU could not be detected by ^{19}F MRS due probably to intrinsic low sensitivity of MRS. (Color bar represents the range of T_1's in milliseconds). Acquisition parameters for MRI for reconstruction of T_1 maps are as follows: echo time = 50 ms; repetition time = 250, 500, 1,000, 2,000, and 4,000 ms; field of view = 16 × 16 × 8 mm; matrix = 128 × 80 × 40 zero-filled to 128 × 128 × 128. Acquisition parameters for T_2^*-weighted MRI are as follows: echo time = 7 ms; repetition time = 100 ms; number of acquisitions = 2; flip angle ≈ 32°. Acquisition parameters are as follows: number of acquisitions = 512; number of points = 2048; sweep width = 10,000; and repetition time = 1,000ms.

We recently demonstrated that following intravenous administration of nanocarriers loaded with an anticancer agent, SPIO particles, and a Gd-based contrast agent, the nanocarriers were successfully delivered to the tumor and released their cargo molecules in the tumor (Fig. 1.7) [43]. As described earlier in this section, biodistribution of nanocarriers is determined by the EPR effect [5].

While the *in vivo* release pattern matched what we observed *in vitro*, the release of 5-FU was undetectable due to the low sensitivity of MRS. As demonstrated in Fig. 1.7, unlike in the *in vitro* phantom study, the small amount of SPIO particles generated a strong negative signal *in vivo* without affecting the T_1 enhancement. Therefore, this technique has the potential to be used for the noninvasive monitoring of drug release from nanocarriers *in vivo* if we employ gadolinium-based contrast agents as surrogate tracers for the drug cargo.

Imaging studies have revealed novel insights into the biological and physiological characteristics of cancers, as well as enabled early detection of cancer. The future of new imaging probes will be to increase their specificity and sensitivity, as well as to include therapy. This goal can be simultaneously achieved by using nanocarriers. Integration of new nanocarrier-based agents and novel imaging techniques will be the vanguard of the next generation of cancer research and treatment.

References

1. Jain R. K. (1997) Delivery of molecular and cellular medicine to solid tumors. *Adv Drug Deliv Rev* **26**, 71–90.
2. Minchinton A. I., and Tannock I. F. (2006) Drug penetration in solid tumours. *Nat Rev Cancer* **6**, 583–592.
3. Torchilin V. P. (2006) Multifunctional nanocarriers. *Adv Drug Deliv Rev* **58**, 1532–1555.
4. Pathak AP, Gimi B, Glunde K, Ackerstaf E, Artemov D, and Bhujwalla ZM (2004) Molecular and functional imaging of cancer: advances in MRI and MRS. *Methods Enzymol* **386**, 3–60.
5. Matsumura Y., and Maeda H. (1986) A new concept for macromolecular therapeutics in cancer chemotherapy: mechanism of tumoritropic accumulation of proteins and the antitumor agent smancs. *Cancer Res* **46**, 6387–6392.
6. Torchilin V. P. (2007) Targeted pharmaceutical nanocarriers for cancer therapy and imaging. *Aaps J* **9**, E128–147.
7. Needham D., Anyarambhatla G., Kong G., and Dewhirst M. W. (2000) A new temperature-sensitive liposome for use with mild hyperthermia: Characterization and testing in a human tumor xenograft model. *Caner Res* **60**, 1197–1201.
8. Nogusa H., Yano T., Okuno S., Hamana H., and Inoue K. (1995) Synthesis of carboxymethylpullulan-peptide-doxorubicin conjugates and their properties. *Chem Pharm Bull (Tokyo)* **43**, 1931–1936.
9. Simões S., Moreira J. N., Fonseca C., Düzgüneş N., and de Lima M. C. (2004) On the formulation of pH-sensitive liposomes with long circulation times. *Adv Drug Deliv Rev* **56**, 947–965.
10. Maeda H., Wu J., Sawa T., Matsumura Y., and Hori K. (2000) Tumor vascular permeability and the EPR effect in macromolecular therapeutics: a review. *J Control Release* **65**, 271–284.

11. Sinko P. J. (2006) 14. Biopharmaceutics. In: Sinko PJ (ed) Martin's Physical Pharmacy and Pharmaceutical Sciences. Lippincott Williams & Wilkins Baltimore, MD, pp 355–396.

12. Moghimi S. M., Muir I. S., Illum L., Davis S. S., and Kolb-Bachofen V. (1993) Coating particles with a block co-polymer (poloxamine-908) suppresses opsonization but permits the activity of dysopsonins in the serum. *Biochim Biophys Acta* **1179**, 157–165.

13. Glunde K., Pathak A. P., and Bhujwalla Z. M. (2007) Molecular-functional imaging of cancer: to image and imagine. *Trends Mol Med* **13**, 287–297.

14. Gupta H., and Weissleder R. (1996) Targeted contrast agents in MR imaging. *Magn Reson Imaging Clin N Am* **4**, 171–184.

15. Artemov D. (2003) Molecular magnetic resonance imaging with targeted contrast agents. *J Cell Biochem* **90**, 518–524.

16. Geraldes C. F., and Laurent S. (2009) Classification and basic properties of contrast agents for magnetic resonance imaging. *Contrast Media Mol Imaging* **4**, 1–23.

17. Albert M. S., Cates G. D., Driehuys B., Happer W., Saam B., Springer C. S., Jr., and Wishnia A. (1994) Biological magnetic resonance imaging using laser-polarized 129Xe. *Nature* **370**, 199–201.

18. Schroder L., Lowery T. J., Hilty C., Wemmer D. E., and Pines A. (2006) Molecular imaging using a targeted magnetic resonance hyperpolarized biosensor. *Science* **314**, 446–449.

19. Fossheim S. L., Fahlvik A. K., Klaveness J., and Muller R. N. (1999) Paramagnetic liposomes as MRI contrast agents: influence of liposomal physicochemical properties on the in vitro relaxivity. *Magn Reson Imaging* **17**, 83–89.

20. Strijkers G. J., Mulder W. J., van Heeswijk R. B., Frederik P. M., Bomans P., Magusin P. C., and Nicolay K. (2005) Relaxivity of liposomal paramagnetic MRI contrast agents. *Magma* **18**, 186–192.

21. Tilcock C., Ahkong Q. F., Koenig S. H., Brown R. D., III, Davis M., and Kabalka G. (1992) The design of liposomal paramagnetic MR agents: effect of vesicle size upon the relaxivity of surface-incorporated lipophilic chelates. *Magn Reson Med* **27**, 44–51.

22. Winter P. M., Caruthers S. D., Kassner A., Harris T. D., Chinen L. K., Allen J. S., Lacy E. K., Zhang H., Robertson J. D., Wickline S. A., and Lanza G. M. (2003) Molecular imaging of angiogenesis in nascent Vx-2 rabbit tumors using a novel alpha(nu)beta3-targeted nanoparticle and 1.5 tesla magnetic resonance imaging. *Cancer Res* **63**, 5838–5843.

23. Morawski A. M., Lanza G. A., and Wickline S. A. (2005) Targeted contrast agents for magnetic resonance imaging and ultrasound. *Curr Opin Biotechnol* **16**, 89–92.

24. Schmieder A. H., Caruthers S. D., Zhang H., Williams T. A., Robertson J. D., Wickline S. A., and Lanza G. M. (2008) Three-dimensional MR mapping of angiogenesis with alpha5beta1(alpha nu beta3)-targeted theranostic nanoparticles in the MDA-MB-435 xenograft mouse model. *Faseb J* **22**, 4179–4189.

25. Folkman J. (1971) Tumor angiogenesis: therapeutic implications. *N Engl J Med* **285**, 1182–1186.

26. Kang H. W., Josephson L., Petrovsky A., Weissleder R., and Bogdanov A., Jr. (2002)

Magnetic resonance imaging of inducible E-selectin expression in human endothelial cell culture. *Bioconjug Chem* **13**, 122–127.

27. Sipkins D. A., Cheresh, D. A., Kazemi, M. R., Nevin, L. M., Bednarski, M. D., and Li, K. C. P. (1998) Detection of Tumor Angiogenesis *in vivo* by AB-targeted Magnetic Resonance Imaging. *Nature Medicine* **4**, 623–626.
28. Sipkins D. A., Gijbels K., Tropper F. D., Bednarski M., Li K. C., and Steinman L. (2000) ICAM-1 expression in autoimmune encephalitis visualized using magnetic resonance imaging. *J Neuroimmunol* **104**, 1–9.
29. Yu X., Song S. K., Chen J., Scott M. J., Fuhrhop R. J., Hall C. S., Gaffney P. J., Wickline S. A., and Lanza G. M. (2000) High-resolution MRI characterization of human thrombus using a novel fibrin-targeted paramagnetic nanoparticle contrast agent. *Magn Reson Med* **44**, 867–872.
30. Sharon N., and Halina L. (2007) *Lectins*. Springer.
31. Mazzetti S., Frigerio S., Gelati M., Salmaggi A., and Vitellaro-Zuccarello L. (2004) Lycopersicon esculentum lectin: an effective and versatile endothelial marker of normal and tumoral blood vessels in the central nervous system. *Eur J Histochem* **48**, 423–428.
32. Paschkunova-Martic I., Kremser C., Mistlberger K., Shcherbakova N., Dietrich H., Talasz H., Zou Y., Hugl B., Galanski M., Solder E., Pfaller K., Holiner I., Buchberger W., Keppler B., and Debbage P. (2005) Design, synthesis, physical and chemical characterisation, and biological interactions of lectin-targeted latex nanoparticles bearing Gd-DTPA chelates: an exploration of magnetic resonance molecular imaging (MRMI). *Histochem Cell Biol* **123**, 283–301.
33. Laitinen L. (1987) Griffonia simplicifolia lectins bind specifically to endothelial cells and some epithelial cells in mouse tissues. *Histochem J* **19**, 225–234.
34. Minamikawa T., Miyake T., Takamatsu T., and Fujita S. (1987) A new method of lectin histochemistry for the study of brain angiogenesis. Lectin angiography. *Histochemistry* **87**, 317–320.
35. Ernst C., and Christie B. R. (2006) Isolectin-IB 4 as a vascular stain for the study of adult neurogenesis. *J Neurosci Methods* **150**, 138–142.
36. Pathak A. P., Kato Y., Zhang J., and Jones M. (2007) A New Lectin-targeted Contrast Agent for MR and Optical Molecular Imaging of Vascular Endothelium. 15th Annual Meeting of the International Society for Magnetic Resonance in Medicine Berlin, Germany.
37. Suto Y., and Shimatani Y. (1995) Dual contrast magnetic resonance imaging with combined use of positive and negative contrast agent in human hepatocellular carcinoma. *Br J Radiol* **68**, 116–120.
38. Weissleder R., Saini S., Stark D. D., Wittenberg J., and Ferrucci J. T. (1988) Dual-contrast MR imaging of liver cancer in rats. *AJR Am J Roentgenol* **150**, 561–566.
39. Kato Y., and Artemov D. (2009) Monitoring of release of cargo from nanocarriers by MRI/MRSI: Significance of T_2/T_2^* effect of iron particles. *Magn Reson Med* **61**, 1059–1065.

40. Dreher M. R., Liu W., Michelich C. R., Dewhirst M. W., Yuan F., and Chilkoti A. (2006) Tumor vascular permeability, accumulation, and penetration of macromolecular drug carriers. *J Natl Cancer Inst* **98**, 335–344.

41. Paran Y., Bendel P., Margalit R., and Degani H. (2004) Water diffusion in the different microenvironments of breast cancer. *NMR Biomed* **17**, 170–180.

42. Shapiro E. M., Skrtic S., Sharer K., Hill J. M., Dunbar C. E., and Koretsky A. P. (2004) MRI detection of single particles for cellular imaging. *Proc Natl Acad Sci U S A* **101**, 10901–10906.

43. Onuki Y., Jacobs I., Artemov D., and Kato, Y. (2010) Noninvasive visualization of *in vivo* release and intratumoral distribution of surrogate MR contrast agent using the dual MR contrast technique. *Biomaterials* **31**, 7132–7138.

Chapter 2

NANO-SIZE SUPERPARAMAGNETIC MAGNETIC RESONANCE CONTRAST AGENTS

Alexei A. Bogdanov, Jr.
Department of Radiology, University of Massachusetts Medical School, Worcester, MA 01655, USA
Alexei.Bogdanov@umassmed.edu

2.1 INTRODUCTION

Proton magnetic resonance imaging (MRI) is frequently performed after a systemic administration of compounds that locally change relaxation times of water proton nuclear spins. These compounds usually comprise chelated cations of paramagnetic metals and exhibit paramagnetic properties unless used as a bolus injection at very high concentrations [30]. These compounds exhibit predominantly longitudinal or spin–lattice (T_1) relaxation time shortening effects *in vivo* and can be either low molecular weight or linked to carrier molecules, e.g., proteins [58], linear polymers [6], nonlinear graft copolymers [7], or dendrimers [67]. In all cases the covalent or noncovalent linking of chelates to larger molecules results in the increase of molar relaxivity (r_1 and r_2), which is a measure of paramagnetic metal efficacy in shortening proton relaxation times.

In contrast, superparamagnetic iron oxide (IO)-based contrast agents exhibit primarily spin–spin or transverse (T_2) relaxation of water protons and high magnetic susceptibility because of arising microinhomogeneities of a local magnetic field [52]. These contrast agents are unique in that they represent the only example of true nano-size diagnostic agents with a history of extensive clinical trials and clinical use.

For the last 25 years a substantial amount of work in developing novel long-lasting and targeted imaging agents was justified by the availability of already clinically approved IO-based nanoparticles that were specifically designed as contrast agents

Nanoimaging
Edited by Beth Goins and William Phillips

www.panstanford.com

for MRI. The recognized advantage of IO is the sensitivity of MRI to the presence of these nanoparticles in the tissue (see below). Unfortunately, during the last five years the prospects for successful development of clinical applications of these nanoparticles substantially worsened. None of the new advanced IO preparations were allowed for clinical use by regulatory boards even though phase II and phase III clinical trials showed promise in terms of potential benefits to cancer patients. Furthermore, some of the existing clinically approved IO-particle-based "negative contrast" MR agents (Ferridex™, Resovist™) are no longer marketed for various reasons, including potential concerns about their safety. Since the benefits of the use of IO nanoparticles in diagnostic radiology could potentially easily outweigh the issue of regulatory concerns in the future, the work on the development of novel nanoparticles has not stopped.

In this regard, a significant effort of many radiology research groups has been dedicated to the task of designing biocompatible, ultrasmall, nano-size particles and macromolecules ranging from 10 to 50 nm in diameter that can be utilized as contrast agents for MRI. The push for designing such small particles was traditionally based on the assumption that nanoparticles, which have hydrodynamic radii below 100 nm, should have a substantially better chance of crossing vascular and tissue barriers and should penetrate the interstitium better than larger particles of several hundred nanometers in diameter. A smaller size of the particles was thought to have a potential of eventually translating into more even distribution in the tissue. High-resolution electron and atomic force microscopies combined with the more advanced techniques of sample preparation resulted in more detailed information about the ultrastructure and morphology of the nanoparticles. The developed nanoparticles, both paramagnetic and superparamagnetic, have several common features. They usually have inner cores containing a plurality of atoms of paramagnetic or superparamagnetic metals, and these cores are coated with a surface layer or layers of stabilizing molecules. The surface coating is amenable to covalent or noncovalent linking to the cores either during the synthesis, or after "naked" nanoparticle fractionation, and is designed to stabilize the core nanoparticles with the formation of stable colloidal solutions. This renders the particles hydrophilic, biocompatible, and potentially less susceptible to the adsorption of proteins. The latter, in turn, is believed to make nanoparticles less detectable for surveying and resident tissue macrophages and other cells of the reticulo-endothelial system (RES). Many of the developed nanoparticle types were at some point regarded as potential platforms for designing targeted contrast agents. The latter is frequently associated with certain limitations in terms of design and chemical composition. The common types of nanoparticle-size MR contrast agents as well as recent innovations and their applications are described below.

2.2 USPIO STRUCTURE AND SYNTHESIS

The preparation of nanoparticles that have long circulation times depends on the initial synthesis conditions as well as on physical separation of the contaminating impurities. Long-circulating nanoparticles are usually subjected to a sizing step by using a series of membrane filtration steps. This is done primarily to remove aggregated particles and for obtaining a more homogenous fraction of nanoparticles in terms of their hydrodynamic diameter distribution. The sizing alone to diameters below 0.1 μm contributes to a more predictable *in vivo* fate after the intravenous administration. The coating of nanoparticles with relatively inert polymers, which usually include polysaccharides and polyethylene glycols, is not only intended to prevent aggregation of nanoparticles during synthesis but is also responsible for the blocking of particle recognition by opsonizing proteins *in vivo*. However, to obtain particles that would be also potentially targetable, nanoparticle surfaces should be modified with an adapter or linker groups. Therefore, in many instances the nanoparticle design is a result of a compromise between a tendency to choose the most indolent surface characteristics and creating a multifunctional surface as a precursor for synthesis of targeted nanoparticles.

Most of the nanoparticles developed as MRI contrast agents are ultrasmall colloids such as stabilized iron oxides, polymeric nanospheres, dendrimers, and apatite particles, which are distributed mainly in the plasma fraction of blood. Their plasma volume distribution, together with their molar longitudinal relaxivity (which is determined by r_2/r_1 ratio), determines their potential efficacy as intravascular MR agents.

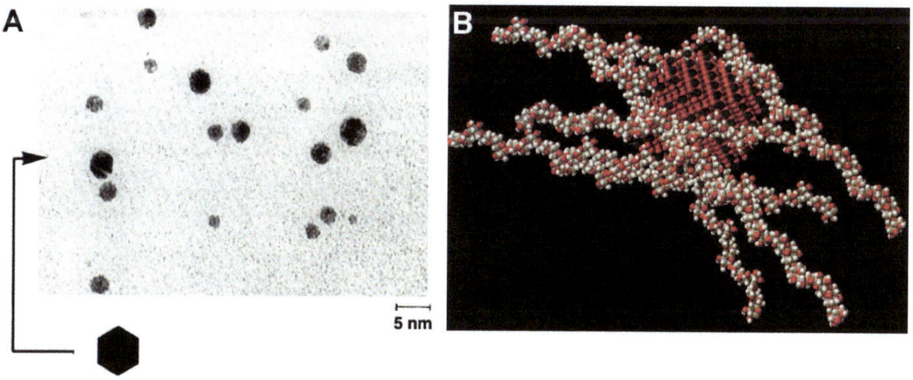

Figure 2.1 Structure of a typical USPIO (MION). (**A**) transmission electron microscopy of MION core ultrasmall iron oxide nanoparticles. MION consists of an electron-dense crystalline core typically 4.6±1.2 nm in diameter according to TEM data. Schematic diagram of hexagonal crystalline core is shown and pointed by an arrow. (**B**) Theoretical model of USPIO nanoparticle indicating that the inner core of MION consists of inverse spinel oxygen packing. Each core contains about 25 stabilizing surface-bound dextran molecules (not all shown). Reproduced with permission from Ref. [56].

The composition of most nanoparticle-based contrast agents includes one or several metal ions usually existing in several oxidation states (e.g., Fe(II), Fe(III), as well as Mn(II) and others), which alongside with oxygen and water inclusions comprise the core of a nanoparticle. These small, metal-containing cores are most commonly constituted of noncrystalline (amorphous) and crystalline mixed metal oxides [11], which usually aggregate with the formation of multi-crystal cores several nanometers in diameter that have the ability to attenuate electron beams. This enables transmission electron microscopy of very diluted IO particle samples (Fig. 2.1A). The core at high magnifications appears hexagonal with a diameter of between 2 and 8 nm (the mean size of IO core is 4.6 nm).Such particles are often referred to as USPIO (ultrasmall iron particles of IO). The aggregation of various stabilized nanoparticles upon drying, frequently presents a problem during sample preparation, and as a result transmission electron microscopy of thin films does not usually yield data comparable to that of dynamic light correlation spectroscopy (or laser light scattering). In addition, the stabilizing layer of molecules usually has low electron density, which complicates data interpretation.

The most common approach to IO synthesis consists of combining the solutions of Fe(II) and Fe(III) chlorides, followed by a slow alkalinization of the medium upon heating in the presence of an excess of a stabilizing macromolecule (e.g., various glucans). Alkalinization with weak bases such as ammonium hydroxide results in the nucleation of mixed iron oxides upon heating. As a result, in an ideal scenario the nanocrystals remain separated as a result of a noncovalently bonded carbohydrate shell consisting of several dextran molecules (poly-alpha-1,6-glucose) (Fig. 1B). The hypothetical structure of these dextran-coated IO nanoparticles shows random positioning of short dextran polymers (e.g., dextran T10). It is known that chemical reduction of terminal reducing sugars in dextrans results in polysaccharides that are more efficient in stabilizing small (<20 nm) IO nanoparticles [45]. It is highly probable that most of the dextran molecules are associated with the IO core nanocrystals as a result of the formation of a plurality of hydrogen bonds involving sugar hydroxyls and Fe(III). The detailed structure of these complexes is still largely unknown [12].

Polysaccharide-coated iron oxides are amenable to chemical stabilization using cross-linked agents, e.g., epichlorohydrin (Fig. 2.2). The epichlorohydrin-treated, IO-associated dextran carry multiple cross-links, which results in a more stable dextran coating. In addition, epoxides can be subjected to ammonolysis with the resultant formation of primary amines. These functionalized particles (amino-CLIOs, i.e., cross-linked iron oxides) have found multiple uses for immobilizing molecules required for cellular targeted imaging applications (see below) [68]. It should be noted that many other polysaccharides, including carboxylated dextran [62], oxidized starch [25] and copolymers of chitosan [61] can be used for stabilizing nanoparticles during their synthesis. Not unexpectedly the magnitude of the T_2^* -weighted signal in the liver depends on the size of nanoparticle cores when equal doses of IO are injected intravenously.

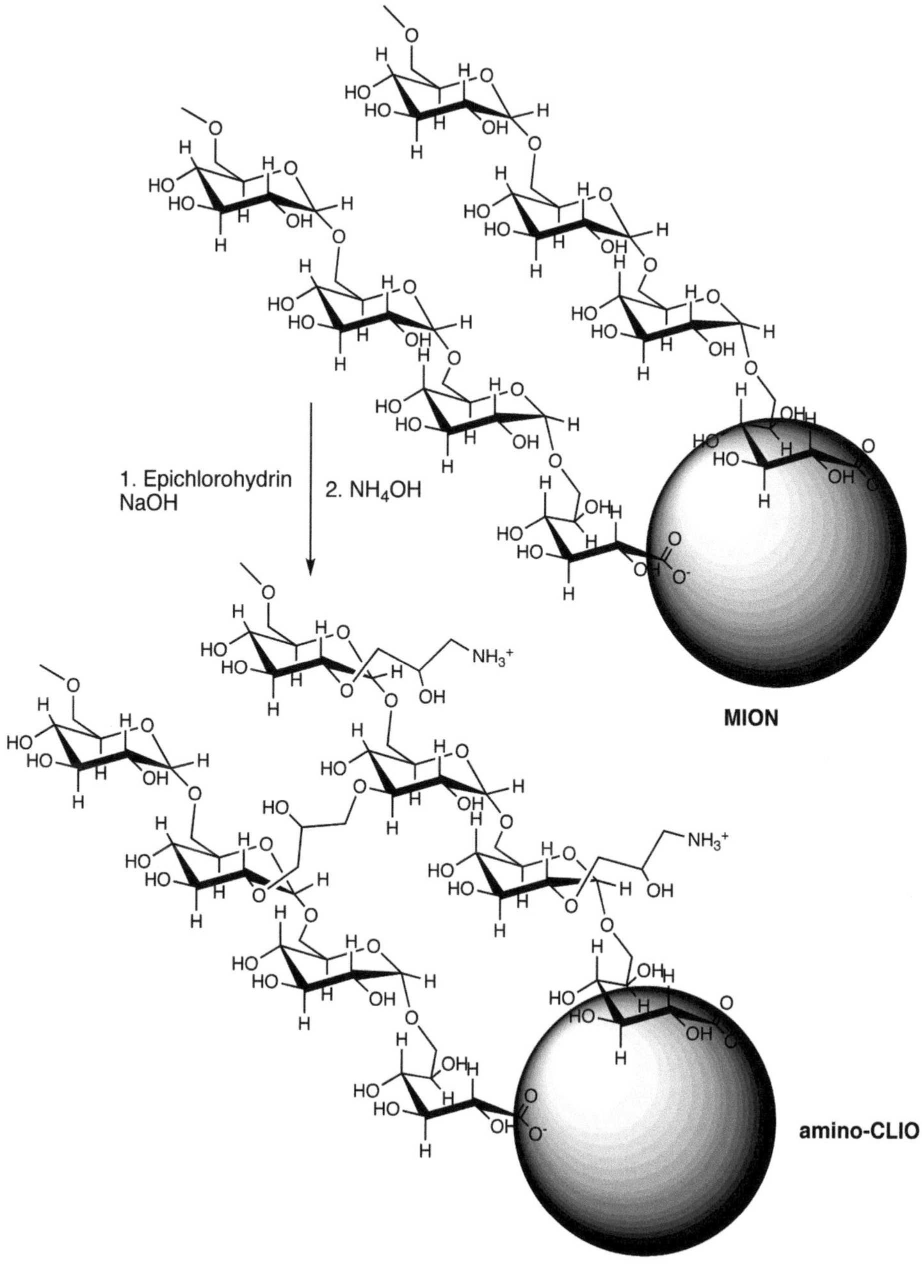

Figure 2.2 A scheme showing synthesis of functionalized cross-linked USPIO (CLIO) using MION as a starting nanoparticle. Dextran coating is cross-linked in the presence of epichlorohydrin with subsequent ammonolysis of free epoxides. The aminated nanoparticle can then be used as a platform for subsequent immobilization of various targeting ligands [20, 23].

However, as shown in a comparative study performed by MR imaging of IO nanoparticle clearance from the livers of rats, IO nanoparticle half-lives in the liver depended exclusively on the properties of nanoparticle coating with oxidized starch-coated small nanoparticles (12 nm in diameter) producing the most lasting MR signal [9].

Apart from the "single-pot" synthesis described above, a separate group of methods for IO nanoparticle synthesis includes includes pre-formed particle coating by using water-soluble dimercaptosuccinic acid [29] or amphiphilic copolymers that can enable stabilization and formation of colloidal solutions of IO nanoparticles [69, 70].

In addition to iron oxides, alternative ultrasmall nanoparticles were developed that consisted of polyethylene glycol (PEG)-stabilized manganese-substituted hydroxylapatite (MnHA-PEG) and that were synthesized by using PEGs of various masses (MW = 140–2000) and approximately 10 nm in diameter [1, 2]. The MnHA-PEG nanoparticles formed 100 nm aggregates, which were partially de-aggregated and cleared from the blood. Unfortunately, concerns about the potential accumulation of non-biodegradable particles after their extravasation precluded further development of these nanoparticles.

2.3 MAGNETIC PROPERTIES OF USPIO AND WATER RELAXATION

Most of the nanoparticle-based contrast agents are designed to have high magnetic moments. This results in strong T_2 and T_2^* effects, i.e., associated with spin–spin (transverse) relaxation effects due to proton spin dephasing leading to increasing R_2 (transverse relaxation rate) and the effects resulting from the presence of local magnetic field inhomogeneities caused by high magnetic moments of nanoparticles. Nanoparticles are more efficient transverse relaxation rate enhancers at higher magnetic field strengths because the dispersion of $1/T_2$ ($1/T_2 = R_2$) (i.e., field dependence) approaches a constant value, whereas the dispersion of $1/T_1$ ($1/T_1 = R_1$) approaches zero. Magnetic moments of nanoparticles are very high because of the tendency of the individual electron spins in paramagnetic cations of the nanoparticle core to align together. In water solutions this results in a very fast water proton spin relaxation due to fluctuations of dipolar magnetic coupling between the nanocrystal magnetization and the spins of water protons. This process of relaxation, which can be described as outer sphere relaxation in most cases, can be used for explaining the relaxation data, in terms of the theory of relaxation of water protons that diffuse in the outer sphere environments of IO nanoparticles [14]. However, relaxation of water protons in the presence of IO (transverse relaxation enhancers) is not always limited by diffusion, and the exchange between protons of hydroxy groups present on the surface of nanoparticles and water protons can also significantly influence the relaxation process by contributing to the overall T_2 mechanism and the rate of T_2 decay (negative MR imaging signal) [15].

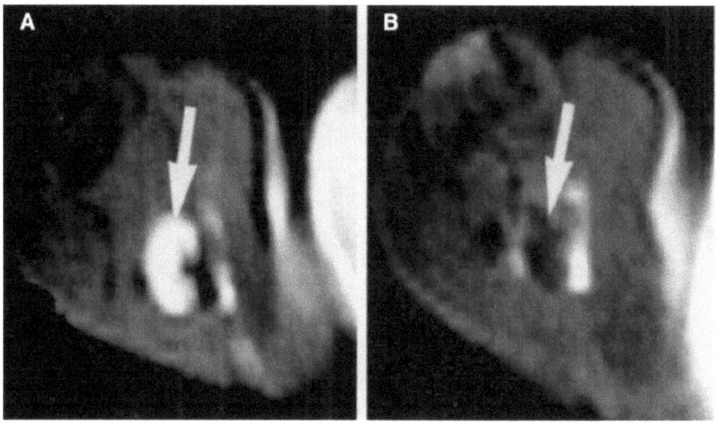

Figure 2.3 MR imaging of lymph nodes at high resolution using 4.7T MRI T_2-weighted images show a reactive left popliteal lymph node in a rat (arrow), which has high signal intensity and low T_2-weighting on (**A**) the precontrast image and very strong T_2-weighted signal (black) on (**B**) the postcontrast image. Reproduced with permission from Ref [60].

Relaxivity of iron oxides can vary greatly depending on the Fe(II)/Fe(III) ratio and nanoparticle core size. However, if the r_2/r_1 ratios of USPIO are between 2 and 10, such nanoparticles can be used for shortening T_2 relaxation times of water essential for T_2-weighted MR imaging.

2.4 BIOLOGICAL PROPERTIES AND BIODISTRIBUTION *IN VIVO*

It has been previously noted elsewhere that the hydrodynamic parameters of IO in blood is larger because of hydration of the particles and that the exact size of IO in blood is "probably" 50–200 nm [63]. It is apparent that the properties of the iron core coating can influence the pharmacokinetic properties of IO. It is known that many IO particles heavily adsorb plasma proteins [40]. The cause of this remarkable affinity to proteins and instability in plasma lies not in the hydrodynamic diameter distribution profiles of the particles (size distribution) that suggest that the average nanoparticle size should not exceed 100 nm, but in the properties of their coating [44]. The adsorption and tight association of plasma proteins outside the loosely bound dextran is a potential primary cause of nanoparticle uptake in the RES as the affinity for macrophage scavenger receptors increase. For example, Feridex (AMI-25), a commercially available preparation of IO, which contains a very broad distribution of particle sizes due to various degrees of aggregation, has been tested extensively as a liver contrast agent. The uptake in large organs of the RES [31], especially in liver macrophages (Kupffer cells) and macrophages of red pulp of the spleen, can lead to a rapid depletion of nanoparticles distributed in plasma with a resultant removal from the circulation. This leads to a strong accumulation of IO particles in the liver and spleen. Early *in vivo* experiments using ^{59}Fe-labeled IO showed that liver accumulated more than 80% of intravenously administered IO nanoparticles, while spleen accumulation varied with an average of 6% [66]. More detailed histology studies showed sequestering of IO in macrophages in both liver and spleen, with a more diffuse distribution in the liver (70 μm between positive cells) than in spleen, where clusters of IO positive cells were separated at the distance of 25 μm [31]. This dense distribution leads to strong T_2^* effects, essentially negating T_1-weighted relaxation of water protons in spleen.

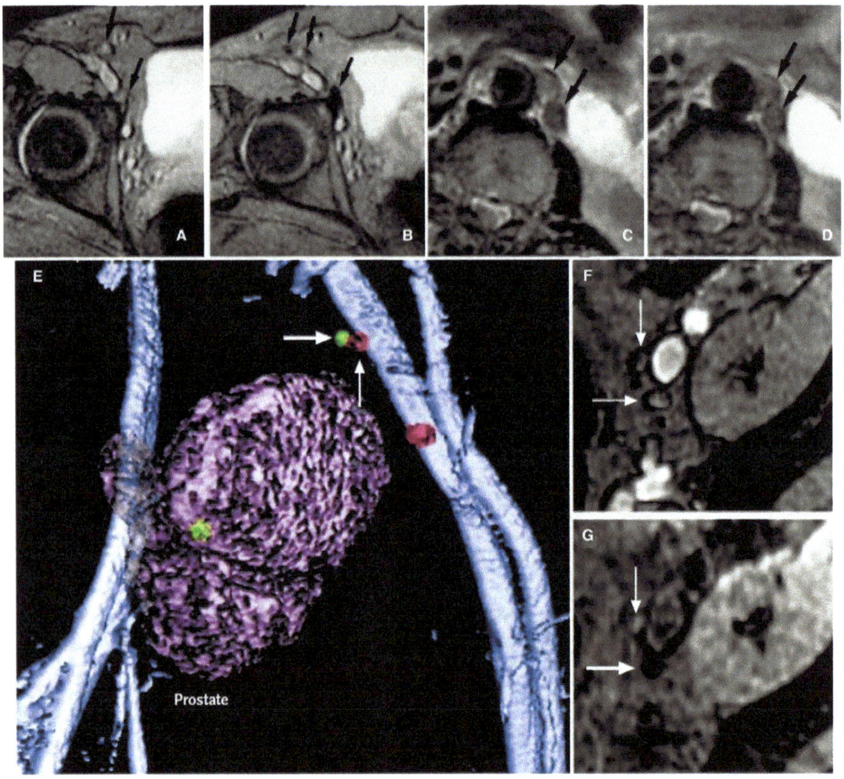

Figure 2.4 (**A, B**) Benign external iliac nodes in a patient with adenocarcinoma of the prostate. (**A**) Precontrast T_2^* -weighted GRE image (500/18, 15' flip angle) does not show a clear delineation of external iliac nodes (arrows). (**B**) Postcontrast T_2^* - weighted GRE image shows the nodes (arrows) clearly, owing to their decrease in SI, a finding that indicates USPIO uptake. (**C, D**) Metastatic retroperitoneal nodes in a patient with transitional cell carcinoma. (**C**) Precontrast T_2-weighted fast SE image (4,000/119) shows two para-aortic nodes (arrows) with intermediate SI. (**D**) Postcontrast T_2-weighted fast SE MR image (4,000/1 19) shows no change in the SI of the nodes (arrows). Histopathologic findings confirmed that they were metastatic nodes. (**E**) A three-dimensional reconstruction of the prostate, iliac vessels, and metastatic (red) and nonmetastatic (green) pelvic lymph nodes, performed in assisting in the planning of surgery and radiotherapy. A malignant node (thick arrow) immediately adjacent to the normal node (thin arrow) posteromedial to the iliac vessels. (**F**) Conventional MRI showing that the signal intensity is identical in the two nodes (arrows). (**G**) MRI with lymphotropic superparamagnetic nanoparticles showing that the signal in the normal node is decreased (thick arrow) but that it is high in the metastatic node (thin arrow). Reproduced from Refs. [5] and [17] with permission.

Early preparations of smaller and size-fractionated IO nanoparticles (USPIO) exhibited much lower liver uptake primarily due to improvements in the synthesis procedure that allowed a whole new area of applications *in vivo,* including lymph node imaging [47, 64]. The detection of metastatic spread of cancer to the lymphatic system using noninvasive imaging is critically important for tumor staging. For example, in prostate carcinoma patients the evidence of metastatic spread to lymph nodes makes radical resection of the prostate a completely inefficient procedure in terms of a potential curative effect and requires alternative treatments that can improve quality of life and save treatment costs. However, differentiating between cancer-affected and normal lymph nodes is problematic because of the frequent inability to reliably differentiate between hyperplastic (benign) and metastatic lymph nodes. The IO nanoparticles with improved coating and size distribution showed the ability to extravasate into the interstitium and accumulate in lymphatic sinuses with the uptake by lymph node resident macrophages [16, 65]. As a result, a remarkable sensitivity, 100%, with 80% specificity was achieved in patients with pelvic and urological cancers spread to their lymph nodes that were injected with a relatively small dose of AMI-227 nanoparticles [5]. These nanoparticles were similar in composition and size distribution to MION-47 [22]. Both AMI-227 and MION-47 consist of an IO (magnetite) core coated with dextran T10. Both preparations have V_2/V_1 ratios of 2.2. The particle size of AM I-227 is 19 nm, and the particle size of MION-47 is 18–24 nm. Early reports of successful imaging of USPIO uptake in the lymph nodes after intravenous injection of commercial USPIO preparations (AMI-227) suggested that hyperplastic nodes exhibited high levels of USPIO uptake and thus were showing a characteristic darkening associated with IO uptake in resident lymph node phagocytes. These hyperplastic nodes showed characteristic negative contrast on T_2- or T_2^*-weighted images (Fig. 2.3A,B). The same trend was observed in an early clinical trial of USPIO in patients. The benign iliac lymph nodes in prostate cancer patients were almost impossible to delineate on precontrast gradient-echo images (Fig. 2.4). However, after the intravenous injection of USPIO the same iliac lymph nodes were clearly visible because of the apparent USPIO uptake in phagocytes manifesting in the characteristic darkening of lymph node tissue (Fig. 2.4A,B). Lymph nodes can become completely or partially replaced by tumor cells as in the reported case of a patient with transitional carcinoma spread to retroperitoneal (paraaortic) lymph nodes (Fig. 2.4C,D). Such replacement of normal lymphatic tissue is frequently a consequence of metastasis of tumors outside the borders of the primary tumors. In these cases lymph nodes no longer show the homogenous hypointensity on T_2-weighted images that is characteristic for normal lymphatic tissue. A more recent multicenter clinical trial study that involved patients with genitourinary cancers showed the importance of 3D-vascular mapping of lymph node location in addition to detection of the uptake of iron oxides in lymph node tissues (Fig. 2.4E) that enable better spatial differentiation between the benign and metastatic lymph nodes (Fig. 2.4F,G). Thus, early clinical trials and their more recent expansion produced very similar results and showed exquisite sensitivity and high specificity of lymph node uptake. These results demonstrated once again that nanoparticles capable of homing to lymph nodes have high importance for potential staging and treatment planning of cancer surgery in clinical practice [36].

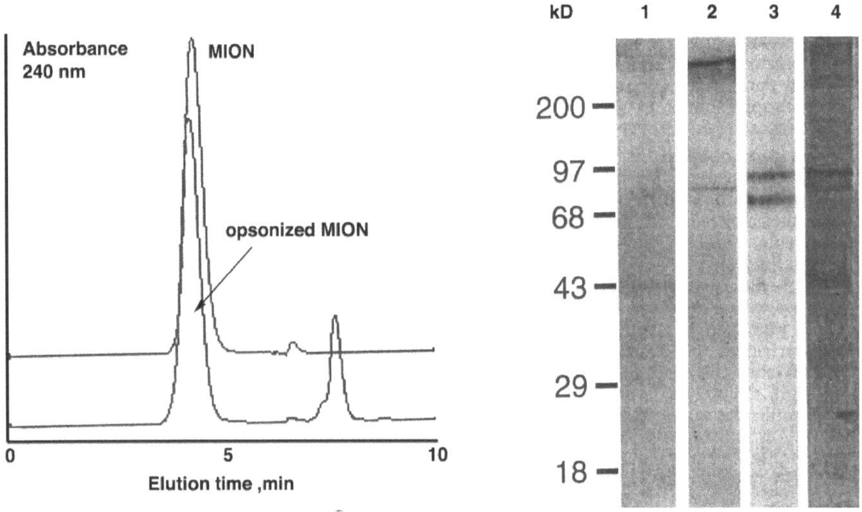

Figure 2.5. (**A**) Size exclusion HPLC of MION (upper trace) and opsonized MION (lower trace). (**B**) Immunoblotting of opsonized MION samples using polyclonal antisera against transferrin (1), vitronectin (2), fibronectin (3), and complement C3 (4). Reproduced from Ref. [40] with permission.

The behavior of USPIO particles after systemic administrations (i.e., long blood half-life and the increased accumulation in lymph nodes) closely correlates with the interface properties of IO nanoparticle coating and its interactions with blood components. We previously performed experiments with ultrasmall IO nanoparticles (also known as USPIO MION-46 and MION-47), which were synthesized in the presence of high concentrations of dextran T10, enabling trapping small IO crystals of several nanometers in diameter in a dense carbohydrate shell [56]. The latter nanoparticles never aggregated in the presence of fresh blood plasma, which is consistent with the extended circulation in the bloodstream. However, gradient density separation showed that after incubations at 37°C, MION-47 contained adhered plasma proteins. Size-exclusion HPLC of these opsonized as well as non-opsonized preparations of MION-47 showed that as a result of the binding of plasma proteins to the surface of MION-47 there was a 25% increase of effective hydrodynamic diameter, which corresponded to a shift corresponding to a 0.75×10^6 Da globular protein to a 1.01×10^6 Da (Fig. 2.5). The immunoblotting of proteins resolved by sodium dodecyl sulfate electrophoresis of the above opsonized nanoparticles showed that fibronectin, vitronectin, and C3 fragments were associated with their coating (Fig. 2.5) [40]. However, no transferrin binding was observed suggesting that these IO nanoparticles were different from larger IO particles that had imperfect coating with dextran and short blood half-lives *in vivo* [8]. The above C3 binding to IO in plasma suggested that (i) there is a potential C3b-mediated, enhanced uptake of MION particles in macrophages and (ii) there is potential toxicity associated with dextran-coated IO.

The discovery that USPIO have unusually long plasma blood half-lives (180–600 min in rats, depending on the dose) enabled early evaluation of these nanoparticles as potential contrast agents for blood pool imaging. Magnetic resonance blood pool imaging (or MR angiography, MRA) generally does not necessarily require any contrast agent administration, because water protons in rapidly flowing blood can be easily differentiated from the stationary ones using special MR pulse sequences. However, for accurate imaging of slow blood flow, contrast agents that are retained in the blood stream without extravasation could potentially benefit the diagnostic study. Contrast-enhanced MRA requires that the selective increase of R_1 (longitudinal) relaxivity of blood would follow the injection of a contrast agent and that this increase in R_1 would be constant (or nearly constant) within the time window sufficient for performing the study by acquiring pulse sequences. Therefore, small-molecular-weight contrast agents (i.e., paramagnetic chelates of Gd) that extravasate quickly are not ideal for such MR studies. The first animal experiments with USPIO in rats indicated that at a dose of 15–50 μmol Fe/kg, aortic signal-to-noise ratio increased three- to fourfold. IO nanoparticles markedly improved the quality of images of the vasculature of the lungs, abdomen, and extremities [13]. During the initial safety studies in human volunteers, significant changes in MR signal intensity of the blood pool was reported for both T_1- and T_2-weighted images [4, 33]. IO nanoparticles formulated for human clinical trials (AMI-227) showed that small doses of iron (1.1–2.6 mg/kg) were useful in contrast-enhanced MR delineation of small arteries. The demonstration of the feasibility of vascular imaging using USPIO-enhanced MR angiography suggested that USPIO could potentially have an independent use in clinical diagnostic imaging as angiographic agents. For example, USPIO increase the specificity and sensitivity of detection of small-diameter-vessel occlusion. The comparison of the data obtained using high-relaxivity IO (Ferridex, AMI-25) particles (Fig. 2.6A) [55] and USPIO (Fig. 2.6B) [3] showed obvious advantages of nanoparticles in terms of small-vessel delineation. However, images with satisfactory quality were obtained using USPIO doses that were higher than regular IO that potentially increases the concerns about their safety for MR angiography applications. Because of the ability of IO nanoparticles to create local inhomogeneities in a magnetic field (causing T_2^* relaxation effects), a substantial loss of the signal-to-noise ratio of the aorta and a decrease in arterial visualization were observed with higher concentrations of the agent, which usually can be avoided by minimizing echo time values during the gradient-echo pulse sequence acquisition. Since such minimization was not always possible with the technology level available in the 1990s, USPIO never received a status of MRA contrast agents because of T_2^* effects at higher doses of iron [28]. However, there is currently a renewed interest in using IO nanoparticles as a potential alternative to Gd-based chelates because of adverse effects in patients with renal disease and kidney dysfunction [42].

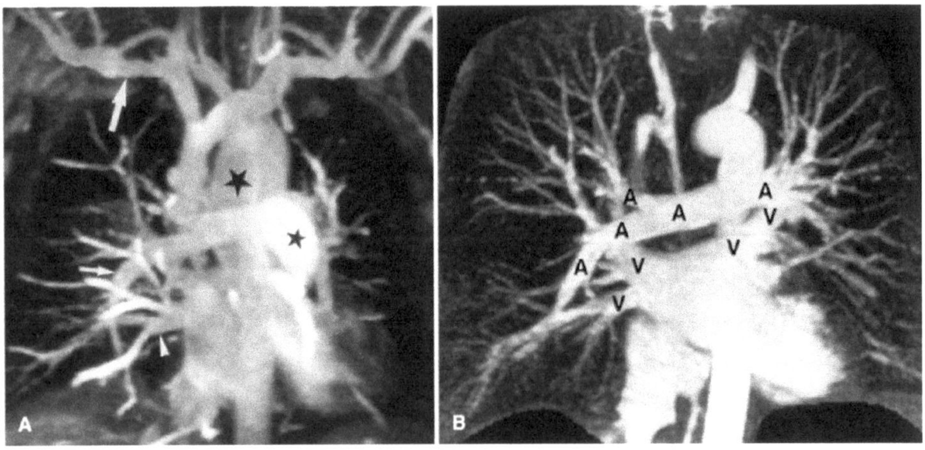

Figure 2.6 Iron oxide-assisted MR angiography in early clinical trials. (A) Maximum intensity projection (MIP) reconstruction image using gradient-echo MR pulse sequence during the infusion of AMI-25 (15 μmol Fe/kg). The pulmonary artery (small arrow) and right intralobar pulmonary vein (arrowhead) are clearly visible. (B) A MIP image of pulmonary MR obtained with 90 μmol Fe/kg NC100150 (experimental USPIO, Nycomed) injection. A denotes arterial branches; V, veins; small star, main pulmonary artery; and large star, ascending aorta. A partial image showing arterial branching. Image artifacts are seen as signal intensity increases and decreases along each vessel. The arrow points to the azygous arch, and IVC stands for inferior vena cava. Overall image quality is judged very satisfactory. Reproduced from Refs.

2.5 IRON OXIDE NANOPARTICLE TARGETING

By applying T2-weighted spin-echo or T_2^* -weighted gradient-echo MRI pulse sequences, low amounts of IO nanoparticles can be detected in tissues. This lead to the widespread use of IO in targeted imaging drug delivery and established IO as a very popular platform for designing nanoparticles targeted to cell surface receptors. The first approach to targeting included polysaccharides chemically different from dextrans in order to cover the particles with carbohydrate-based receptor ligands that would be associated with a given population of cells after systemic administration *in vivo*. For the purpose of retargeting IO nanoparticles (which normally are taken up by lymph node phagocytes) to an asialoglycoprotein receptor on hepatocytes, IO nanoparticles were synthesized in the presence of arabinogalactan instead of the commonly used dextran. This resulted in a better in vivo MRI contrast between asialoglycoprotein-negative cancer and normal liver tissues [48]. However, substituting dextran with other polysaccharides, including arabinogalactan, results in particles with low stability in solutions. As an alternative, asialofetuin conjugation to preformed dextran-covered MION USPIO [53] resulted in nanoparticles that were very rapidly removed from circulation in rats and accumulated to a greater extent in liver hepatocytes when compared with nonconjugated MION. In cell culture experiments the specificity of asialofetuin-conjugated particle binding was proven by using galactose as a competitive inhibitor [21, 46]. It appears that after relaxation patterns in the liver are responsible for contrast phenomena and that they can be selectively modified as a result of hepatocyte targeting. It has been reported that in the case of nontargeted MION or larger PION particles, water proton relaxation in the liver was similar suggesting that PION is in fact an aggregated form of MION. Water relaxation in the liver was consistent with rapid water exchange between vascular and extravascular compartments because the particles were primarily internalized in liver Kupffer cells, which was not the case for asialofetuin-conjugated particles, in which relaxation was not exchange-limited to the same extent [32]. However, achieving the complete lack of uptake in Kupffer cells is practically impossible without eliminating macrophages with liposomal apoptosis-inducing drugs [59]. Therefore, even in the case of highly efficient retargeting to hepatocytes, the overall distribution of IO nanoparticles would reflect the uptake of a fraction of injected nanoparticles in liver macrophages.

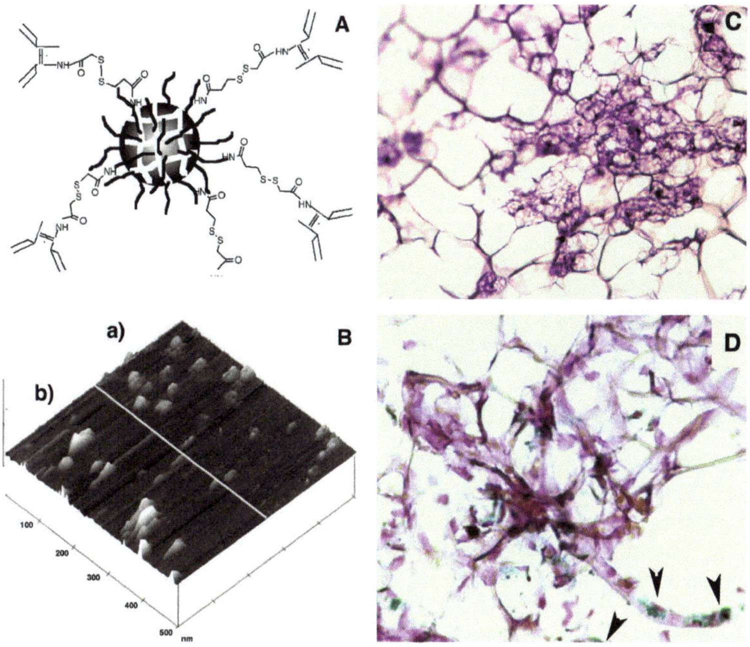

Figure 2.7 HUVEC-targeted CLIO nanoparticles. (A) A schematic drawing showing a cross-linked iron oxide nanoparticle (CLIO) with several $F(ab')_2$ molecules tethered to the surface via S–S bonds. (B) Results of atomic force microscopy of samples adsorbed on the mica surface. (a) CLIO particles modified with 2-pyridyldithiopropionyl groups (average diameter 30 nm) and (b) CLIO conjugated to H18/7 $F(ab')_2$ (average diameter 50 nm). (C, D) *Ex vivo* analysis of Matrigel implants (20 days after implantation) isolated from mice 3 hours after intravenous administration of targeted CLIO conjugates at the dose equivalent to 3 mg/kg of Fe that followed intraperitoneal treatment with IL-1beta. The tissues were stained with Perl's reagent and hematoxylin/eosin. (C) Control (HUVEC-free) Matrigel implants; (D) HUVEC-containing implants. Prussian blue stain was detected only in HUVEC-containing implants. Magnification×15,000. Reproduced from Ref. [24] with permission

A shorter T_2 in the normal tissue due to nanoparticle accumulation as opposed to lesser accumulation in tumor cells can yield better delineation of tumor margins and enable better surgical planning. The recent finding that SPIO-enhanced imaging of pancreas can assist in confirming diagnosis of pancreatic cancer in humans in the case of clinical doubt [18] validates early efforts directed at obtaining preparations of IO targeted to the exocrine pancreas. The initial attempt to link cholecystokinin noncovalently to nanoparticles as a potential targeting ligand for delivery of MION resulted in a 20% decrease of relaxation times in the pancreas in rats [49]. The noncovalent, high-affinity avidin–biotin interaction between biotinylated secretin and streptavidin-linked MION enabled attachment of two secretin molecules per nanoparticle on average. These targeted nanoparticles showed a marginally but significantly higher binding to normal acinar cells and pancreatic tissue *in vivo* than IO nanoparticles that carried no immobilized secretin [57].

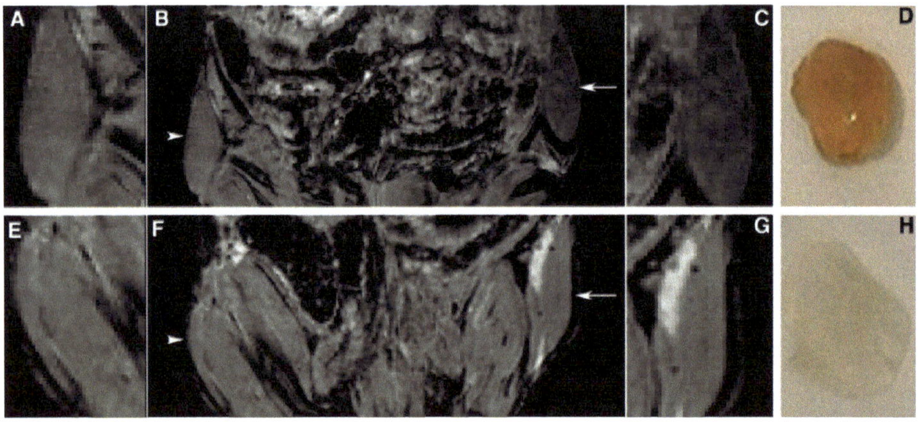

Figure 2.8 7T MR imaging of vessels after adoptive transfer of HUVEC *in vivo*. T_2^*-weighted gradient-echo MR images of mice bearing Matrigel implants (**A–C** and **E–G**). Implants containing Matrigel alone are shown on the left (**A**, enlarged, shown on **B** and **F** by arrowheads), and HUVEC-containing implants are shown on the right (**E**, enlarged, shown on **B** and **F** by arrows). Images were obtained 3 hours after the injection either of anti-E-selectin CLIO-F(ab')$_2$ (images **A–C**) or of nonspecific CLIO-F(ab')$_2$ nanoparticles (images **E–G**). Shown also are light photography images of HUVEC seeded implant (**D**) and a control implant (**H**), surgically removed from a euthanized mouse injected with E-selectin specific CLIO-F(ab')$_2$. Reproduced from Ref. [24] with permission.

A significant step toward the potential clinical applications of targeted IO was made by demonstrating the ability of IO to bind with high selectivity only to cells that bear specific markers of activation (E-selectin in the case of endothelial cells), and not to nonactivated cells [23]. By using an approach utilizing aminated cross-linked IO nanoparticles (CLIO-F(ab')$_2$, conjugates with an average hydrodynamic diameter of 40.6 nm were synthesized (Fig. 2.7A,B). These endothelium-targeted IO nanoparticles were shown to bind to live human endothelial umbilical vein cells (HUVECs) in culture only after treating cells with IL-1 beta, which induces expression of E-selectin. E-selectin-expressing HUVECs showed a 100–200 times higher binding of targeted particles than control cells. The binding resulted in a high superparamagnetism of HUVECs with the transverse water proton relaxation time (T_2) decreasing to 30–40 ms in cell precipitates of activated cells and resulted in a highly specific T_2-weighted ^1H_2O signal decrease. The same type of targeted CLIO was used *in vivo* in a model of adoptive HUVEC transfer in athymic mice [24]. Imaging of E-selectin expression in functional blood vessels was accomplished by implanting HUVECs in a Matrigel solution, which served as a temporary neovascularization scaffold after the solidification of Matrigel. The formation of human endothelial cell-containing vessels in mice was confirmed using immunohistology and staining of sections for iron after the administration of HUVEC-targeted CLIO nanoparticles (Fig. 2.7C,D). CLIO-F(ab')$_2$ conjugate probes were administered via an intravenous injection following the induction of E-selectin expression by IL-1 beta. A specific hypointensity on MR imaging was observed only if the animals were preinjected with IL-1 beta (Fig. 2.8). Image signal intensity did not change in control animals that either harbored Matrigel alone, imaged after the CLIO conjugate administration in the absence of IL-1 beta treatment, or were injected with CLIO linked to the idiotype-matched control F(ab')$_2$. The demonstration of E-selectin expression imaging in a model system justified further development of MR-targeted agents for monitoring endothelial cells in angiogenesis and atherosclerosis. The latter common and complex disease of blood vessel walls are especially relevant to USPIO-assisted diagnostic imaging due to the ability of USPIO to target activated macrophages populating inflamed, unstable atherosclerotic plaques [10]. Active macrophage mediated macropinocytosis or phagocytosis of "naked" or lipid-modified USPIO in the regions of plaque instability can potentially serve as a diagnostic marker for local vulnerability of vascular wall and can highlight areas of prospective intravascular intervention. The targeted delivery into phagocytes and retargeting of nanoparticles to other cells of the RES is facilitated by the presence of the discontinuous endothelium lining and fenestrated basal membrane of liver and spleen sinusoids. As shown above, targeting of endothelial cells of the major blood vessels can be even more straightforward in terms of the efficacy of delivery. Specific delivery of nanoparticles to endothelial caveolae can potentially enhance trans-endothelial delivery to perivascular space and potentially to lymph nodes [43]. As a rule, all other nanoparticle delivery attempts with the exception of targeting of some tumor cell-specific receptors have been unsuccessful.

One of the reasons is that both USPIO linked to targeting molecules and those conjugated to control molecules usually have similar long circulation times in the bloodstream and accumulate in tumor interstitium because of similar cross-vascular concentration gradients. For example, experimental gliomas (GFP-expressing variant of rat 9L gliosarcoma [39]) readily accumulate nontargeted, fluorochrome-tagged USPIO, as demonstrated by registration of tumor-specific and USPIO-specific optical images [26] (Fig. 2.9). However, the binding to the cells and uptake is certainly higher in the case of targeted USPIO that potentially translates into either higher binding to cells or combined elevated binding and intracellular accumulation in the cells. Both can result in changes of iron relaxivity and contrast effects on MR images. Transferrin receptor is frequently overexpressed in cancer and plays a central role in providing necessary intracellular pool of iron. The expression of this receptor is tightly regulated. Iron enters mammalian cells as a result of binding of blood plasma transferrin (Tf) to the transferrin receptor, which is internalized, via receptor-mediated endocytosis in endosomes where Fe(II) is released from the carrier apo-transferrin as a result of local acidification of endosomes [51]. Since Tf-Fe(II) is required for cellular proliferation it was early on assumed that Tf can be used as a targeting ligand for contrast agent delivery to tumor cells because many tumor cells express higher numbers of Tf receptors [27]. Indeed, quantitative measurement of Tf receptor overexpression in a small number of human breast cancer biopsy samples (n = 27) has shown that 74% of tumors had certain levels of Tf receptor overexpression that could be potentially detectable by using targeted IO delivery [19]. By covalent linking of transferrin to the surface of nanoparticles, a class of IO was obtained that had transferrin protein-to-nanoparticle ratio. This resulted in a 40% decrease in the T_2-weighted signal in rat mammary tumors compared with a 10% reduction in the case of albumin-linked IO [27]. Experiments in genetically engineered 9L rat gliosarcoma cells showed that cells expressing stabilized (3'-UTR deleted) Tf receptor mRNA demonstrated significantly higher levels of iron uptake compared with wild-type 9L cells [37, 38]. By linking IO to Tf it was possible to deliver iron into cells in a relatively chemically inert form in the amounts that would not be physiologically possible if only Tf-bound iron was used as a sole source of Fe(II)/Fe(III). The use of cross-linked and aminated IO nanoparticles (CLIOs) enabled conjugating Tf using the heterobifunctional linking agent N-succinimidyl 3-(2-pyridyldithio)propionate (SPDP), which resulted in conjugated nanoparticles linked to protein via a cleavable bond [20]. It has been demonstrated that these nanoparticles are superior to Tf-linked MION. The latter were obtained as a result of oxidative activation of a nanoparticle dextran-coat with the subsequent reduction of Schiff's bonds after a co-incubation with Tf [20]. This conjugation method resulted in a fourfold higher number of Tf molecules linked per IO nanoparticle, causing a 10-fold increase of binding and uptake in cell culture experiments.

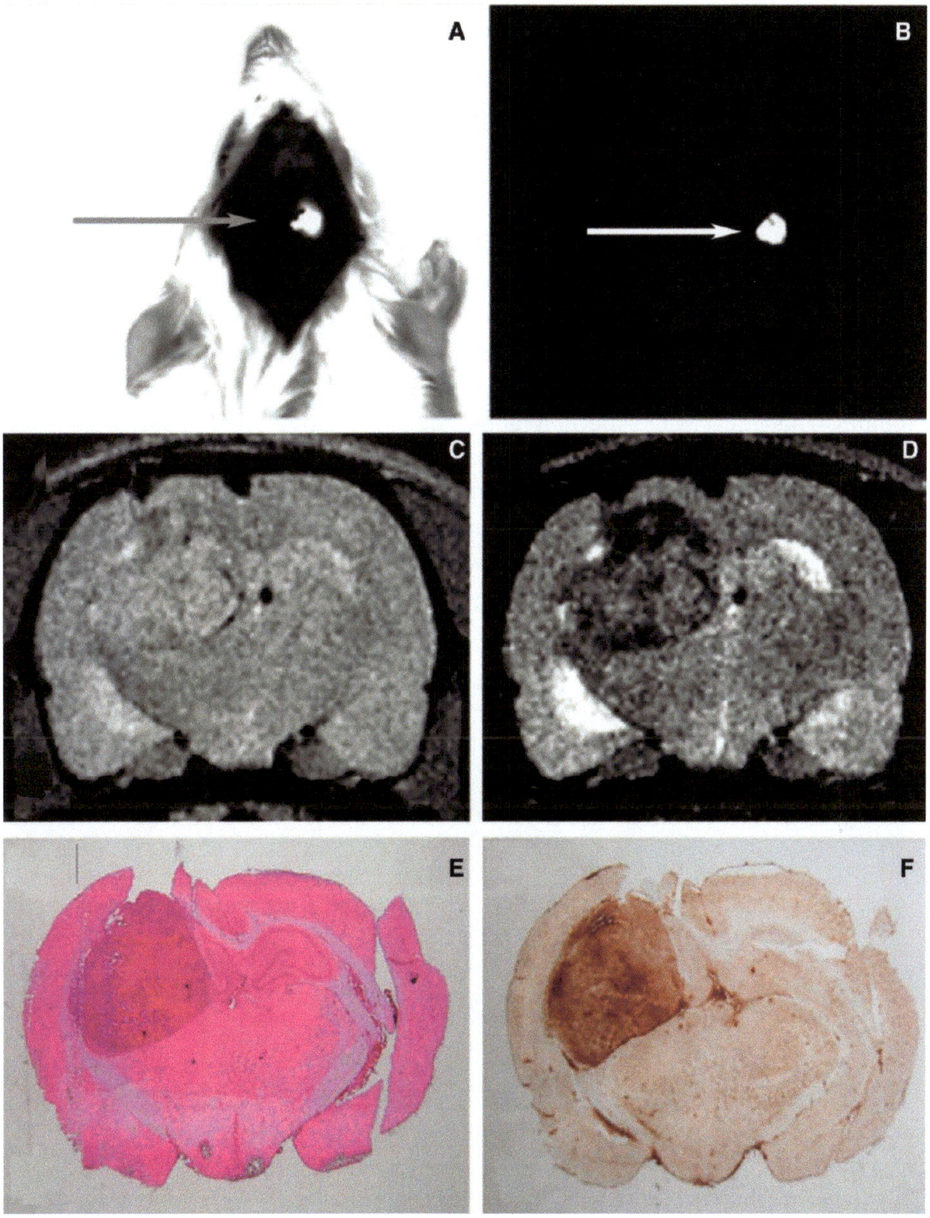

Figure 2.9 Delineation of a GFP-expressing 9L glioma tumor by optical imaging of Cy5.5-CLIO nanoparticles. (**A**) GFP channel, (**B**) Cy5.5-channel, (**C**) proton density, and (**D**) T_2-weighted MR images at 4.7T. Tumor uptake of iron is evident on T_2-weighted images as regions of low signal intensity, whereas tumor is isointense to surrounding tissue using proton density images. (**E**) Histological section corresponding to MRI slices in **C** and **D**. (**F**) Diaminobenzidine-amplified Prussian blue staining of histological section corresponding to MRI slices in **A** and **B**. Reproduced from Ref [26] with permission.

The notable examples of successes in experimental tumor receptor targeting were commonly achieved in rodent brain tumor models. Brain tumors have extensive vascular supply, disrupted blood brain barrier within the tumor margins, which enables efficient "passive" delivery of long-circulating nanoparticles (such as USPIO), regardless of whether the expression of tumor-associated molecules is present or absent. For example, in a model of non–small cell lung carcinoma, metastasis to the brain IO nanoparticle-conjugated anti-ganglioside L6 monoclonal antibody showed the accumulation in the peripheral regions of tumors that had a higher density of blood vessels and proliferating carcinoma cells [50]. Both carcinoma-targeted and control USPIO that were conjugated to nonspecific idiotype-matched IgG2a showed an uptake in tumors several hours after the administration. However, only the L6-linked USPIO showed a ring-like enhancement of peripheral areas of the brain tumors. This effect was especially apparent on T_1- (and not T_2-) weighted images (Fig. 2.10). The strong T_1 effect was presumably due to the lack of internalization of USPIO since L6 is a noninternalizing antibody. The nanoparticles used in the study also had a relatively low R_2/R_1 ratio. The internalization of iron oxide nanoparticles in cells is usually accompanied by a decrease in T_1-weighted contrast due to limitations in water diffusion and strong T_2^* effects due to USPIO clustering inside the intracellular organelles.

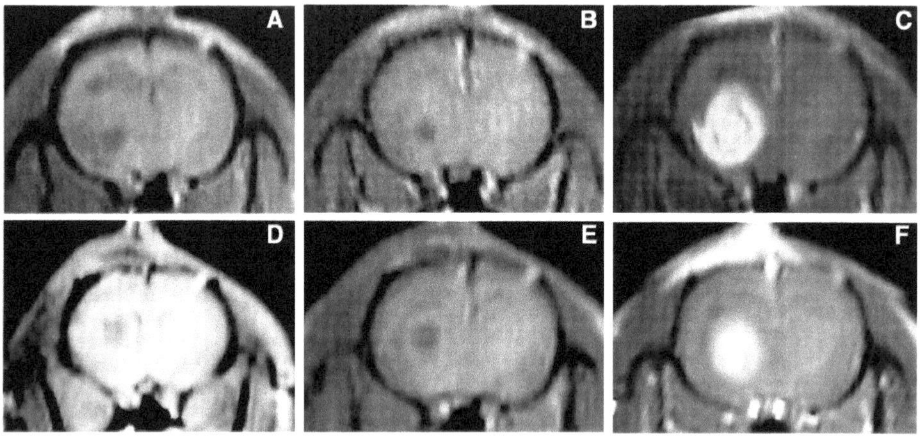

Figure 2.10 T_1-weighted MR images (1.5T, SE TR/TE=300/6 ms) of LX-1 (human small cell lung carcinoma) tumors implanted in the brain as a model of lung cancer metastasis. (**A–C**) The animal was injected with L6-USPIO targeted nanoparticles. (**D–F**) The animal for a control group injected with control IgG2a-USPIO. The dose of USPIO was 10 mg Fe/kg. (**A,D**) Precontrast images; (**B,E**) 30 minutes post injection; (**C,F**) 6 hours post injection of USPIO. Reproduced from Ref. [50] with permission.

More recent reports seem to suggest that the plurality of small-molecular-weight affinity molecules either engineered from full-size IgG or completely unrelated to antibodies can potentially replace monoclonal IgG as targeting molecules for targeted delivery of IO nanoparticles to tumors *in vivo*. For example, more than 10 molecules of recombinant bivalent single-chain antibodies, di-scFv-c [41], were linked to maleimide functionalized IO nanoparticles for the purpose of imaging and therapy of MUC-1-expressing cancers. Overexpression of aberrant mucins (e.g., undeglycosylated uMUC-1) is typical for many human adenocarcinomas, including prostate, pancreatic, and breast cancer. Mucin-1 binding peptides (EPPT) that specifically recognize uMUC-1 were covalently linked to Cy5.5-linked nanoparticles for the purpose of monitoring tumor response to breast cancer therapy using MRI or optical imaging in mice [34, 35].

Chlorotoxin, a small protein that exhibits certain selectivity of binding to glioma compared with normal cells, was recently investigated as a potential tumor-targeting ligand [61]. The toxin was covalently linked to chitosan–PEG-coated IO nanoparticles, and these nanoparticles showed the ability to accumulate in the cerebellum of transgenic animals with spontaneously developing medulloblastomas. The accumulation of these tumors, which are unrelated to glioblastoma, was semi-quantitated by measuring changes in R_2 (transverse relaxation rates) values, which increased several times over the R_2 value of the control frontal lobe or over the R_2 measured when chlorotoxin-free IO nanoparticles were used for injecting in transgenic animals.

Another molecule alternative to antibody targeting was tested by linking a recombinant peptide corresponding to the 17 kD N-terminal fragment of urokinase-type plasminogen activator (uPA) to 18 nm IO nanoparticles stabilized with a triblock-copolymer carrying free carboxyl groups [69, 70]. Such uPA receptor–targeted imaging nanoparticles were tested for molecular MRI of mammary and pancreatic tumors in mice. Since uPA receptor expression in tumor stromal cells is an established negative predictor of treatment outcomes in some adenocarcinomas, the imaging assisted with such targeted nanoparticles would have clinical significance if methods of selective nanoparticle delivery to stromal cells existed.

The imaging of more rare cells present in tumors presents a serious challenge. However, in the case of large numbers of cell death present in the tissue imaging using annexin V [54] or C2 domain of synaptotagmin [71] conjugated to USPIO can be a viable option because of the high sensitivity of MRI to the presence of IO nanoparticles in the tissues.

2.6 CONCLUSIONS

Ultrasmall IO nanoparticles were in the focus of investigative radiology and molecular MR imaging research for more than two decades. These nanoparticles were eventually developed into efficient, highly homogenous diagnostic pharmaceutical agents that can be manufactured using a scalable and GLP-compatible process. These iron-based imaging agents developed specifically for systemic administration into the bloodstream went through several multiphase clinical trials and were shown to be excellent blood perfusion agents with a high specificity in imaging lymph nodes and organs of the reticulo-endothelial system. Some concerns about the potential clinical benefits associated with the use of USPIO in humans and safety concerns are still preventing USPIO from gaining more widespread acceptance by clinical radiologists. However, USPIO still hold promise for potential use for MR imaging in cancer and especially for detecting the spread of cancer to lymph nodes. This is due to the very high sensitivity of MR imaging to magnetic field microinhomogeneities created by USPIO particles with very high magnetic moments. With the advancement of new MR imaging protocols enabling USIPO imaging at lower doses and the development of novel methods of image processing, the use of USPIO in diagnostic imaging is destined to expand in the future.

References

1. Adzamli K., Dorshow R. B., Hynes M. R., Nosco D. L., and Adams M. D. (1997) Characterization of polyethyleneglycol-stabilized, manganese-substituted hydroxylapatite (MnHA-PEG). A potential MR blood pool agent. *Acta Radiol Suppl* **412**, 73–78.
2. Adzamli K., Dorshow R. B., Hynes M. R., Nosco D. L., Beaty-Nosco J. A., Fallis S., Galen K. P., Levine J. A., Nema S., Polta J., and Adams M. D. (1996) Manganese-substituted hydroxyapatites coated with polyethylene glycol diphosphonate: potential blood-pool agents for magnetic resonance imaging. *Acad Radiol* **3**, S349–S352.
3. Ahlstrom K. H., Johansson L. O., Rodenburg J. B., Ragnarsson A. S., Akeson P., and Borseth A. (1999) Pulmonary MR angiography with ultrasmall superparamagnetic iron oxide particles as a blood pool agent and a navigator echo for respiratory gating: pilot study. *Radiology* **211**, 865–869.
4. Anzai Y., Prince M. R., Chenevert T. L., Maki J. H., Londy F., London M., and McLachlan S. J. (1997) MR angiography with an ultrasmall superparamagnetic iron oxide blood pool agent. *J Magn Reson Imaging* **7**, 209–214.
5. Bellin M. F., Roy C., Kinkel K., Thoumas D., Zaim S., Vanel D., Tuchmann C., Richard F., Jacqmin D., Delcourt A., Challier E., Lebret T., and Cluzel P. (1998) Lymph node metastases: safety and effectiveness of MR imaging with ultrasmall superparamagnetic iron oxide particles--initial clinical experience. *Radiology* **207**, 799–808.
6. Berthezene Y., Vexler V., Price D. C., Wisner-Dupon J., Moseley M. E., Aicher K. P., and Brasch R. C. (1992) Magnetic resonance imaging detection of an experimental pulmonary perfusion deficit using a macromolecular contrast agent. Polylysine-gadolinium-DTPA40 [published erratum appears in *Invest Radiol* (1992) **27**(8), 582]. *Invest Radiol* **27**, 346–351.
7. Bogdanov A. A., Weissleder R., Frank H. W., Bogdanova A. V., Nossif N., Schaffer B. K., Tsai E., Papisov M. I., and Brady T. J. (1993) A new macromolecule as a contrast agent for MR-angiography - preparation, properties, and animal studies. *Radiology* **187**, 701–706.

8. Bogdanov A. J., Papisov M., Weissleder R., Shen T., and Brady T. (1992) Opsonization of dextran-stabilized iron oxides with plasma proteins. 11th Annual Scientific Meeting Of Society of Magnetic Resonance in Medicine. SMRM, Berkley CA Berlin, Germany, p. 864.

9. Briley-Saebo K. C., Johansson L. O., Hustvedt S. O., Haldorsen A. G., Bjornerud A., Fayad Z. A., and Ahlstrom H. K. (2006) Clearance of iron oxide particles in rat liver: effect of hydrated particle size and coating material on liver metabolism. *Invest Radiol* **41**, 560–571.

10. Briley-Saebo K. C., Mani V., Hyafil F., Cornily J. C., and Fayad Z. A. (2008) Fractionated Feridex and positive contrast: in vivo MR imaging of atherosclerosis. *Magn Reson Med* **59**, 721–730.

11. Di Marco M., Port M., Couvreur P., Dubernet C., Ballirano P., and Sadun C. (2006) Structural characterization of ultrasmall superparamagnetic iron oxide (USPIO) particles in aqueous suspension by energy dispersive X-ray diffraction (EDXD). *J Am Chem Soc* **128**, 10054–10059.

12. Di Marco M., Sadun C., Port M., Guilbert I., Couvreur P., and Dubernet C. (2007) Physicochemical characterization of ultrasmall superparamagnetic iron oxide particles (USPIO) for biomedical application as MRI contrast agents. *Int J Nanomed* **2**, 609–622.

13. Frank H., Weissleder R., and Brady T. J. (1994) Enhancement of MR angiography with iron oxide: preliminary studies in whole-blood phantom and in animals. *Am J Roentgenol* **162**, 209–213.

14. Gillis P. and Koenig S. H. (1987) Transverse relaxation of solvent protons induced by magnetized spheres: application to ferritin, erythrocytes, and magnetite. *Magn Reson Med* **5**, 323–345.

15. Gossuin Y., Roch A., Muller R. N., and Gillis P. (2002) An evaluation of the contributions of diffusion and exchange in relaxation enhancement by MRI contrast agents. *J Magn Reson* **158**, 36–42.

16. Guimaraes R., Clement O., Bittoun J., Carnot F., and Frija G. (1994) MR lymphography with superparamagnetic iron nanoparticles in rats: pathologic basis for contrast enhancement. *Am J Roentgenol* **162**, 201–207.

17. Harisinghani M. G., Barentsz J., Hahn P. F., Deserno W. M., Tabatabaei S., van de Kaa C. H., de la Rosette J., and Weissleder R. (2003) Noninvasive detection of clinically occult lymph-node metastases in prostate cancer. [see comment] [erratum appears in *N Engl J Med* (2003) **349**(10), 1010]. *N Engl J Med* **348**, 2491–2499.

18. Heredia V., Altun E., Bilaj F., Ramalho M., Hyslop B. W., and Semelka R. C. (2008) Gadolinium- and superparamagnetic-iron-oxide-enhanced MR findings of intrapancreatic accessory spleen in five patients. *Magn Reson Imaging* **26**, 1273–1278.

19. Hogemann-Savellano D., Bos E., Blondet C., Sato F., Abe T., Josephson L., Weissleder R., Gaudet J., Sgroi D., Peters P. J., and Basilion J. P. (2003) The transferrin receptor: a potential molecular imaging marker for human cancer. *Neoplasia* **5**, 495–506.

20. Hogemann D., Josephson L., Weissleder R., and Basilion J. P. (2000) Improvement of MRI probes to allow efficient detection of gene expression. *Bioconjug Chem* **11**, 941–946.

21. Huang G., Diakur J., Xu Z., and Wiebe L. I. (2008) Asialoglycoprotein receptor-targeted superparamagnetic iron oxide nanoparticles. *Int J Pharm* **360**, 197–203.

22. Jung C., Weissleder R., Josephson L., Bengele H., and Brady T. (1996) Physical properties of MION-46 and AMI-227. Society of Magnetic Resonance Annual Meeting, New York, 1681.

23. Kang H. W., Josephson L., Petrovsky A., Weissleder R., and Bogdanov A., Jr. (2002) Magnetic resonance imaging of inducible E-selectin expression in human endothelial cell culture. *Bioconjugate Chem* **13**, 122–127.

24. Kang H. W., Torres D., Wald L., Weissleder R., and Bogdanov A. A., Jr. (2006) Targeted imaging of human endothelial-specific marker in a model of adoptive cell transfer. *Lab Invest* **86**, 599–609.

25. Kellar K. E., Fujii D. K., Gunther W. H., Briley-Saebo K., Bjornerud A., Spiller M., and Koenig S. H. (2000) NC100150 Injection, a preparation of optimized iron oxide nanoparticles for positive-contrast MR angiography. *J Magn Reson Imaging* **11**, 488–494.

26. Kircher M. F., Mahmood U., King R. S., Weissleder R., and Josephson L. (2003) A multimodal nanoparticle for preoperative magnetic resonance imaging and intraoperative optical brain tumor delineation. *Cancer Res* **63**, 8122–8125.

27. Kresse M., Wagner S., Pfefferer D., Lawaczeck R., Elste V., and Semmler W. (1998) Targeting of ultrasmall superparamagnetic iron oxide (USPIO) particles to tumor cells in vivo by using transferrin receptor pathways. *Magn Reson Med* **40**, 236–242.

28. Kroft L. J. and de Roos A. (1999) Blood pool contrast agents for cardiovascular MR imaging. *J Magn Reson Imaging* **10**, 395–403.

29. Lee J. H., Huh Y. M., Jun Y. W., Seo J. W., Jang J. T., Song H. T., Kim S., Cho E. J., Yoon H. G., Suh J. S., and Cheon J. (2007) Artificially engineered magnetic nanoparticles for ultra-sensitive molecular imaging. *Nat Med* **13**, 95–99.

30. Maeda M., Itoh S., Kimura H., Iwasaki T., Hayashi N., Yamamoto K., Ishii Y., and Kubota T. (1993) Tumor vascularity in the brain: evaluation with dynamic susceptibility-contrast MR imaging. *Radiology* **189**, 233–238, 1993 Oct.

31. Magin R. L., Bacic G., Niesman M. R., Alameda J. C., Jr., Wright S. M., and Swartz H. M. (1991) Dextran magnetite as a liver contrast agent. *Magn Reson Med* **20**, 1–16.

32. Mandeville J. B., Moore J., Chesler D. A., Garrido L., Weissleder R., and Weisskoff R. M. (1997) Dynamic liver imaging with iron oxide agents: effects of size and biodistribution on contrast. *Magn Reson Med* **37**, 885–890.

33. Mayo-Smith W. W., Saini S., Slater G., Kaufman J. A., Sharma P., and Hahn P. F. (1996) MR contrast material for vascular enhancement: value of superparamagnetic iron oxide [see comments]. *Am J Roentgenol* **166**, 73–77.

34. Medarova Z., Pham W., Kim Y., Dai G., and Moore A. (2006) *In vivo* imaging of tumor response to therapy using a dual-modality imaging strategy. *Int J Cancer* **118**, 2796–2802.

35. Medarova Z., Rashkovetsky L., Pantazopoulos P., and Moore A. (2009) Multiparametric monitoring of tumor response to chemotherapy by noninvasive imaging. *Cancer Res* **69**, 1182–1189.

36. Misselwitz B. (2006) MR contrast agents in lymph node imaging. *Eur J Radiol* **58**, 375–382.

37. Moore A., Basilion J. P., Chiocca E. A., and Weissleder R. (1998) Measuring transferrin receptor gene expression by NMR imaging. *Biochim Biophys Acta* **1402**, 239–249.

38. Moore A., Josephson L., Bhorade R. M., Basilion J. P., and Weissleder R. (2001) Human transferrin receptor gene as a marker gene for MR imaging. *Radiology* **221**, 244–250.

39. Moore A., Marecos E., Simonova M., Weissleder R., and Bogdanov A. J. (1998) Novel gliosarcoma cell line expressing green fluorescent protein: a model for quantitative assessment of angiogenesis. *Microvasc Res* **56**, 145–153.

40. Moore A., Weissleder R., and Bogdanov A., Jr. (1997) Uptake of dextran-coated monocrystalline iron oxides in tumor cells and macrophages. *J Magn Reson Imaging* **7**, 1140–1145.

41. Natarajan A., Xiong C. Y., Gruettner C., DeNardo G. L., and DeNardo S. J. (2008) Development of multivalent radioimmunonanoparticles for cancer imaging and therapy. *Cancer Biother Radiopharm* **23**, 82–91.

42. Neuwelt E. A., Hamilton B. E., Varallyay C. G., Rooney W. R., Edelman R. D., Jacobs P. M., and Watnick S. G. (2009) Ultrasmall superparamagnetic iron oxides (USPIOs): a future alternative magnetic resonance (MR) contrast agent for patients at risk for nephrogenic systemic fibrosis (NSF)? *Kidney Int* **75**, 465–474.

43. Oh P., Borgstrom P., Witkiewicz H., Li Y., Borgstrom B. J., Chrastina A., Iwata K., Zinn K. R., Baldwin R., Testa J. E., and Schnitzer J. E. (2007) Live dynamic imaging of caveolae pumping targeted antibody rapidly and specifically across endothelium in the lung. *Nat Biotechnol* **25**, 327–337.

44. Papisov M. I., Bogdanov A., Schaffer B., Nossiff N., Shen T., Weissleder R., and Brady T. J. (1993) Colloidal magnetic-resonance contrast agents: effect of particle surface on biodistribution. *J Magn Magn Mater* **122**, 383–386.

45. Paul K. G., Frigo T. B., Groman J. Y., and Groman E. V. (2004) Synthesis of ultrasmall superparamagnetic iron oxides using reduced polysaccharides. *Bioconjug Chem* **15**, 394–401.

46. Reimer P., Bader A., and Weissleder R. (1998) Preclinical assessment of hepatocyte-targeted MR contrast agents in stable human liver cell cultures. *J Magn Reson Imaging* **8**, 687–689.

47. Reimer P., Kwong K. K., Weisskoff R., Cohen M. S., Brady T. J., and Weissleder R. (1992) Dynamic signal intensity changes in liver with superparamagnetic MR contrast agents. *J Magn Reson Imag* **2**, 177–181.

48. Reimer P., Weissleder R., Lee A. S., Wittenberg J., and Brady T. J. (1990) Receptor imaging: application to MR imaging of liver cancer. *Radiology* **177**, 729–734.

49. Reimer P., Weissleder R., Shen T., Knoefel W. T., and Brady T. J. (1994) Pancreatic

receptors: initial feasibility studies with a targeted contrast agent for MR imaging. *Radiology* **193**, 527–531.

50. Remsen L. G., McCormick C. I., Roman-Goldstein S., Nilaver G., Weissleder R., Bogdanov A., Hellstrom I., Kroll R. A., and Neuwelt E. A. (1996) MR of carcinoma-specific monoclonal antibody conjugated to monocrystalline iron oxide nanoparticles: the potential for noninvasive diagnosis. *Am J Neuroradiol* **17**, 411–418.

51. Richardson D. R. and Ponka P. (1997) The molecular mechanisms of the metabolism and transport of iron in normal and neoplastic cells. *Biochim Biophys Acta* **1331**, 1–40.

52. Rosen B. R. and Brady T. J. (1993) Future uses of MR imaging agents. *J Comput Assist Tomogr* **17**(Suppl 1), S36-S42.

53. Schaffer B. K., Linker C., Papisov M., Tsai E., Nossiff N., Shibata T., Bogdanov A., Jr., Brady T. J., and Weissleder R. (1993) MION-ASF: biokinetics of an MR receptor agent. *Magn Reson Imaging* **11**, 411–417.

54. 54. Schellenberger E. A., Bogdanov A. J., Hogemann D., Tait J., Weissleder R., and Josephson L. (2002) Annexin V-CLIO : a nanoparticle for detecting apoptosis by MRI. *Mol Imaging* **1**, 102–107.

55. 55. Schmitz S. A., Albrecht T., and Wolf K. J. (1999) MR angiography with superparamagnetic iron oxide: feasibility study. *Radiology* **213**, 603–607.

56. Shen T., Weissleder R., Papisov M., Bogdanov A., Jr., and Brady T. J. (1993) Monocrystalline iron oxide nanocompounds (MION): physicochemical properties. *Magn Reson Med* **29**, 599–604.

57. Shen T. T., Bogdanov A., Jr., Bogdanova A., Poss K., Brady T. J., and Weissleder R. (1996) Magnetically labeled secretin retains receptor affinity to pancreas acinar cells. *Bioconjug Chem* **7**, 311–316.

58. Spanoghe M., Lanens D., Dommisse R., Van der Linden A., and Alderweireldt F. (1992) Proton relaxation enhancement by means of serum albumin and poly-L-lysine labeled with DTPA-Gd3+: relaxivities as a function of molecular weight and conjugation efficiency. *Magn Reson Imaging* **10**, 913–917.

59. van Rooijen N., Sanders A., and van den Berg T. K. (1996) Apoptosis of macrophages induced by liposome-mediated intracellular delivery of clodronate and propamidine. *J Immunol Methods* **193**, 93–99.

60. Vassallo P., Matei C., Heston W. D., McLachlan S. J., Koutcher J. A., and Castellino R. A. (1994) AMI-227-enhanced MR lymphography: usefulness for differentiating reactive from tumor-bearing lymph nodes. *Radiology* **193**, 501–506.

61. Veiseh O., Sun C., Fang C., Bhattarai N., Gunn J., Kievit F., Du K., Pullar B., Lee D., Ellenbogen R. G., Olson J., and Zhang M. (2009) Specific targeting of brain tumors with an optical/magnetic resonance imaging nanoprobe across the blood-brain barrier. *Cancer Res* **69**, 6200–6207.

62. Wang Y. X., Hussain S. M., and Krestin G. P. (2001) Superparamagnetic iron oxide contrast agents: physicochemical characteristics and applications in MR imaging. *Eur Radiol* **11**, 2319–2331.

63. Weinmann H. J., Ebert W., Misselwitz B., and Schmitt-Willich H. (2003) Tissue-specific MR contrast agents. *Eur J Radiol* **46**, 33–44.

64. Weissleder R., Elizondo G., Wittenberg J., Lee A., Josephson L., and Brady T. (1990) Ultrasmall superparamagnetic iron oxide: an intravenous contrast agent for assessing lymph nodes with MR imaging. *Radiology* **175**, 414–498.

65. Weissleder R., Heautot J. F., Schaffer B. K., Nossiff N., Papisov M. I., Bogdanov A., Jr., and Brady T. J. (1994) MR lymphography: study of a high-efficiency lymphotrophic agent. *Radiology* **191**, 225–230.

66. Weissleder R., Stark D. D., Engelstad B. L., Bacon B. R., Compton C. C., White D. L., Jacobs P., and Lewis J. (1989) Superparamagnetic iron oxide: pharmacokinetics and toxicity. *Am J Roentgenol* **152**, 167–173.

67. Wiener E. C., Brechbiel M. W., Brothers H., Magin R. L., Gansow O. A., Tomalia D. A.,and Lauterbur P. C. (1994) Dendrimer-based metal chelates: a new class of magnetic resonance imaging contrast agents. *Magn Reson Med* **31**, 1–8.

68. Wunderbaldinger P., Josephson L., and Weissleder R. (2002) Tat peptide directs enhanced clearance and hepatic permeability of magnetic nanoparticles. *Bioconjug Chem* **13**, 264–268.

69. Yang L., Mao H., Cao Z., Wang Y. A., Peng X., Wang X., Sajja H. K., Wang L., Duan H., Ni C., Staley C. A., Wood W. C., Gao X., and Nie S. (2009) Molecular imaging of pancreatic cancer in an animal model using targeted multifunctional nanoparticles. *Gastroenterology* **136**, 1514–1525, e1512.

70. Yang L., Peng X. H., Wang Y. A., Wang X., Cao Z., Ni C., Karna P., Zhang X., Wood W. C., Gao X., Nie S., and Mao H. (2009) Receptor-targeted nanoparticles for in vivo imaging of breast cancer. *Clin Cancer Res* **15**, 4722–4732.

71. Zhao M., Beauregard D. A., Loizou L., Davletov B., and Brindle K. M. (2001) Non-invasive detection of apoptosis using magnetic resonance imaging and a targeted contrast agent. *Nat Med* **7**, 1241–1244.

Chapter 3

IN VIVO IMAGING OF IMMUNOTHERAPY USING NANOPARTICLES

Christopher M. Long[a,b,d] and Jeff W. M. Bulte[a–d]

[a] *Russell H. Morgan Department of Radiology and Radiological Science, Division of MR Research*
[b] *Department of Biomedical Engineering,*
[c] *Department of Chemical & Biomolecular Engineering, and*
[d] *Cellular Imaging Section and Vascular Biology Program, Institute for Cell Engineering, The Johns Hopkins University School of Medicine, Baltimore, MD 21205, USA*
jwmbulte@mri.jhu.edu

3.1 INTRODUCTION

Nanotechnology is proving to be an invaluable resource in the multidisciplinary biomedical field and has become a mainstream topic of interest for many researchers. Nanotechnology can be utilized for broad applications across all divisions of biological sciences. One of the best examples is the blossoming field of molecular imaging, which utilizes various imaging systems, including fluorescent, bioluminescent, radionuclide, and magnetic resonance imaging, and has provided fundamental improvements in our understanding of multiple *in vivo* molecular networks. One of those molecular networks currently under intense study is our immune system. Advances in optics and light microscopy have allowed us to obtain a much more detailed picture of the complex interactions that occur between individual cells of the immune system. However, most of the past studies were limited to *ex vivo* and *in vitro* models. What was missing from all of those past studies was a dynamic, high-resolution window into the complex interaction of immune cells in their natural physiological context. Recent advances in molecular and cellular imaging have allowed scientists to begin to study the immune system in a serial manner without perturbing the physiological environment [47, 48, 79]. One of the fields beginning to take advantage of these advances is the field of immunotherapy.

Nanoimaging
Edited by Beth Goins and William Phillips

www.panstanford.com

Most immunotherapies are based on two main strategies, active and passive. Active immunotherapy relies on the activation of the patient's immune system, as is the case in traditional vaccine development. Vaccines are currently used for disease infection, but in the case of cancer, vaccines in development are being designed to specifically attack tumor cells already growing within a patient's body. Passive immunotherapy, on the other hand, bypasses the patient's innate immune system and relies on *in vitro* stimulated immune cells or immune components to carry the burden as effectors. *Ex vivo* expansion of lymphocytes or antibody production are utilized, and the products are administered to the patient in order to directly take effect.

The goal of immunotherapy is to harness the powers of the immune system to either attack foreign antigens, as is the case for tumor immunotherapy, or inhibit the response to a foreign antigen, which would be beneficial in an organ transplant setting. Techniques such as vaccination, adoptive transfer of T cells, and administration of monoclonal antibodies and other immune-mediating small molecules have all been used in an attempt to control immune responses to known antigens. In each of these cases, the molecular and cellular events responsible for an immune response have not been completely defined, and only recently have correlative studies been used to help understand the mechanisms and extent of immune activation following each technique.

Real-time monitoring of immune responses to immunotherapy would be incredibly beneficial but has proven to be quite difficult. Until recently, none of the cellular activities could be monitored *in vivo* and serial studies within individuals were not possible [33, 55, 95]. Advances within the field of molecular imaging have pushed past some of these boundaries and the use of novel nanoparticles has created multiple modalities with which a basic scientist can now attempt to monitor the efficacy of an experimental immunotherapy. The future holds the promise of not only monitoring the trafficking of immune cells throughout the body but also quantifying the molecular activity and viability of the cells involved in the immune response [50, 51].

In this chapter, we will review the role nanoparticles play in aiding the study of cellular and molecular function following immunotherapy. Each of the main imaging modalities currently being used for cellular and molecular imaging will be discussed with a heavy emphasis on magnetic resonance imaging (MRI) as this is our lab's main area of interest. The practice of immunotherapy is currently far from being optimized, and hopefully with the aid of novel nanoparticles and breakthrough imaging techniques the field will be advanced to the point of standard of care for a variety of ailments.

3.2 THE ROLE OF MRI IN MONITORING THE EFFICACY OF IMMUNOTHERAPY

3.2.1 MRI Technology

MRI is a technique that relies upon unpaired nuclear spins, called magnetic dipoles, which align themselves when placed into a magnetic field. It just so happens that one of the most abundant elements in our body, hydrogen, has such a dipole, and when placed within the strong magnetic field present in an MRI scanner, a small portion of the hydrogen atoms will align. An external radiofrequency coil can be used to create a temporary pulse of energy that will change the alignment of the spins. This momentary perturbation will be righted as the magnetic dipoles return to their baseline orientation, which is detected as an electromagnetic flux or current.

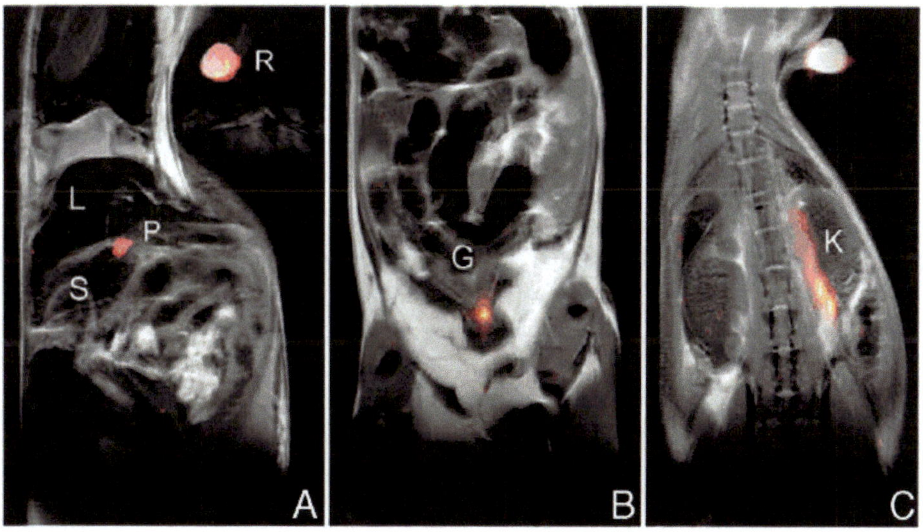

Figure 3.1 In vivo MRI showing transferred T cells labeled with ^{19}F perfluoropolyether nanoemulsions homing to the pancreas. Panels a–c are ^{19}F pseudo-color "hot spot" images superimposed on the ^{1}H anatomical images through the mouse torso. (a) Image of ^{19}F-labeled, in vitro activated diabetogenic T cells (5×10^6) transferred intraperitoneally into an NOD SCID mouse 48 hours earlier. The image shows specific T cells (pseudo-color) homing to the pancreas (P). The lungs, spleen (S), and liver (L) are labeled. (b) A negative control image of an NOD SCID mouse that received cell-free ^{19}F nanoparticles in PBS at an equivalent ^{19}F dose of 1×10^7 labeled T cells. For this control, ^{19}F signal (pseudo-color) can be detected only near the gut (G). (c) A negative control image of a NOD SCID mouse that received activated, labeled MHC-mismatched T cells. No signal is present in or around the pancreas, liver, or spleen, but cells (pseudo-color) can be seen at a site near the kidney (K). Reproduced, with permission, from Ref. [78].

Images in MRI are obtained using the varying rates at which the dipoles within the field of view relax to their baseline orientation. The varying physiochemical environments (i.e., tissue water concentration and the chemically bound state of the protons) will play a role in the different relaxation times present within the field of view and contribute to the image contrast. By varying the times of dipole excitation and recording, one can create images with different types of contrast. Two relaxation time constants are most commonly used when describing the various images obtained in MR, namely, T_1 and T_2. T_1 characterizes the relaxation of the nuclear spin to its longitudinal equilibrium, whereas T_2 measures the loss of coherency among adjacent nuclear spins. Inevitably there will be imperfections within the magnetic field, and these inhomogeneities act to shorten the natural T_2 of a tissue, thereby leading to another time constant, T_2^*. Current contrast agents act on one or more of these time constants in order to produce an identifiable signal within the cells of interest.

MRI has two unique advantages over techniques using radionuclides or optical probes: improved spatial resolution and the ability to simultaneously monitor molecular and anatomical imaging. The spatial resolution of MRI is on the order of micrometers as opposed to millimeters for most radionuclide or *in vivo* optical studies. This added resolution is critical in elucidating the interaction between individual immune cells during evaluation of immunotherapies. The ability to obtain both functional and anatomical information in one scan saves time as well as improves upon the accuracy of the study since there is no need to rely on image registration between various modalities.

Two drawbacks for the use of MRI in evaluating immunotherapies is the length of each study and the expertise or supervision needed to conduct a study. In order to achieve in-plane resolutions on the order of 100 μm, most studies require at least 15–20 minutes for each scan. This time constraint combined with the ability to properly scan only one mouse at a time significantly reduces the number of animals that can be analyzed in any given experimental cycle. The complexity and cost of a high-field animal MR scanner limits the numbers of scientists who are capable of conducting experiments with this technology. In addition, a significant amount of training is needed before one can safely and competently begin using the machine.

3.2.2 MR Contrast Agents

A variety of contrast agents are utilized to manipulate the signal received during an MR scan. Most agents target one or more of the time constants mentioned above in order to help distinguish labeled cells from surrounding tissue. The goal in developing nanoparticles for cell labeling is to maintain high biocompatibility and low toxicity while maximizing the sensitivity of the agent. Additionally, the contrast agent should be highly specific to the cells of interest and not contribute to background signal. Finally, since dividing immune cells is the focus of many immunotherapies, the contrast agent should not be significantly diluted upon cell division in order to maintain long-term cell tracking [31]. Of course no current agent meets all of these criteria perfectly, but there are a variety of agents that have proven themselves quite useful for the time being.

3.2.2.1 Superparamagnetic nanoparticles

The first such agents are superparamagnetic particles of iron oxide (SPIO). These particles have been used extensively in cell imaging and cell tracking studies because of a very strong effect on T_2^* as well as the T_2 time constant [17, 49, 72-73, 89, 93]. Iron oxide particles come in a variety of different sizes ranging from the ultrasmall superparamagnetic particles of iron oxide (USPIO) nanoparticles (~35 nm diameter) to the regular SPIO particles (50–200 nm diameter) and even busting out of the nanoparticle category and into the micrometer size (MPIO). Each one of these particles is used in conjunction with a T_2- or T_2^* -sensitive pulse sequence where the agents create local hypointensities (black signals) within the MR image. Attempts to increase solubility while maintaining biological inertness have led to coating the iron oxide particles with varying substrates such as dextran, carboxydextran, and polystyrene [17, 73, 93].

The success that has been obtained with MRI cell tracking in animals has led to its introduction in the clinic for melanoma patients [24], brain [99] and spinal cord [19] injury patients, and type 1 diabetic patients [82]. It should be noted, however, that a drawback of using an SPIO contrast agent is that it is not easily quantifiable in contrast to fluorinated agents (see below) [3, 78]. The contrast obtained through iron oxides is not a direct measurement of the agent but an indirect picture of how the iron affects the surrounding magnetic field. This limitation may prove to be a large one as scanning resolutions make their way down to single cell resolutions, which is, at present, far from feasible in the clinical setting. Another concern with iron oxides is the fact that negative contrast, loss of signal, can sometimes make it difficult to positively identify cell locations. In some cases, *a priori* knowledge of cell biodistribution is necessary in order to evaluate cellular trafficking. However, clinical MRI cell tracking has now been applied in several hospital settings and overall appears to have a promising future [16].

3.2.2.2 Paramagnetic MR contrast agents

Paramagnetic agents primarily play a role in shortening the T_1 relaxation time and include metals such as manganese and gadolinium. In contrast to the SPIO agents, which produce a dark signal in T_2^* images wherever present, the paramagnetic agents actually produce a bright signal. This positive contrast effect is quite useful and would be ideal if it were not for the fact that paramagnetic metals, with their unpaired electrons, are naturally toxic to cells and have only been used in chelated forms such as gadolinium diethylenetriamine pentaacetic acid (Gd-DTPA) or gadolinium tetraazacyclododecanetetraacetic acid (Gd-DOTA). Additionally, the sensitivity of gadolinium-based contrast agents is substantially lower than that of iron oxide nanoparticles [71]. In order to attain sufficient contrast for the monitoring of labeled cells, it is necessary to entrap on the order of 10^7–10^8 Gd complexes per cell [20]. Some groups have overcome this limitation in sensitivity by creating nanoparticles loaded with gadolinium that may be taken up by cells with natural phagocytic activity [4, 27, 36, 67]. The majority of these studies have used liposomes, which will dissolve upon entering the cellular compartment as sensitivity is enhanced when the gadolinium has more access to free water.

3.2.2.3 Fluorinated nanoparticles

Although using contrast agents that affect either T_1 or T_2^* relaxation times of surrounding tissues is a very sensitive method of creating contrast, it can, at times, be very difficult to discriminate between labeled cells and the surrounding tissue. Additionally, quantification of cell numbers in regions of interest (ROIs) can be challenging [78]. To combat these drawbacks, some labs have begun to use fluorine-19 (^{19}F) as a contrast agent in nanoparticle formulations [3, 69, 78] (Fig. 3.1). ^{19}F is a completely distinct element from hydrogen and can be imaged and quantified separately [15]. This fact, along with the fact that fluorine is only found in trace amounts throughout the body, make it an ideal contrast agent for localizing labeled cells. However, the bonus of being atomically distinct from hydrogen does come at a cost. The fluorine agents act similarly to nuclear and bioluminescent probes in the sense that the functional data must then by overlain upon similarly obtained anatomical data. Additionally, the sensitivity of fluorine contrast agents is still a strong limiting factor in their widespread use. As only a minute fraction of dipoles align with even some of our highest field strength magnets, it becomes difficult to draw signal from the limited amount of fluorine present within each cell or group of cells.

3.3 APPLICATIONS OF MAGNETIC RESONANCE IMAGING

As mentioned in the introduction of this chapter, there are two different approaches to effectively manipulating the immune response: active and passive immunotherapies. MRI has been used to study both approaches, but the first MR studies relied upon the naturally phagocytic cell types of the body and, therefore, began by studying the active arm of immunotherapies.

3.3.1 Dendritic Cell Imaging

The foundation for studying phagocytic cells, such as dendritic cells, stemmed from groups who were studying Kupffer cells in the liver by simply injecting nanoparticle contrast agents, both positive and negative agents, and analyzing the uptake within Kupffer cells in the liver [29, 64]. Groups then extended this work to study another phagocytic cell type, the dendritic cell (DC). DCs are the body's most efficient antigen-presenting cell, and *ex vivo* manipulated DCs are currently being explored in the clinic for use in active immunotherapies [9, 30, 58]. In order for these cells to do their job, they must migrate either from the site of antigen delivery or from the site in which they were injected via afferent lymphatics to draining lymph nodes. In these lymph nodes, the DCs present captured antigens to T cells and B cells, which then proliferate and act as the effector cells of the immune system. The migratory pattern of DCs presents a perfect intermediate in which *in vivo* cell tracking could be used to study the effectiveness of active immunotherapies. Not only can antigen uptake be monitored but the route of administration, optimal dose, and frequency for *ex vivo* labeled cells can be interrogated as a parameter in need of optimization. Most clinical studies utilizing *ex vivo* generated DCs have injected the cells through a variety of routes, including intravenous, subcutaneous, intradermal, intranodal, or intralymphatic. However, the most effective administration pathway for DC vaccines has yet to be determined [11]. Multiple groups have begun to study this issue using MRI as their imaging modality of choice. Ahrens *et al.* begun the work in this area by proving that DCs could be efficiently labeled using SPIO nanoparticles and serially tracked using *in vivo* MRI [2]. These initial studies did not follow the DCs as they migrated throughout the lymphatics but, rather, imaged them in the quadriceps of mice post injection. Baumjohann *et al.* then took this study one step further, realizing that if these cells could be visualized *in vivo* using MRI, one may be able to monitor the migration of SPIO-labeled DCs until they reached their terminal location within a draining lymph node [10] (Fig. 3.2). This study utilized subcutaneous footpad injections of SPIO-labeled DCs and serially tracked the accumulation of these labeled DCs in the draining popliteal lymph nodes. Ahrens's group has also attempted similar studies using their novel perfluoropolyether (PFPE) nanoparticle composition, which relies upon hot spot fluorine imaging [3]. Each of these studies provided confidence in MRI's ability to aid in the intelligent design of DC vaccines.

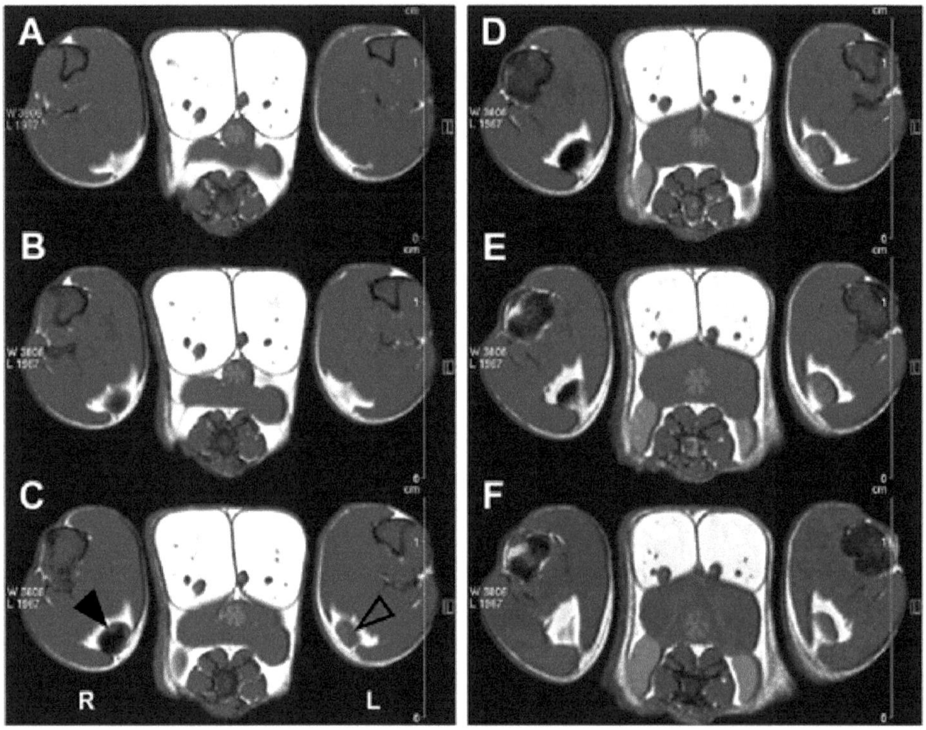

Figure 3.2 SPIO-labeled DCs home to the draining lymph node. BM-DCs (BALB/c) were labeled with SPIO or remained unlabeled, and 1 × 10^7 mature cells were injected into the right (R) or left (L) footpad of a syngeneic mouse, respectively, followed by a second injection of cells 24 hours later. Another 24 hours later the mouse was killed, and high-resolution MRI of popliteal lymph node regions was performed. A series of six 0.5 mm thick axial slices ranging from proximal to distal is shown in A–F. The right lymph node (R) shows strong signal reduction (filled arrow) where SPIO-DCs were injected into the footpad. In contrast, no signal reduction (open arrow) can be observed in the left lymph node (L) where unlabeled DCs were injected. Reproduced, with permission, from Ref. [10].

The last few studies discussed, however, were entirely conducted in preclinical settings. MRI cell tracking has not been limited to preclinical studies, as there have been few clinical studies to date [16]. The first clinical study using this technology was to evaluate the administration of DCs for use as a cancer vaccine in melanoma patients [24] (Fig. 3.3). Dextran-coated SPIO nanoparticles were used in conjunction with indium-111 (^{111}In) oxine in order to determine the accuracy of ultrasound-guided intranodal injections. MR imaging, with its added spatial resolution, was found to be more sensitive than gamma scintigraphy in detecting lymph nodes containing DCs. Most importantly, the study showed that even with ultrasound guidance, four out of the eight patients had been mis-injected with their cellular therapy. This study, as well as its follow-up, proved that MR imaging is ideally suited to track *ex vivo* labeled DC therapies [85].

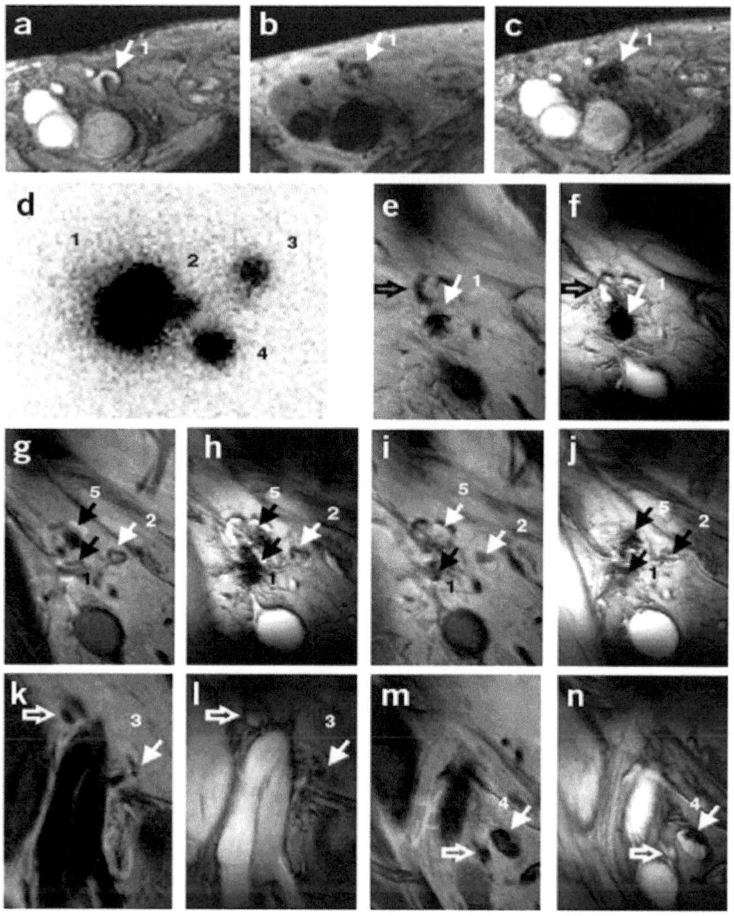

Figure 3.3 Scintigraphy and MRI of stage III melanoma patients following intranodal injection of autologous DCs, primed with synthetic melanoma antigens and labeled with Endorem™ (Feridex®) and ^{111}In-oxine. (a) Gradient-echo (GE) MR image before vaccination, showing a right inguinal lymph node with a hyperintense signal area (1). (b) Spin-echo (SE) MR image obtained from the same lymph node after vaccination. (c) GE MR images after vaccination in the same position as (b) show a decreased signal intensity of lymph node 1. (d–n) Monitoring of *in vivo* migration of SPIO- and ^{111}In-labeled DCs with MRI and scintigraphy after injection in a right inguinal lymph node. (d) Scintigraphy two days after vaccination shows migration of DCs from the injection lymph node (1) to three nearby lymph nodes (2–4). (e–n) Five image pairs of a GE and SE image two days after vaccination, showing migration of DCs from the injection lymph node 1 (e and f) to four nearby lymph nodes (g–n). Open arrows indicate lymph nodes that do not contain Endorem-labeled DCs. On the SE images, these nodes are dark gray; on the GE images, they are white. Closed arrows indicate lymph nodes that are positive for Endorem-labeled cells on the GE images and that have a decreased signal intensity compared with the SE images. The lymph node that was identified by the scintigraphy as the injection lymph node (lymph node 1 in d) actually consisted of two distinct lymph nodes, as evidenced by MRI (lymph nodes 1 and 5). Overall, MRI demonstrated a better accuracy for determining the number and location of lymph nodes that contained injected or migrating labeled DCs. Reproduced, with permission, from Ref. [24].

In our own studies, we have chosen to take a different approach to imaging DCs. Rather than rely on DCs that have been removed from their natural environment and labeled *in vitro*, we have chosen to label the vaccine itself with iron oxide nanoparticles and utilize endogenous DCs [48]. This approach, called magnetovaccination, relies upon the phagocytic nature of immature DCs and their tendency to sample their surroundings, especially when stimulated by cytokines to do so. This *in situ* targeting of DCs is seen by many to be superior to *ex vivo* labeled DCs as it exploits the inherent migratory capacity of DCs *in vivo* [28, 32].

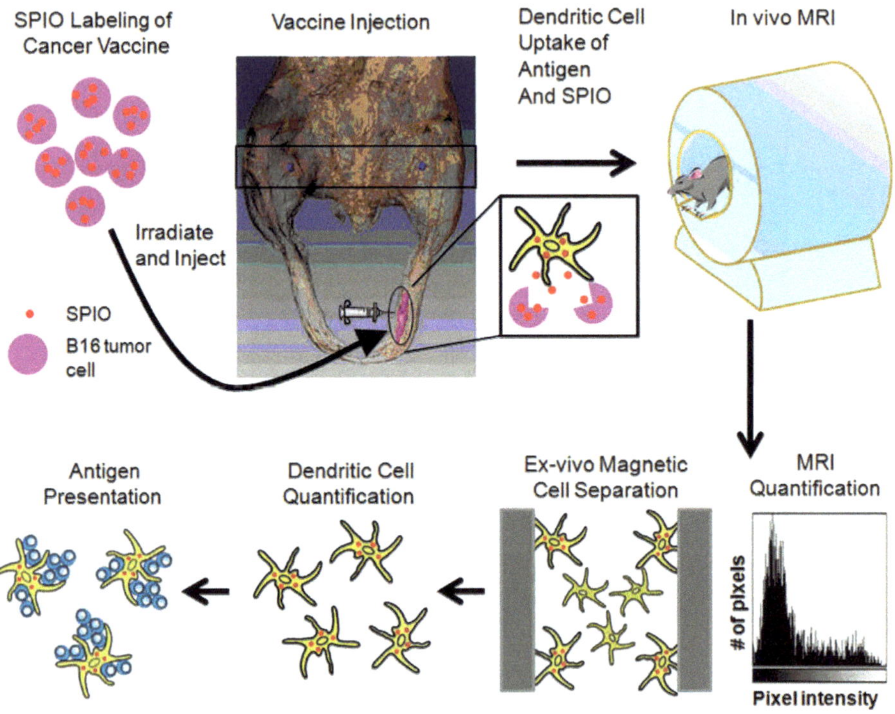

Figure 3.4 Magnetovaccination technique. (Clockwise from top left) B16 melanoma cells are incubated with SPIO overnight and washed, irradiated, and combined with irradiated B78H1-GM-CSF cells. Vaccine cells are then injected into the footpads of C57BL/6 mice. Dendritic cells survey the site of the vaccination and are labeled *in vivo* by capture of SPIO associated with apoptotic material released from labeled tumor cells, and traffic to popliteal lymph nodes where axial images (black rectangle) are later obtained by MRI (top right). After *in vivo* imaging, the iron signal is quantified and the lymph nodes are resected and separated magnetically following single-cell suspension. Magnetically separated cells are then counted and their phenotype determined by FACS. The magnet-positive cells are then co-incubated with antigen-specific T cells and proliferation assayed by [^{3}H]-thymidine incorporation. Reproduced, with permission, from Ref. [48].

A granulocyte-macrophage colony-stimulating factor (GM-CSF) tumor cell-based vaccine (GVAX) is a cancer vaccine currently under clinical investigation at our institution. This vaccine is produced from irradiated tumor cells that have been transduced to produce GM-CSF. After injection, the cells begin to attract monocytes that aid in their maturation into DCs. Over time, the irradiated cells begin to die and release their cellular contents. This process usually provides antigen, which is carried by DCs through the lymphatics to draining lymph nodes. Our magnetovaccine simply piggybacks on this process to load DCs with SPIO nanoparticles as well as antigen, which allows us to visualize the trafficking and accumulation of DCs within lymph nodes.

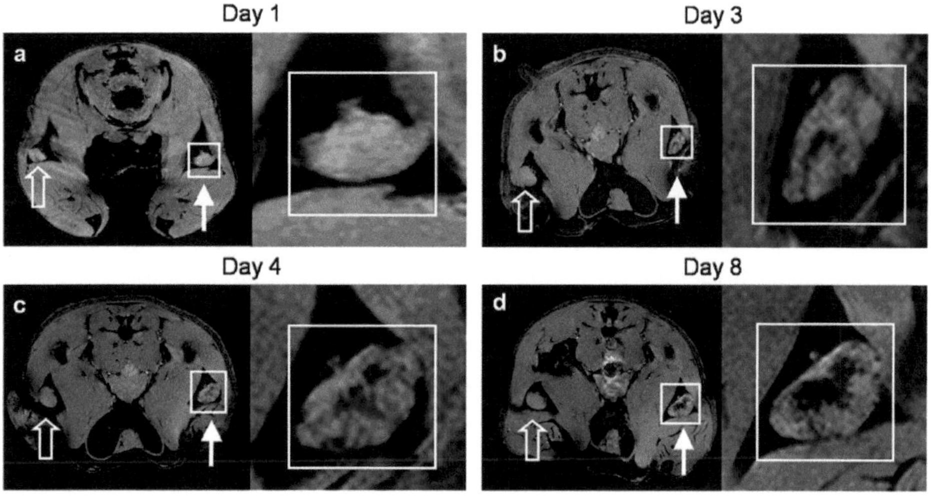

Figure 3.5 *In vivo* MRI DCs that have taken up SPIO *in vivo* after intradermal injection of GM-CSF tumor cell vaccine into footpads of mice. Right panels for each day represent magnifications of corresponding insets in a, b, c, and d, respectively. Open arrows indicate draining popliteal lymph nodes from footpads receiving unlabeled GM-CSF vaccines. Closed arrows indicate draining popliteal lymph nodes from footpads receiving SPIO-labeled GM-CSF vaccines. On multi-gradient-echo images, SPIO-containing lymph nodes have decreased signal intensity. (a) Day 1: Neither popliteal lymph nodes show any evidence of hypo-intensities. (b) Day 3: Decreased signal intensity apparent in the lymph node corresponding to the popliteal lymph node of SPIO-labeled vaccine. (c and d). On day 4 and day 8, respectively, the signal decrease persists and actually increases in popliteal lymph nodes. Reproduced, with permission, from Ref. [48].

In our studies, mice were injected with a magnetovaccine subcutaneously in the footpad (Figs. 3.4, 3.5). The draining popliteal lymph nodes were then serially monitored by MRI in order to evaluate the kinetics of DC trafficking. After the mice had been imaged, they were sacrificed and their popliteal lymph nodes removed. Single-cell suspensions were then run over a magnetic column in order to separate out the cells that were labeled with SPIO nanoparticles. After this process was complete, flow cytometry was run in order to evaluate the cell populations that had stuck to the magnetic column. In this manner, we were able to quantify the DCs that had positively interacted with our tumor vaccine. After utilizing image thresholding to create a "black pixel" quantification, DC numbers for each lymph node were correlated with each lymph node image. A clear correlation was found between DC number and black pixel counts in each node. This process proved sensitive enough to evaluate the wide biological variability present following this vaccine. Additionally, immune adjuvants were evaluated, and again, *in vivo* MR imaging proved valuable in determining the efficacy of the immunotherapy.

Each of the methodologies described above represents one of the many steps needed in order to exploit non-invasive imaging techniques for the purpose of systematically evaluating the many variables present in creating an effective active immunotherapy. Whether the future lies in *ex vivo* antigen-pulsed DC vaccines or whole-cell vaccine adjuvants, noninvasive MR imaging will definitely be there to help evaluate the cutting-edge technology.

3.3.2 T Cells

Although active immunotherapies will no doubt be major players in the future, they are not alone in the field of immunotherapy. Passive immunotherapies, especially the adoptive transfer of T cells, are starting to play a larger role especially with the advances in molecular biology that allow the creation of transgenic T-cell receptors. Now that scientists can create T cells capable of targeting any given antigen *in vitro*, the next big question is, Are those cells trafficking to the proper location and targeting the correct antigens *in vivo*? *In vivo* MR imaging should be able to aid in answering this question, especially in the arena of tumor immunotherapy, where T cells should ideally home to the sites of tumors, proliferate, and eradicate the tumors.

For many years now, scientists have been isolating tumor-specific T cells from cancer patients, culturing them and re-injecting them in hopes of eradicating a patient's tumor burden [34, 90, 97]. The results have been highly variable, which has led some researchers to turn to MRI and utilize its high spatial resolution in order to provide a noninvasive method of tracking these adoptively transferred T cells. Although T cells are not naturally phagocytic, like macrophages and DCs, groups have found creative ways to label them with iron oxides and other contrast agents. Among the many techniques used, transfection seems to be one of the methods of choice [7-8, 18, 88]. However, many studies have also been conducted using HIV-derived Tat peptides that are attached to a cross-linked iron oxide (CLIO) [39-40, 43, 86]. Kircher *et al.* used these techniques of T-cell labeling and MR imaging in an attempt to display the utility of MRI cell tracking in evaluating passive immunotherapies. The group developed a quantitative approach of monitoring antigen-specific T-cell recruitment in tumors and then followed up their initial study with another describing possible mechanisms behind tumor immune escape [40, 86]. Although these studies and others using ^{19}F contrast agents have aided the study of passive immunotherapies, there still exists a daunting challenge in serially tracking these adoptively transferred cells as they ideally proliferate within tumors and lymphatic tissues [78].

3.4 THE ROLE OF OPTICAL IMAGING IN MONITORING IMMUNOTHERAPY EFFICACY

3.4.1 Optical Imaging Technology

Optical imaging (OI) techniques have already been developed for *in vitro* and *ex vivo* applications in molecular and cellular biology. The extension of this technology toward noninvasive *in vivo* imaging represents an exciting avenue of monitoring biological processes. OI typically employs a light source, such as a monochromatic laser, that excites a fluorophore, whose emission is then captured by a detector, most commonly a highly sensitive charged couple device. Fluorescence and bioluminescence imaging are the modalities most frequently employed for *in vivo* cell tracking, and both techniques can often be conducted independently on the same OI system [59]. Optical imaging modalities such as confocal microscopy, two-photon microscopy, and diffuse optical tomography are often used for *in vivo* imaging but are not capable of monitoring dynamic cell migration and so are not useful for *in vivo* cell-tracking studies [35, 79].

Limitations of optical imaging-based cell-tracking techniques include limited depth of penetration, limited quantification, and poor spatial resolution due to scatter [59]. Advances have been made in instrumentation and reconstruction algorithms in order to generate 3D imaging systems [92]. Ntziachristos *et al.* have even developed a tomographic imaging system called fluorescence-mediated tomography (FMT), which generates a 3D quantitative distribution map of near-infrared (NIR) fluorochromes in mouse models [60]. Fluorescent probes with emission wavelengths in the NIR spectra (700–1000 nm) have enhanced the feasibility of tracking cells in small animal models because of their ability to penetrate increased depths of tissue in relation to probes of longer wavelengths [59]. Each of these improvements, combined with the relatively low cost and utilization of multiple probes with different spectral characteristics, makes OI a viable imaging modality for monitoring immunotherapies *in vivo*.

3.4.2 Optical Imaging Contrast Agents

Desirable features for an optical imaging contrast agent do not differ all that much from those of an MR imaging agent, in the sense that one desires biocompatibility, tolerable toxicity profile, low dilution with cell division, limited transference to other cells, and a low background signal with a prominent contrast-to-noise ratio. Some characteristics unique to an ideal optical imaging agent are bright signal, which depends on the extinction coefficient and quantum yield of the agent, photostability, and pH insensitivity.

Fluorescent dyes make up a large class of agents used in optical imaging, with NIR cyanine dyes representing the most prominent group [46]. Many have been incorporated into nanoparticles for labeling or simply added to cell suspensions, which allow for binding to cell membranes. A relatively new and promising nanoparticle for use in optical imaging is the quantum dot (QD) [54]. These inorganic fluorophores comprise colloidal semiconductor cores (gold [Au], cadmium [Cd], silicon [S], selenium [Se], tellurium [Te], zinc [Zn]) surrounded by a chemical coating that allows for protein conjugation [41]. QDs have a narrow emission and broad excitation spectrum, which reduces autofluorescence. Their sensitivity can reach levels of single-molecule detection, and since their emission spectra are so narrow, multiple QDs can be visualized with a single light source [38, 76]. Additionally, QDs demonstrate reduced photobleaching and blinking, which may allow them to remain fluorescent in tissues *in vivo* for months [54]. Obviously, this last feature has its advantages but is also a concern as issues still remain about the cytotoxicity of these agents.

3.4.3 Applications of Optical Imaging

To date, most applications of optical imaging for cell tracking have involved either cells directly labeled with fluorophores or transduced to express fluorescent proteins [75]. As mentioned earlier, optical imaging methods such as confocal and two-photon microscopy are not typically considered for cell-tracking studies, because one must have *a priori* knowledge of where the cells will travel and are limited by depth of light penetration in tissue. In any case, various groups have used these imaging techniques in conjunction with QD nanoparticles in order to monitor DC accumulation in lymph nodes [70]. In particular, one study was able to show that QDs were not only useful as an imaging platform but could also be used to deliver antigen for priming an immune response [70].

Moore *et al.* have shown that optical imaging is a useful adjunct to MRI as studied for the role of CD8 T cells in the progression of diabetes in NOD mice. In these studies, T cells were labeled with an antigen-specific superparamagnetic iron oxide label (CLIO-NRP-V7 nanoparticle), and followed by MRI *in vivo*. There was progressive accumulation of T cells within the pancreas that was later correlated with OI studies conducted postmortem [56]. Our own lab has begun to use OI to fully characterize the cells involved in the immune response following our magnetovaccination technique. We have utilized QD nanoparticles in conjunction with our magnetovaccination in order to dual-label antigen-presenting DCs. With the help of flow cytometry and MRI we are beginning to work out both the kinetics and the morphology of the cells involved in the vaccine immune response. Our work, as well as other immune cell studies using fluorescently labeled cells, shows that optical imaging definitely has a role in cell tracking but may currently need the help of multimodal nanoparticles in order to fully find its role in cell tracking.

3.5 THE ROLE OF NUCLEAR IMAGING IN MONITORING IMMUNOTHERAPY EFFICACY

Molecular imaging strategies using positron emission tomography (PET) or single-photon emission computed tomography (SPECT) hold great promise for detecting *in vivo* molecular changes in a noninvasive fashion. The functional information gained may greatly facilitate advancements in the field. Not only is PET noninvasive, but this imaging modality also provides a quantitative and tomographic assay of biological processes [21].

3.5.1 Nuclear Imaging Technology

PET records high-energy γ rays emitted from within a subject. Molecules labeled with a positron-emitting isotope are monitored indirectly by measuring the two 511,000 eV γ rays that are produced at ~180° from each other following the annihilation of a nearby electron by the emitted positron. Detection of γ rays is achieved through scintigraphic instrumentation, which converts γ-ray energy into visible light. The coincidence detection of γ rays in PET defines both the line of response in space and the direction of flight. The knowledge of the direction of flight allows one to correct for signal attenuation of the emitted radiation. By doing so, quantitative information about tracer localization and concentration can be obtained. Clinical PET scanners have a spatial resolution between 3 and 5 mm, while commercially available micro-PET scanners can achieve resolutions of ~1.5 mm in small animals. Additionally, the sensitivity of PET is relatively high, i.e., in the range of 10^{-11}–10^{-12} mole/L, and is independent of the location or depth of the emission source.

Gamma-emitting isotopes can also be used for imaging and rely on gamma cameras that are rotated around a subject to create a tomographic image. This imaging technique is known as SPECT, where the direction of flight has to be determined by geometric collimation since emission of γ rays is isotropic. The use of collimators reduces the sensitivity of SPECT, which is typically at least a log order less sensitive than PET. However, the collimator design is always a compromise between spatial resolution and sensitivity, with micro-pinhole apertures achieving resolutions as low as 200 μm, an order of magnitude higher than that achieved with PET. SPECT also allows for multiple probes to be imaged simultaneously as long as each is labeled with a different radionuclide.

3.5.2 Nuclear Imaging Tracers

Positron-emitting isotopes frequently used include oxygen-15 (^{15}O), nitrogen-13 (^{13}N), carbon-11 (^{11}C), and fluorine-18 (^{18}F), with the latter frequently being used as a substitute for hydrogen. Other less commonly used positron emitters include oxygen-14 (^{14}O), copper-64 (^{64}Cu), copper-62 (^{62}Cu), iodine-124 (^{124}I), bromium-76 (^{76}Br), rubidium-82 (^{82}Rb), and gallium-68 (^{68}Ga). Most of these isotopes are produced in a cyclotron, but some may be produced using a generator. Many of the positron-emitting isotopes used have relatively short half-lives (e.g., ^{18}F has a half-life of 110 minutes). Common single-photon-emitting isotopes include technium-99m (^{99m}Tc), indium-111 (^{111}In), iodine-123 (^{123}I), and iodine-131 (^{131}I).

Since the mid-1980s, cell labeling with radionuclides has been based on the use of ^{99m}Tc-labeled hexamethyl-propylene amine oxime (^{99m}Tc-HMPAO) and ^{111}In-oxine, joined later by ^{18}F 2-fluoro-2-deoxy-D-glucose (^{18}F-FDG) in the early 1990s [5, 12, 66]. T cells incorporate significant amounts of ^{18}F-FDG and have been used to track these lymphocytes [14, 61]. However, most malignancies are also associated with increased glucose metabolic activity, which places a limitation on the utility of FDG for tracking T cells when used in tumor immunotherapy applications.

3.5.3 Applications of Nuclear Imaging

The burgeoning field of cancer immunotherapy and the limited number of PET probes, which can reliably differentiate between metabolically active tumor cells and rapidly dividing immune cells, underlines the need to develop new molecular imaging strategies [84]. The earliest attempts involved investigators directly labeling T cells *ex vivo* with radioactive probes and adoptively transferring those cells into the patient [1, 13, 14, 62]. This strategy precluded long-term monitoring due to apoptosis or proliferation of the labeled cells *in vivo,* which lead to detrimental levels of probe dilution. To overcome this limitation, adoptively transferred T cells have been monitored using radiolabeled antibodies specific for a cell surface marker on the T cells [52]. In this study, anti-Thy1.2 antibodies were labeled with ^{64}Cu and injected on days 4, 7, and 14 post T-cell transfer. The authors were able to use this technique to show that not all tumor-specific T cells trafficked to the tumor but that some of them preferentially homed to major lymphoid organs throughout the body.

Next, investigators attempted an indirect imaging route where immune cells were genetically engineered, *ex vivo*, with the herpes simplex virus (HSV) thymidine kinase (tk) reporter gene, reinfused into the host, and a PET probe specific for the reporter gene later used [25, 26, 42, 80]. This method has proven to be successful, but it still has several key issues that need to be addressed. The first is that of sensitivity. Dilution does not play a role with this method, but diffuse distribution of lymphocytes throughout the body means that PET must be able to detect relatively small numbers of cells and at a high enough resolution to determine accurate anatomic location. The second issue is one of specificity. The reporter genes used must have limited endogenous expression in order to provide an adequate signal-to-noise ratio. The last two issues are that of toxicity and immunogenicity. The reporter gene proteins must not induce cell death or lead to a host response against the transferred cells. As of this writing and to the best of our knowledge, only one clinical study has been performed with the HSV-tk reporter gene [96]. The case report described a glioma patient who received autologous cytotoxic T cells transfected with the HSV-tk, and PET imaging signal superimposed on anatomical MRI images and demonstrated that these cells had accumulated within the tumor (Fig. 3.6).

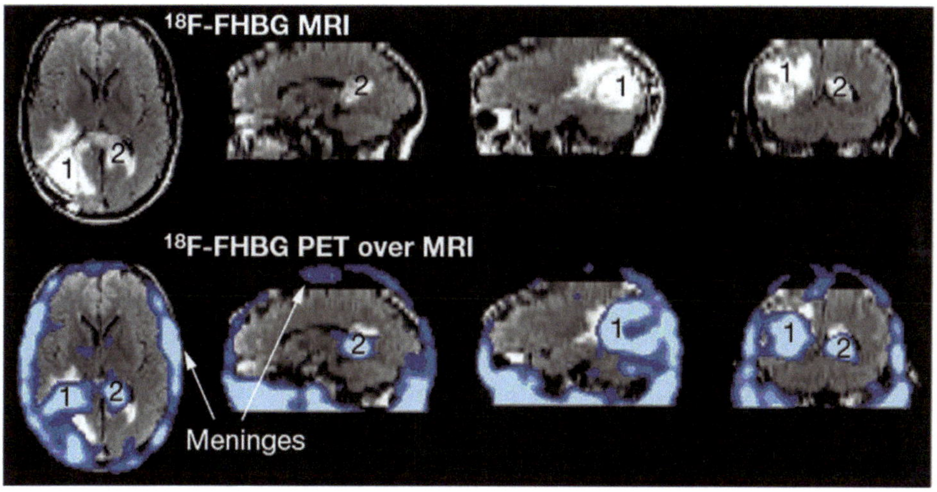

Figure 3.6 MRI and PET over MRI-superimposed brain images of a glioma patient who received infusions of autologous cytolytic T cells transfected with the HSV-tk reporter gene. Images were acquired ~2 hours after injection of the ^{18}F–FHBG substrate. A surgically resected tumor (1) was evident in the left corner of his brain and a new, nonresected tumor in the center (2), near the corpus callosum. The infused cells had localized at the site of tumor 1 and trafficked to tumor 2. Reproduced, with permission, from Ref. [96].

Radionuclide imaging has even been used to evaluate dendritic cell migration post DC vaccine [23, 65, 68, 83]. Ridolphi *et al.* compared the migration of mature DCs and immature DCs while also evaluating various injection sites in order to determine the optimal administration site for their clinical trial of an antitumor DC vaccine [68]. The DCs for their study were labeled with ^{99m}Tc-HMPAO or ^{111}In-oxine, and the draining lymph nodes were evaluated up to 72 hours after the injections. In another vaccine study for melanoma subjects, Quillien *et al.* also investigated administration routes for DCs by using ^{111}In-oxine-labeled DCs, which were injected either into lymphatic vessels intradermally or into lymph nodes directly [65]. In the most recent study of DC vaccines for melanoma patients, Trakatelli *et al.* used a variety of immune adjuvants to increase the immunostimulatory capacity of DCs and monitored their trafficking using ^{111}In-oxine [83].

Although the last group of studies discussed involved terminally differentiated cells aggregated in one location, T-cell studies do not necessarily benefit from the same kinetics and distribution. On the contrary, in an ideal vaccine situation, T cells would continue to divide and spread throughout the subject's body, thus leading to decreased cellular signal. Nanoparticles may help in this arena as the structure of nanoparticles allows for a broad range of radiolabeled chemistries to be used to attach various PET nuclides and multifunctional targeting moieties. They have the ability to deliver large numbers of imaging agents per each targeted molecular recognition event, which allows for high-sensitivity imaging even through *in vivo* labeling [94].

3.6 OUTLOOK

Molecular and cellular imaging will continue to play a key role in the future of immunotherapy development. These techniques can provide an in-depth understanding of the fundamental aspects of immunotherapy, which will, in turn, help to refine therapeutic strategies. However, among molecular imaging techniques, no single imaging modality stands out as the perfect player. No single modality is capable of providing all the necessary information needed to completely study an immune reaction. For instance, optical imaging is difficult to quantify and has severely limited tissue penetration, MRI has superb resolution but low sensitivity, and PET has very high sensitivity but poor resolution. The combination of multiple molecular imaging techniques for use in monitoring immunotherapies *in vivo* could, therefore, offer synergistic advantages over any modality alone. However, one cannot simply add two different imaging probes together and hope to achieve ideal results. Unless those two agents just happened to behave identically when added to cells and an *in vivo* environment, the information obtained would not be easily registered between modalities. Therefore, multimodal nanoparticle agents seem to be an obvious answer to the limitations imposed by each imaging modality.

The obvious benefit of combining imaging modalities has not gone unnoticed by researchers in the field of molecular imaging. Studies have already been conducted with nanoparticle contrast agents that can be detected with optical imaging and MRI [22, 77, 87]. These agents integrate the high sensitivity of optical imaging with the high spatial resolution and anatomical information provided by MRI. Fluorochromes are also being fused with radiotracers for use in nuclear imaging and optical imaging. These probes would offer improved depth penetration, 3D resolution, and quantification [44, 63]. Although MRI has great spatial resolution, the sensitivity is not nearly as high as that of radionuclide PET imaging, which has a weakness in spatial resolution. Multimodal imaging probes may overcome each of these individual shortcomings by providing complementary information about cellular and molecular activities, especially in the *in vivo* setting.

Most relevant to the theme of this chapter is the recent development of bifunctional nanoparticles used for dual-imaging *in vivo* studies. Combinations such as MR with fluorescence, radioisotope with fluorescence, or gold with MR have been used for cell-tracking purposes [6, 57, 74, 81, 91]. Bifunctional particles have even been used specifically for the purpose of tracking immune cells such as cytotoxic T cells as they traffic to tumors [40]. However, dual-imaging modalities may still prove to be insufficient as many investigators are looking to gain an insight into cellular metabolism, gene function, as well as *in vivo* cellular location. In order to monitor all of these activities, a sensitive and highly accurate imaging agent is necessary. Recent studies have begun to look into the use of triple and quadruple imaging probes that rely on the strengths of many different imaging modalities in the same agent [37, 43].

In addition to multimodal nanoparticles, the future of *in vivo* imaging of vaccines will rely upon the ability to track multiple cell types simultaneously. Optical imaging has had this ability for some time now, with fluorophores of different wavelengths being utilized to distinguish many different cell types. MRI has only recently gained this ability with the advent of chemical exchange saturation transfer (CEST) reporter genes, multispectral magnetofluorescent nanocomposites, and newly microfabricated magnetizable elements [45, 53, 98]. These agents, along with varying radioisotopes and nuclear imaging, should make it possible to image multiple cell types *in vivo* without the limited depth penetration of optical imaging.

It is only a matter of time before we witness the impact that multimodal nanoparticles in conjunction with advances in imaging technology will make on cellular therapeutics. Advanced scanners such as the combined PET/CT and the newly developed PET/MRI will eventually allow molecular and cellular imaging to become a mainstay in clinical diagnostics, clinical research, and drug development. The combination of molecular immunology and molecular imaging with dedicated instruments and nanoparticle contrast agents will undoubtedly permit new discoveries related to how the immune system responds to immunotherapies and may help predict therapeutic responses to specific treatments.

References

1. Adonai N., Nguyen K. N., Walsh J., Iyer M., Toyokuni T., Phelps M. E., McCarthy T., McCarthy D. W., and Gambhir S. S. (2002) Ex vivo cell labeling with 64Cu-pyruvaldehyde-bis(*N*4-methylthiosemicarbazone) for imaging cell trafficking in mice with positron-emission tomography. *Proc Natl Acad Sci U S A* **99**, 3030-3035.

2. Ahrens E. T., Feili-Hariri M., Xu H., Genove G., and Morel P. A. (2003) Receptor-mediated endocytosis of iron-oxide particles provides efficient labeling of dendritic cells for *in vivo* MR imaging. *Magn Reson Med* **49**, 1006-1013.

3. Ahrens E. T., Flores R., Xu H., and Morel P. A. (2005) In vivo imaging platform for tracking immunotherapeutic cells. *Nat Biotechnol* **23**, 983-987.

4. Aime S., Cabella C., Colombatto S., Geninatti Crich S., Gianolio E., and Maggioni F. (2002) Insights into the use of paramagnetic Gd(III) complexes in MR-molecular imaging investigations. *J Magn Reson Imaging* **16**, 394-406.

5. Allan R. A., Sladen G. E., Bassingham S., Lazarus C., Clarke S. E., and Fogelman I. (1993) Comparison of simultaneous 99mTc-HMPAO and 111In oxine labelled white cell scans in the assessment of inflammatory bowel disease. *Eur J Nucl Med* **20**, 195-200.

6. Alric C., Taleb J., Le Duc G., Mandon C., Billotey C., Le Meur-Herland A., Brochard T., Vocanson F., Janier M., Perriat P., Roux S., and Tillement O. (2008) Gadolinium chelate coated gold nanoparticles as contrast agents for both X-ray computed tomography and magnetic resonance imaging. *J Am Chem Soc* **130**, 5908-5915.

7. Anderson S. A., Shukaliak-Quandt J., Jordan E. K., Arbab A. S., Martin R., McFarland H., and Frank J. A. (2004) Magnetic resonance imaging of labeled T-cells in a mouse model of multiple sclerosis. *Ann Neurol* **55**, 654-659.

8. Arbab A. S., Rad A. M., Iskander A. S., Jafari-Khouzani K., Brown S. L., Churchman J. L., Ding G., Jiang Q., Frank J. A., Soltanian-Zadeh H., and Peck D. J. (2007) Magnetically-labeled sensitized splenocytes to identify glioma by MRI: a preliminary study. *Magn Reson Med* **58**, 519-526.

9. Banchereau J., Schuler-Thurner B., Palucka A. K., and Schuler G. (2001) Dendritic cells as vectors for therapy. *Cell* **106**, 271-274.

10. Baumjohann D., Hess A., Budinsky L., Brune K., Schuler G., and Lutz M. B. (2006) *In vivo* magnetic resonance imaging of dendritic cell migration into the draining lymph nodes of mice. *Eur J Immunol* **36**, 2544-2555.

11. Baumjohann D. and Lutz M. B. (2006) Non-invasive imaging of dendritic cell migration *in vivo*. *Immunobiology* **211**, 587-597.

12. Becker W., Schomann E., Fischbach W., Borner W., and Gruner K. R. (1988) Comparison of 99Tcm-HMPAO and 111In-oxine labelled granulocytes in man: first clinical results. *Nucl Med Commun* **9**, 435-447.

13. Bernhard H., Neudorfer J., Gebhard K., Conrad H., Hermann C., Nahrig J., Fend F., Weber W., Busch D. H., and Peschel C. (2008) Adoptive transfer of autologous, HER2-specific, cytotoxic T lymphocytes for the treatment of HER2-overexpressing breast cancer. *Cancer Immunol Immunother* **57**, 271-280.

14. Botti C., Negri D. R., Seregni E., Ramakrishna V., Arienti F., Maffioli L., Lombardo C., Bogni A., Pascali C., Crippa F., Massaron S., Remonti F., Nerini-Molteni S., Canevari S., and Bombardieri E. (1997) Comparison of three different methods for radiolabelling human activated T lymphocytes. *Eur J Nucl Med* **24**, 497-504.

15. Bulte J. W. (2005) Hot spot MRI emerges from the background. *Nat Biotechnol* **23**, 945-946.

16. Bulte J. W. (2009) In vivo MRI cell tracking: clinical studies. *AJR Am J Roentgenol* **193**, 314-325.

17. Bulte J. W. and Kraitchman D. L. (2004) Iron oxide MR contrast agents for molecular and cellular imaging. *NMR Biomed* **17**, 484-499.

18. Bulte J. W., Laughlin P. G., Jordan E. K., Tran V. A., Vymazal J., and Frank J. A. (1996) Tagging of T cells with superparamagnetic iron oxide: uptake kinetics and relaxometry. *Acad Radiol* **3**(Suppl 2), S301-303.

19. Callera F. and de Melo C. M. (2007) Magnetic resonance tracking of magnetically labeled autologous bone marrow CD34+ cells transplanted into the spinal cord via lumbar puncture technique in patients with chronic spinal cord injury: CD34+ cells' migration into the injured site. *Stem Cells Dev* **16**, 461-466.

20. Crich S. G., Biancone L., Cantaluppi V., Duo D., Esposito G., Russo S., Camussi G., and Aime S. (2004) Improved route for the visualization of stem cells labeled with a Gd-/Eu-chelate as dual (MRI and fluorescence) agent. *Magn Reson Med* **51**, 938-944.

21. Czernin J. and Phelps M. E. (2002) Positron emission tomography scanning: current and future applications. *Annu Rev Med* **53**, 89-112.

22. Daldrup-Link H. E., Rudelius M., Metz S., Piontek G., Pichler B., Settles M., Heinzmann U., Schlegel J., Oostendorp R. A., and Rummeny E. J. (2004) Cell tracking with gadophrin-2: a bifunctional contrast agent for MR imaging, optical imaging, and fluorescence microscopy. *Eur J Nucl Med Mol Imaging* **31**, 1312-1321.

23. De Vries I. J., Krooshoop D. J., Scharenborg N. M., Lesterhuis W. J., Diepstra J. H., Van Muijen G. N., Strijk S. P., Ruers T. J., Boerman O. C., Oyen W. J., Adema G. J., Punt C. J., and Figdor C. G. (2003) Effective migration of antigen-pulsed dendritic cells to lymph nodes in melanoma patients is determined by their maturation state. *Cancer Res* **63**, 12-17.

24. de Vries I. J., Lesterhuis W. J., Barentsz J. O., Verdijk P., van Krieken J. H., Boerman O. C., Oyen W. J., Bonenkamp J. J., Boezeman J. B., Adema G. J., Bulte J. W., Scheenen T. W., Punt C. J., Heerschap A., and Figdor C. G. (2005) Magnetic resonance tracking of dendritic cells in melanoma patients for monitoring of cellular therapy. *Nat Biotechnol* **23**, 1407-1413.

25. Doubrovin M. M., Doubrovina E. S., Zanzonico P., Sadelain M., Larson S. M., and O'Reilly R. J. (2007) In vivo imaging and quantitation of adoptively transferred human antigen-specific T cells transduced to express a human norepinephrine transporter gene. *Cancer Res* **67**, 11959-11969.

26. Dubey P., Su H., Adonai N., Du S., Rosato A., Braun J., Gambhir S. S., and Witte O. N. (2003) Quantitative imaging of the T cell antitumor response by positron-emission tomography. *Proc Natl Acad Sci U S A* **100**, 1232-1237.

27. Esposito G., Geninatti Crich S., and Aime S. (2008) Efficient cellular labeling by CD44 receptor-mediated uptake of cationic liposomes functionalized with hyaluronic acid and loaded with MRI contrast agents. *ChemMedChem* **3**, 1858-1862.

28. Esslinger C., Chapatte L., Finke D., Miconnet I., Guillaume P., Levy F., and MacDonald H. R. (2003) In vivo administration of a lentiviral vaccine targets DCs and induces efficient CD8(+) T cell responses. *J Clin Invest* **111**, 1673-1681.

29. Ferrucci J. T. and Stark D. D. (1990) Iron oxide-enhanced MR imaging of the liver and spleen: review of the first 5 years. *AJR Am J Roentgenol* **155**, 943-950.

30. Figdor C. G., de Vries I. J., Lesterhuis W. J., and Melief C. J. (2004) Dendritic cell immunotherapy: mapping the way. *Nat Med* **10**, 475-480.

31. Frangioni J. V. and Hajjar R. J. (2004) In vivo tracking of stem cells for clinical trials in cardiovascular disease. *Circulation* **110**, 3378-3383.

32. Furumoto K., Soares L., Engleman E. G., and Merad M. (2004) Induction of potent antitumor immunity by in situ targeting of intratumoral DCs. *J Clin Invest* **113**, 774-783.

33. Grakoui A., Bromley S. K., Sumen C., Davis M. M., Shaw A. S., Allen P. M., and Dustin M. L. (1999) The immunological synapse: a molecular machine controlling T cell activation. *Science* **285**, 221-227.

34. Hanson H. L., Donermeyer D. L., Ikeda H., White J. M., Shankaran V., Old L. J., Shiku H., Schreiber R. D., and Allen P. M. (2000) Eradication of established tumors by CD8+ T cell adoptive immunotherapy. *Immunity* **13**, 265-276.

35. Hardy J., Edinger M., Bachmann M. H., Negrin R. S., Fathman C. G., and Contag C. H. (2001) Bioluminescence imaging of lymphocyte trafficking in vivo. *Exp Hematol* **29**, 1353-1360.

36. Himmelreich U., Aime S., Hieronymus T., Justicia C., Uggeri F., Zenke M., and Hoehn M. (2006) A responsive MRI contrast agent to monitor functional cell status. *Neuroimage* **32**, 1142-1149.

37. Hwang do W., Ko H. Y., Kim S. K., Kim D., Lee D. S., and Kim S. (2009) Development of a quadruple imaging modality by using nanoparticles. *Chemistry* **15**, 9387-9393.

38. Jaiswal J. K., Mattoussi H., Mauro J. M., and Simon S. M. (2003) Long-term multiple color imaging of live cells using quantum dot bioconjugates. *Nat Biotechnol* **21**, 47-51.

39. Josephson L., Tung C. H., Moore A., and Weissleder R. (1999) High-efficiency intracellular magnetic labeling with novel superparamagnetic-Tat peptide conjugates. *Bioconjug Chem* **10**, 186-191.

40. Kircher M. F., Allport J. R., Graves E. E., Love V., Josephson L., Lichtman A. H., and Weissleder R. (2003) In vivo high resolution three-dimensional imaging of antigen-specific cytotoxic T-lymphocyte trafficking to tumors. *Cancer Res* **63**, 6838-6846.

41. Kirchner C., Liedl T., Kudera S., Pellegrino T., Munoz Javier A., Gaub H. E., Stolzle S., Fertig N., and Parak W. J. (2005) Cytotoxicity of colloidal CdSe and CdSe/ZnS nanoparticles. *Nano Lett* **5**, 331-338.

42. Koehne G., Doubrovin M., Doubrovina E., Zanzonico P., Gallardo H. F., Ivanova A., Balatoni J., Teruya-Feldstein J., Heller G., May C., Ponomarev V., Ruan S., Finn R., Blasberg R. G., Bornmann W., Riviere I., Sadelain M., O'Reilly R. J., Larson S. M., and Tjuvajev J. G. (2003) Serial in vivo imaging of the targeted migration of human HSV-TK-transduced antigen-specific lymphocytes. *Nat Biotechnol* **21**, 405-413.

43. Lewin M., Carlesso N., Tung C. H., Tang X. W., Cory D., Scadden D. T., and Weissleder R. (2000) Tat peptide-derivatized magnetic nanoparticles allow in vivo tracking and recovery of progenitor cells. *Nat Biotechnol* **18**, 410-414.

44. Li C., Wang W., Wu Q., Ke S., Houston J., Sevick-Muraca E., Dong L., Chow D., Charnsangavej C., and Gelovani J. G. (2006) Dual optical and nuclear imaging in human melanoma xenografts using a single targeted imaging probe. *Nucl Med Biol* **33**, 349-358.

45. Lim Y. T., Noh Y. W., Cho J. H., Han J. H., Choi B. S., Kwon J., Hong K. S., Gokarna A., Cho Y. H., and Chung B. H. (2009) Multiplexed Imaging of Therapeutic Cells with Multispectrally Encoded Magnetofluorescent Nanocomposite Emulsions. *J Am Chem Soc*

46. Lin Y., Weissleder R., and Tung C. H. (2002) Novel near-infrared cyanine fluorochromes: synthesis, properties, and bioconjugation. *Bioconjug Chem* **13**, 605-610.

47. Long C. M. and Bulte J. W. (2009) In vivo tracking of cellular therapeutics using magnetic resonance imaging. *Expert Opin Biol Ther* **9**, 293-306.

48. Long C. M., van Laarhoven H. W., Bulte J. W., and Levitsky H. I. (2009) Magnetovaccination as a novel method to assess and quantify dendritic cell tumor antigen capture and delivery to lymph nodes. *Cancer Res* **69**, 3180-3187.

49. Martina M. S., Fortin J. P., Menager C., Clement O., Barratt G., Grabielle-Madelmont C., Gazeau F., Cabuil V., and Lesieur S. (2005) Generation of superparamagnetic liposomes revealed as highly efficient MRI contrast agents for in vivo imaging. *J Am Chem Soc* **127**, 10676-10685.

50. Massoud T. F. and Gambhir S. S. (2003) Molecular imaging in living subjects: seeing fundamental biological processes in a new light. *Genes Dev* **17**, 545-580.

51. Massoud T. F. and Gambhir S. S. (2007) Integrating noninvasive molecular imaging into molecular medicine: an evolving paradigm. *Trends Mol Med* **13**, 183-191.

52. Matsui K., Wang Z., McCarthy T. J., Allen P. M., and Reichert D. E. (2004) Quantitation and visualization of tumor-specific T cells in the secondary lymphoid organs during and after tumor elimination by PET. *Nucl Med Biol* **31**, 1021-1031.

53. McMahon M. T., Gilad A. A., DeLiso M. A., Berman S. M., Bulte J. W., and van Zijl P. C. (2008) New "multicolor" polypeptide diamagnetic chemical exchange saturation transfer (DIACEST) contrast agents for MRI. *Magn Reson Med* **60**, 803-812.

54. Michalet X., Pinaud F. F., Bentolila L. A., Tsay J. M., Doose S., Li J. J., Sundaresan G., Wu A. M., Gambhir S. S., and Weiss S. (2005) Quantum dots for live cells, in vivo imaging, and diagnostics. *Science* **307**, 538-544.

55. Monks C. R., Freiberg B. A., Kupfer H., Sciaky N., and Kupfer A. (1998) Three-dimensional segregation of supramolecular activation clusters in T cells. *Nature* **395**, 82-86.

56. Moore A., Grimm J., Han B., and Santamaria P. (2004) Tracking the recruitment of diabetogenic CD8+ T-cells to the pancreas in real time. *Diabetes* **53**, 1459-1466.

57. Nahrendorf M., Zhang H., Hembrador S., Panizzi P., Sosnovik D. E., Aikawa E., Libby P., Swirski F. K., and Weissleder R. (2008) Nanoparticle PET-CT imaging of macrophages in inflammatory atherosclerosis. *Circulation* **117**, 379-387.

58. Nestle F. O., Banchereau J., and Hart D. (2001) Dendritic cells: On the move from bench to bedside. *Nat Med* **7**, 761-765.

59. Ntziachristos V., Bremer C., and Weissleder R. (2003) Fluorescence imaging with near-infrared light: new technological advances that enable in vivo molecular imaging. *Eur Radiol* **13**, 195-208.

60. Ntziachristos V., Tung C. H., Bremer C., and Weissleder R. (2002) Fluorescence molecular tomography resolves protease activity in vivo. *Nat Med* **8**, 757-760.

61. Paik J. Y., Lee K. H., Byun S. S., Choe Y. S., and Kim B. T. (2002) Use of insulin to improve [18 F]fluorodeoxyglucose labelling and retention for in vivo positron emission tomography imaging of monocyte trafficking. *Nucl Med Commun* **23**, 551-557.

62. Pittet M. J., Grimm J., Berger C. R., Tamura T., Wojtkiewicz G., Nahrendorf M., Romero P., Swirski F. K., and Weissleder R. (2007) In vivo imaging of T cell delivery to tumors after adoptive transfer therapy. *Proc Natl Acad Sci U S A* **104**, 12457-12461.

63. Ponomarev V., Doubrovin M., Serganova I., Vider J., Shavrin A., Beresten T., Ivanova A., Ageyeva L., Tourkova V., Balatoni J., Bornmann W., Blasberg R., and Gelovani Tjuvajev J. (2004) A novel triple-modality reporter gene for whole-body fluorescent, bioluminescent, and nuclear noninvasive imaging. *Eur J Nucl Med Mol Imaging* **31**, 740-751.

64. Pouliquen D., Le Jeune J. J., Perdrisot R., Ermias A., and Jallet P. (1991) Iron oxide nanoparticles for use as an MRI contrast agent: pharmacokinetics and metabolism. *Magn Reson Imaging* **9**, 275-283.

65. Quillien V., Moisan A., Carsin A., Lesimple T., Lefeuvre C., Adamski H., Bertho N., Devillers A., Leberre C., and Toujas L. (2005) Biodistribution of radiolabelled human dendritic cells injected by various routes. *Eur J Nucl Med Mol Imaging* **32**, 731-741.

66. Read E. J., Keenan A. M., Carter C. S., Yolles P. S., and Davey R. J. (1990) In vivo traffic of indium-111-oxine labeled human lymphocytes collected by automated apheresis. *J Nucl Med* **31**, 999-1006.

67. Ribot E. J., Miraux S., Delville M. H., Bouchaud V., Lacomme S., Gontier E., Bouzier-Sore A. K., Franconi J. M., Thiaudiere E., and Voisin P. (2009) Study of the MR relaxation of microglia cells labeled with Gd-DTPA-bearing nanoparticles. *Contrast Media Mol Imaging* **4**, 109-117.

68. Ridolfi R., Riccobon A., Galassi R., Giorgetti G., Petrini M., Fiammenghi L., Stefanelli M., Ridolfi L., Moretti A., Migliori G., and Fiorentini G. (2004) Evaluation of in vivo labelled dendritic cell migration in cancer patients. *J Transl Med* **2**, 27.

69. Ruiz-Cabello J., Walczak P., Kedziorek D. A., Chacko V. P., Schmieder A. H., Wickline S. A., Lanza G. M., and Bulte J. W. (2008) In vivo "hot spot" MR imaging of neural stem cells using fluorinated nanoparticles. *Magn Reson Med* **60**, 1506-1511.

70. Sen D., Deerinck T. J., Ellisman M. H., Parker I., and Cahalan M. D. (2008) Quantum dots for tracking dendritic cells and priming an immune response in vitro and in vivo. *PLoS One* **3**, e3290.

71. Sengar R. S., Spokauskiene L., Steed D. P., Griffin P., Arbujas N., Chambers W. H., and Wiener E. C. (2009) Magnetic resonance imaging-guided adoptive cellular immunotherapy of central nervous system tumors with a T1 contrast agent. *Magn Reson Med* **62**, 599-606.

72. Shapiro E. M., Sharer K., Skrtic S., and Koretsky A. P. (2006) In vivo detection of single cells by MRI. *Magn Reson Med* **55**, 242-249.

73. Shapiro E. M., Skrtic S., Sharer K., Hill J. M., Dunbar C. E., and Koretsky A. P. (2004) MRI detection of single particles for cellular imaging. *Proc Natl Acad Sci U S A* **101**, 10901-10906.

74. Shen J., Zhong X. M., Duan X. H., Cheng L. N., Hong G. B., Bi X. B., and Liu Y. (2009) Magnetic resonance imaging of mesenchymal stem cells labeled with dual (MR and fluorescence) agents in rat spinal cord injury. *Acad Radiol* **16**, 1142-1154.

75. Simon G. H., Daldrup-Link H. E., Kau J., Metz S., Schlegel J., Piontek G., Saborowski O., Demos S., Duyster J., and Pichler B. J. (2006) Optical imaging of experimental arthritis using allogeneic leukocytes labeled with a near-infrared fluorescent probe. *Eur J Nucl Med Mol Imaging* **33**, 998-1006.

76. Smith A. M., Ruan G., Rhyner M. N., and Nie S. (2006) Engineering luminescent quantum dots for in vivo molecular and cellular imaging. *Ann Biomed Eng* **34**, 3-14.

77. Sosnovik D. and Weissleder R. (2005) Magnetic resonance and fluorescence based molecular imaging technologies. *Prog Drug Res* **62**, 83-115.

78. Srinivas M., Morel P. A., Ernst L. A., Laidlaw D. H., and Ahrens E. T. (2007) Fluorine-19 MRI for visualization and quantification of cell migration in a diabetes model. *Magn Reson Med* **58**, 725-734.

79. Stoll S., Delon J., Brotz T. M., and Germain R. N. (2002) Dynamic imaging of T cell-dendritic cell interactions in lymph nodes. *Science* **296**, 1873-1876.

80. Su H., Forbes A., Gambhir S. S., and Braun J. (2004) Quantitation of cell number by a positron emission tomography reporter gene strategy. *Mol Imaging Biol* **6**, 139-148.

81. Talanov V. S., Regino C. A., Kobayashi H., Bernardo M., Choyke P. L., and Brechbiel M. W. (2006) Dendrimer-based nanoprobe for dual modality magnetic resonance and fluorescence imaging. *Nano Lett* **6**, 1459-1463.

82. Toso C., Vallee J. P., Morel P., Ris F., Demuylder-Mischler S., Lepetit-Coiffe M., Marangon N., Saudek F., James Shapiro A. M., Bosco D., and Berney T. (2008) Clinical magnetic resonance imaging of pancreatic islet grafts after iron nanoparticle labeling. *Am J Transplant* **8**, 701-706.

83. Trakatelli M., Toungouz M., Blocklet D., Dodoo Y., Gordower L., Laporte M., Vereecken P., Sales F., Mortier L., Mazouz N., Lambermont M., Goldman S., Coulie P., Goldman M., and Velu T. (2006) A new dendritic cell vaccine generated with interleukin-3 and interferon-beta induces CD8+ T cell responses against NA17-A2 tumor peptide in melanoma patients. *Cancer Immunol Immunother* **55**, 469-474.

84. Tumeh P. C., Radu C. G., and Ribas A. (2008) PET imaging of cancer immunotherapy. *J Nucl Med* **49**, 865-868.

85. Verdijk P., Scheenen T. W., Lesterhuis W. J., Gambarota G., Veltien A. A., Walczak P., Scharenborg N. M., Bulte J. W., Punt C. J., Heerschap A., Figdor C. G., and de Vries I. J. (2007) Sensitivity of magnetic resonance imaging of dendritic cells for in vivo tracking of cellular cancer vaccines. *Int J Cancer* **120**, 978-984.

86. Vianello F., Papeta N., Chen T., Kraft P., White N., Hart W. K., Kircher M. F., Swart E., Rhee S., Palu G., Irimia D., Toner M., Weissleder R., and Poznansky M. C. (2006) Murine B16 melanomas expressing high levels of the chemokine stromal-derived factor-1/CXCL12 induce tumor-specific T cell chemorepulsion and escape from immune control. *J Immunol* **176**, 2902-2914.

87. Vuu K., Xie J., McDonald M. A., Bernardo M., Hunter F., Zhang Y., Li K., Bednarski M., and Guccione S. (2005) Gadolinium-rhodamine nanoparticles for cell labeling and tracking via magnetic resonance and optical imaging. *Bioconjug Chem* **16**, 995-999.

88. Walczak P., Kedziorek D. A., Gilad A. A., Lin S., and Bulte J. W. (2005) Instant MR labeling of stem cells using magnetoelectroporation. *Magn Reson Med* **54**, 769-774.

89. Walczak P., Zhang J., Gilad A. A., Kedziorek D. A., Ruiz-Cabello J., Young R. G., Pittenger M. F., van Zijl P. C., Huang J., and Bulte J. W. (2008) Dual-modality monitoring of targeted intraarterial delivery of mesenchymal stem cells after transient ischemia. *Stroke* **39**, 1569-1574.

90. Wang R. F. and Rosenberg S. A. (1999) Human tumor antigens for cancer vaccine development. *Immunol Rev* **170**, 85-100.

91. Wang W., Ke S., Kwon S., Yallampalli S., Cameron A. G., Adams K. E., Mawad M. E., and Sevick-Muraca E. M. (2007) A new optical and nuclear dual-labeled imaging agent targeting interleukin 11 receptor alpha-chain. *Bioconjug Chem* **18**, 397-402.

92. Weissleder R. (2001) A clearer vision for in vivo imaging. *Nat Biotechnol* **19**, 316-317.

93. Weissleder R., Elizondo G., Wittenberg J., Rabito C. A., Bengele H. H., and Josephson L. (1990) Ultrasmall superparamagnetic iron oxide: characterization of a new class of contrast agents for MR imaging. *Radiology* **175**, 489-493.

94. Welch M. J., Hawker C. J., and Wooley K. L. (2009) The Advantages of nanoparticles for PET. *J Nucl Med* **50**, 1743-1746.

95. Wulfing C., Sjaastad M. D., and Davis M. M. (1998) Visualizing the dynamics of T cell activation: intracellular adhesion molecule 1 migrates rapidly to the T cell/B cell interface and acts to sustain calcium levels. *Proc Natl Acad Sci U S A* **95**, 6302-6307.

96. Yaghoubi S. S., Jensen M. C., Satyamurthy N., Budhiraja S., Paik D., Czernin J., and Gambhir S. S. (2009) Noninvasive detection of therapeutic cytolytic T cells with 18F-FHBG PET in a patient with glioma. *Nat Clin Pract Oncol* **6**, 53-58.

97. Yee C., Thompson J. A., Byrd D., Riddell S. R., Roche P., Celis E., and Greenberg P. D. (2002) Adoptive T cell therapy using antigen-specific CD8+ T cell clones for the treatment of patients with metastatic melanoma: in vivo persistence, migration, and antitumor effect of transferred T cells. *Proc Natl Acad Sci U S A* **99**, 16168-16173.

98. Zabow G., Dodd S., Moreland J., and Koretsky A. (2008) Micro-engineered local field control for high-sensitivity multispectral MRI. *Nature* **453**, 1058-1063.

99. Zhu J., Zhou L., and XingWu F. (2006) Tracking neural stem cells in patients with brain trauma. *N Engl J Med* **355**, 2376-2378.

Chapter 4

RADIOLABELED LIPOSOMES AS THERANOSTIC AGENTS

William T. Phillips,[a] Beth Goins,[a] Keitaro Sou,[c] and Ande Bao[a,b]
[a] *Department of Radiology and* [b] *Otolaryngology–Head and Neck Surgery, University of Texas Health Science Center, San Antonio, TX 78229, USA*
Phillips@uthscsa.edu
[c] *Center for Advanced Biomedical Sciences, Waseda University, Tokyo 162-8580, Japan*
k.so@kurenai.waseda.jp

4.1 INTRODUCTION TO LIPOSOMAL NANOPARTICLES (NANOLIPOSOMES)

Liposomes are spontaneously forming lipid bilayers that enclose an aqueous space. Soon after their initial discovery more than 40 years ago, liposomes were recognized as promising drug carriers for diagnostic and therapeutic agents [13, 50]. When liposomes form spontaneously, they initially form as multi-lipid bilayer (multi-lamellar) structures that are in the micrometer size range. To reduce these multi-lamellar micro-size liposomes into nanoparticles generally requires the application of energy-intensive shear forces to the multi-lamellar liposomes. The most widely used methods are filter extrusion and high-pressure emulsification. These processes can reduce liposomes to unilamellar nanoparticles in the size range of 80–250 nm [7]. Liposomes in this nanoparticle size range, particularly those that are smaller than 100 nm, often have special properties and functional relationships that are not found in larger liposomes formulations, and therefore these smaller liposomes can be referred to as nanoliposomes. For instance, nanoliposomes have been shown to have a long circulation time after intravenous injection not found with larger liposome formulations [7]. This makes nanoliposomes ideal agents as intravenously administered drug carriers because the longer circulation time generally results in increased uptake by the targeted tissue [38]. It is likely for this reason that

Nanoimaging
Edited by Beth Goins and William Phillips

www.panstanford.com

several clinically approved anticancer and antifungal products are nanoliposomal formulations with size ranges that are smaller than 100 nm [18, 40]. In addition, nanoliposomes have been shown to have special properties for *in vivo* nanoassembly [83] and aggregation for trapping in lymph nodes and intracavitary spaces [87, 106, 108] as well as convection-enhanced drug delivery (CED) [17, 63, 70, 115] when administered directly into interstitial tissue and tumors.

Nanoliposomes are composed of naturally occurring components and have a myriad of possible compositions and modifications, making them extremely flexible drug carriers. Incremental continuous progress in the field of liposome technology has greatly increased the viability of liposomes as clinically useful drug nanocarriers. These achievements include the development of (i) methods to produce homogeneously sized liposomes in the nanometer size range, (ii) methods to stably encapsulate sufficient quantities of therapeutic agents within the liposomes, (iii) methods for large-scale liposome manufacture, (iv) methods to prolong the circulation time of liposomes in the blood [3, 107, 135], (v) methods to actively target liposomes *in vivo* by surface modification with targeted ligands [4, 35, 51, 118], and (vi) effective methods of noninvasively tracking and quantifying the distribution of liposomes in the body [16, 37, 43, 75, 109]. This chapter will review the uses of radioactive nanoliposomes for *in vivo* imaging and therapeutic applications. Examples of practical imaging applications of radioactive nanoliposomes for research and potential clinical application will also be reviewed. Finally, nanoliposomes labeled with radionuclides that have both diagnostic monitoring and therapeutic functions (theranostic) [82, 101] will be discussed. Although the methods will specifically focus on scintigraphic imaging of liposome nanoparticle formulations, many of the scintigraphic imaging approaches for localizing and quantifying nanoliposome distribution *in vivo* can also be utilized for the tracking and imaging of most any other type of nanoparticle.

4.2 ADVANTAGES OF SCINTIGRAPHIC IMAGING FOR *IN VIVO* NANOPARTICLE IMAGING

Scintigraphic imaging, also known as gamma photon imaging, is a noninvasive imaging technology that involves the detection of gamma photons emitted from radioactive molecules. It is commonly used in clinical patients following administration of a specific radionuclide-labeled molecule or particle for diagnostic purposes. It has a very high sensitivity for detecting radiolabeled molecules in the body that is not possible with standard X-ray, X-ray computed tomography (CT) imaging, or magnetic resonance imaging (MRI). These high sensitivities of detection in the human body are possible because the relative high energies of the gamma photons can penetrate completely through almost any depth of human tissue and exit from the body to be detected by specialized detectors outside the patient. Because such small quantities of radionuclides can be detected (<0.01–1 nanogram of actual matter), scintigraphic imaging is ideally suited for the imaging of physiologic alterations in living animals of all sizes. Technological advances over the last 30 years have resulted in great improvements in the resolution and sensitivity of cameras used for scintigraphic imaging [94, 136, 143]. Recently, dedicated imaging systems have been developed for small animal imaging. In addition, dual imaging systems that acquire anatomical CT images along with the scintigraphic images have been developed for both clinical and small animal imaging systems [26, 79, 144].

4.3 TYPES OF SCINTIGRAPHIC IMAGING

There are two types of scintigraphic imaging, which are based on either single gamma photon detection or dual-annihilation photon detection.

4.3.1 Single-Photon Imaging

Single-photon imaging requires lead collimation, which is located in front of a detector to localize the source of the gamma photons. Single-photon imaging can be either a two-dimensional planar projection imaging or a tomographic three-dimensional imaging known as single-photon emission computed tomography (SPECT) imaging. Commonly used single-photon-emitting radionuclides for nanoparticle imaging are technetium-99m (^{99m}Tc) with a 6 hour half-life, indium-111 (^{111}In) with a half-life of 67 hours, and gallium-67 (^{67}Ga) with a half-life of 78 hours. These radionuclides are all widely available from local clinical radiopharmacies.

4.3.2 PET Imaging

A second type of imaging known as positron emission tomography (PET) imaging is based on the detection of two annihilation photons that are emitted simultaneously at approximately 180° from each other. These annihilation photons are generated following an annihilation reaction between an emitted positron from the positron-emitting radionuclide and a shell electron in the region of the emitted positron. These dual-photon emissions can be localized along a line by the simultaneous detection of the two 180° emitted photons by crystal scintillation detectors that are part of a ring of detectors surrounding the body. Multiple emissions can be iteratively localized in the body and provide a three-dimensional tomographic image of the distribution of the positron radionuclide. Collimators are not required for this type of imaging, although fairly thick crystals are required to detect the high-energy 511 keV photons that are emitted from all positron emission radioisotopes. PET imaging is always acquired for display of tomographic images, and planar projection imaging is not generally feasible. The most commonly used radionuclides for PET imaging have a relatively short half-life, such as oxygen-15 (2 minute half-life), carbon-11 (20 minute half-life), and fluorine-18 (110 minute half-life). Although some work has been done with shorter-lived PET agents and liposome imaging [6, 137], for most nanoparticle imaging applications these PET isotopes have too short a half-life to track the nanoparticle *in vivo*. PET radionuclides with longer half-lives, such as copper-64 (^{64}Cu) (12.7 hour half-life) and iodine-124 (^{124}I) (4.2 day half-life), are more promising for tracking the distribution of long circulating nanoparticles such as liposomes. Recently, several methods have been described for imaging liposomes with ^{64}Cu [100, 119].

4.3.3 Nanoparticle Scintigraphic Imaging

For either PET or single-photon imaging of nanoparticles, the nanoparticle is radiolabeled using specifically developed methodology and the distribution and uptake monitored with detectors. The emitted photon travels through body tissue and exits the body, whereupon it is detected by a crystal inside the gamma camera. Blood clearance profile of radiolabeled nanoparticles can be modified by adjusting particle sizes or attaching certain molecules, such as polyethylene glycol (PEG). Liposomes can also be molecularly targeted to specifically localize in a disease process or organ of interest [118, 134].

4.4 METHODS OF LABELING NANOLIPOSOMES WITH RADIONUCLIDES FOR SCINTIGRAPHIC IMAGING

Commonly used radionuclides for the radiolabeling of liposomes for imaging studies with a gamma camera are ^{99m}Tc, ^{111}In, and ^{67}Ga. ^{99m}Tc has the most ideal properties of the readily available single-photon-emission diagnostic imaging radionuclides due to its optimal imaging characteristics, which include an ideal photon energy of 140 keV and the fact that it is relatively inexpensive because it can be eluted daily from a commercially available molybdenum-99 (^{99}Mo)/^{99m}Tc generator. Because ^{67}Ga, ^{111}In, and ^{123}I are cyclotron products, these agents are more expensive and not always available in every nuclear medicine department.

Several detailed reviews have been recently written describing the methodology for radiolabeling liposomes [48, 73, 76]. Important factors to consider in regard to liposome labeling methods are ease of preparation and *in vitro* and *in vivo* stability. Ideally, previously manufactured liposomes should be labeled just prior to the initiation of experiments. This situation is ideal because the radionuclides used in clinical imaging and research studies have half-lives of 6 hours to 3 days. The labeling of previously manufactured lipsomes is known as "after loading" or "remote labeling," in which the preformed liposomes are labeled just prior to the start of the experiment. Also after radiolabeling, the radionuclide should stay firmly associated with the liposomes as excessive instability will obviously lead to complications in interpreting biodistribution results. Stability of the radiolabeled liposomes can be assessed *in vitro* by incubating the radiolabeled liposomes in serum at body temperature (37°C) and assessing stability. Serum incubation generally mirrors *in vivo* stability.

Many different methods have been developed for radiolabeling liposomes. A simple method of labeling liposomes is to incubate the premanufactured liposomes with a lipophilic radiolabel so that the radionuclide becomes associated with the lipid bilayer. This approach generally yields very unstable radiolabeling, which is undesirable [81].

Alternatively, two after-loading approaches have proven to yield radiolabeled liposomes that have a high efficiency and good radiochemical stability. These include methods of (i) trapping of the radionuclide in the enclosed aqueous phase of liposomes and (ii) binding the radionuclide to a lipid-conjugated chelator that is located in the lipid bilayer of preformed liposomes. These approaches have been used to label liposomes with the single-photon agents ^{67}Ga [95], ^{111}In, ^{123}I, and ^{99m}Tc [8, 14, 16, 20, 37, 39, 54, 60–61, 65, 75, 84, 91, 95, 110, 112]. These approaches to labeling liposomes have been covered in a more detailed review of labeling technology in previous publications by the authors [44, 105]. Two common methods that are used by our group to label liposomes will be reviewed in more detail to provide a better understanding of the imaging studies that will be described in the following sections.

4.4.1 HMPAO Liposome Radiolabeling Method

The first method labels liposomes with ^{99m}Tc by using a lipophilic complex, ^{99m}Tc-hexamethylpropyleneamine oxime (^{99m}Tc-HMPAO) [109]. With this method, lipophilic ^{99m}Tc-HMPAO enters the liposome, where it interacts with previously encapsulated glutathione and is converted to the hydrophilic form, and thus is trapped in the liposome. This method has been used successfully by many investigators [5, 28, 34, 45, 64, 77, 98, 109, 126, 133]. It has been documented to be stable *in vitro* and *in vivo* [102].

4.4.2 BMEDA Liposome Radiolabeling Method

Another after-loading method for radiolabeling liposomes with ^{99m}Tc-*N*,*N*-bis (2-mercaptoethyl)-*N'*,*N'*-diethyl-ethylenediamine (^{99m}Tc-BMEDA) has been developed [14]. The lipophilic ^{99m}Tc-BMEDA crosses the liposomal membrane, where it interacts with either (i) glutathione or (ii) a pH gradient to become trapped in the hydrophilic portion of the liposomes, as shown in Fig. 4.1. This method has been shown to have good *in vitro* and *in vivo* stability with a variety of preformed liposome formulations. Compared with ^{99m}Tc-HMPAO, the BMEDA labeling method has the advantage of being effective for labeling liposomes with beta-particle-emitting therapeutic radionuclides rhenium-186 (^{186}Re) or rhenium-188 (^{188}Re) [15]. It also has the advantage that it can be used to directly label commercially available liposome formulations such as liposomal doxorubicin (Doxil™) via use of the pH gradient mechanism used to load drugs into the liposomes [16]. BMEDA suitable for labeling with ^{99m}Tc or $^{186/188}$Re for research purposes can be obtained from a commercial vendor (ABX, Radeburg, Germany).

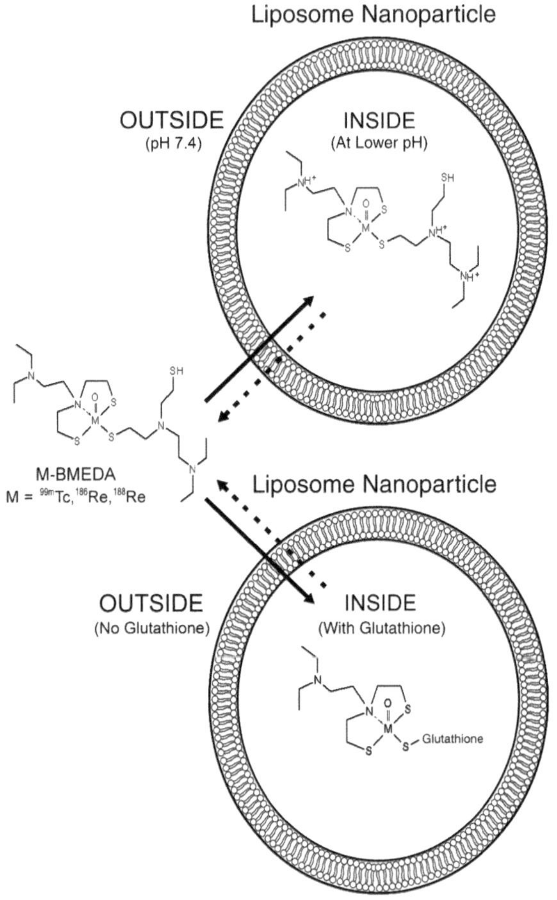

Figure 4.1 Schematic diagram for radiolabeling liposome nanoparticles with the BMEDA after-loading method via pH gradient or glutathione mechanisms. *Top panel*: Lipophilic radiolabeled BMEDA at pH 7.4 can traverse the liposome bilayer and become stably trapped in the aqueous interior following protonation in lower pH environment. *Bottom panel*: Lipophilic radiolabeled BMEDA at pH 7.4 can transverse the liposome bilayer and become stably trapped in the aqueous interior following conversion to a more hydrophilic form with interaction with glutathione.

4.5 ADVANTAGES OF SCINTIGRAPHIC IMAGING FOR DEVELOPMENT OF LIPOSOME PRODUCTS

The *in vivo* imaging of nanoliposome formulations has many advantages as an important aid during the development of new clinical applications of nanoliposomes. Imaging can provide feedback on the percentage of nanoliposomes that accumulate in various organs and at targeted disease sites such as tumors and inflammation. Scintigraphic imaging provides a rapid way to assess the effect of formulation changes, and it is very efficient in terms of decreasing the number of animals used for these assessments as the same animal can be imaged at multiple time points after administration of the radiolabeled nanoliposomes. Liposome scintigraphic imaging can provide a quantitative assessment of the uptake of liposome formulations in different critical organs of the body, which can be helpful in determining the organs that are at most risk for toxicity.

4.6 SPECIFIC USES OF SCINTIGRAPHIC IMAGING FOR LIPOSOME DRUG DEVELOPMENT

Scintigraphic imaging has proved useful for the development of several nanoliposome formulations over the last 20 years. Reviews have been previously written describing in detail a variety of uses of scintigraphic imaging for development of nanoliposome formulations [46, 47, 73, 104]. In the following section some specific examples of the use of scintigraphic imaging for nanoliposome studies will be provided with emphasis on recent research. Because nanoliposomes as nanoparticles have been investigated for a longer period of time than most recently developed nanoparticles, these studies can provide a model for the type of scintigraphic imaging that can be carried out with other nanoparticles.

4.6.1 Studies of Noninvasive Determination of Drug Concentrations

The tracking of radiolabeled liposomes by scintigraphic imaging can be used as a method to noninvasively and quantitatively determine the quantity of drug taken up by the targeted site. Kleiter *et al.* have recently shown a validation of this approach [64]. Liposomes were labeled with ^{99m}Tc using the HMPAO method in order to determine if noninvasive imaging could be used to predict actual doxorubicin tissue levels that accumulated in a rat fibrosarcoma tumor following the administration of long circulating PEG liposomes containing doxorubicin (Doxil). Scintigraphic images were obtained at 18 hours following injection of the ^{99m}Tc-labeled liposomes, and immediately following imaging the tumors were removed for radioactivity counting and for analytical measurement of the doxorubicin concentration. A significant positive correlation between the intratumoral radioactivity and the amount of doxorubicin in the tumor was found. This study provides good support for the use of this labeling and imaging methodology to noninvasively track the distribution of drugs delivered by liposome carriers. This study was also interesting in that the hyperthermia method of high-intensity ultrasound directed at a portion of the tumors was used in order to increase liposome uptake and drug levels in the tumor. Noninvasive scintigraphic imaging of the radiolabeled liposomes clearly demonstrated an increased uptake in the hyperthermia-treated tumors compared with untreated tumors. The hyperthermia increased the uptake of the radiolabel associated with the liposomes by approximately fourfold, and it increased the uptake of measured doxorubicin concentrations in the tumor by a slightly less increment that ranged from 2.6 to 3 times [64].

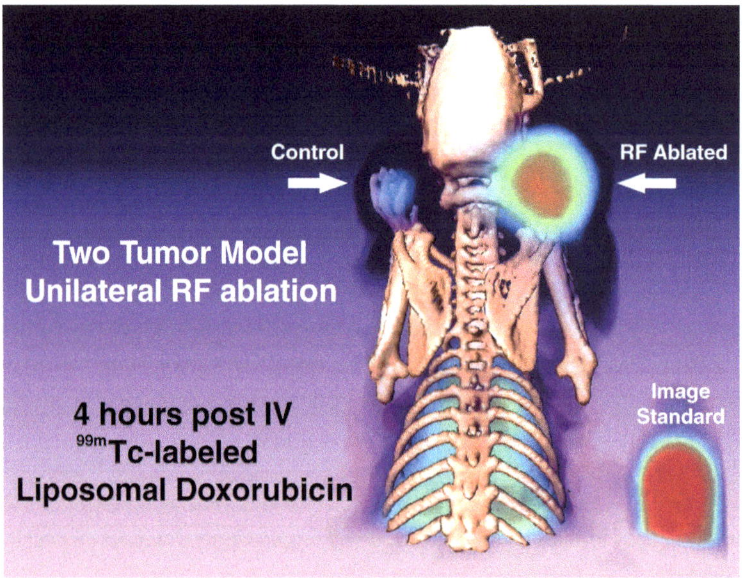

Figure 4.2 SPECT/CT image of rat with head and neck xenograft tumors located on both sides of neck. The tumor on right side was subjected to RF ablation. Images were acquired 4 hours after intravenous injection of ^{99m}Tc-Doxil. There was significantly higher uptake of ^{99m}Tc-Doxil in ablated tumor compared with unablated tumor.

A similar study was performed by our group in which we radiolabeled commercially available liposomal doxorubicin (Doxil) with ^{99m}Tc using the BMEDA radiolabeling method to investigate the increase in tumor uptake following tumor hyperthermia that was administered by insertion of an electrified needle that delivers radiofrequency-induced hyperthermia [57]. This type of radiofrequency (RF) ablation is frequently used clinically to treat liver tumors [71, 86, 121], and its use in combination with liposomal doxorubicin is currently under investigation [1, 2, 49]. In our study, two head and neck xenograft tumors were present on both sides of the neck, but only one tumor was treated with RF ablation. Immediately following treatment of one tumor by RF ablation, rats were injected with ^{99m}Tc-radiolabeled liposomes. Serial scintigraphic images were obtained at 1, 4, 24, and 48 hours following administration of the liposomes. The head and neck xenograft tumor treated with RF ablation was shown to have approximately threefold increased uptake of radiolabeled liposomal doxorubicin compared with nontreated tumors. As can be seen in Fig. 4.2, the treated tumor clearly had much greater uptake of the radiolabeled liposomes than the untreated tumor. The image shown in Fig. 4.2 is a combined fused image of the whole rat CT image, which was windowed to show only the bones of the rat, that was then fused with the SPECT scintigraphic image. Clearly visualizing the bones on the CT image along with the scintigraphic image allows for a much better appreciation of the relative size of the uptake in the tumor and also provides better spatial localization of the tumor in relationship to the other parts of the body. Region of interest analysis revealed the activity to be approximately three times greater in the treated tumor. Following imaging of the rats at 44 hours, tumors were removed for precise quantification of tissue levels of doxorubicin.

Noninvasive image quantification was found to correlate very well with actual measured tissue doxorubicin levels, as shown in Fig. 4.3. These results confirmed the previous finding of Kleiter *et al.* [64] and demonstrates the feasibility of performing noninvasive imaging studies in human subjects to specifically determine chemodosimetry in patients.

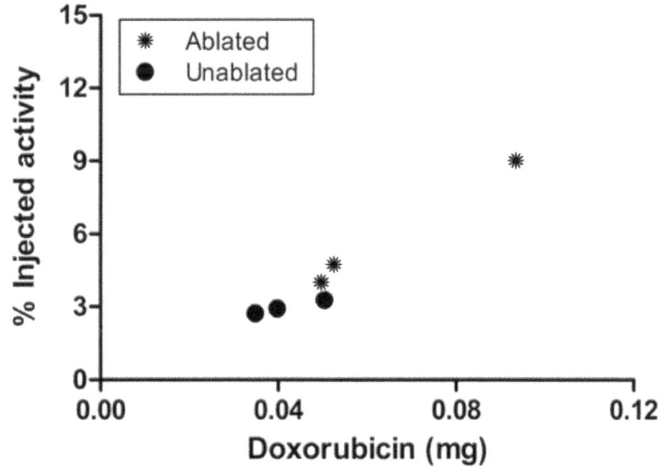

Figure 4.3 Percentage of ^{99m}Tc-Doxil activity determined by planar scintigraphic images acquired at 44 hours as a function of the quantity of doxorubicin (milligrams per gram of tissue) extracted from the ablated and unablated tumors. There was significant positive correlation ($r = 0.9830$, $p = 0.0004$).

4.6.2 Infection and Inflammation Targeting of Liposomes

Nanoliposomes have been shown to accumulate heavily at sites of inflammation and infection by nonspecific passive targeting [5, 43, 72, 90]. Microscopic studies have shown that the liposomes accumulate in inflamed tissue by enhanced extravasation into leaky infected and inflamed tissue, where they become localized in macrophages, and to a lesser extent neutrophils and endothelial cells in the region of the infection/ inflammation [74]. It is likely that any nanoparticle that has a sufficiently prolonged circulation time will accumulate in inflamed or infected tissue due to enhanced capillary leakage at the site of the infection. Nanoliposomes labeled with ^{99m}Tc have been shown to be effective diagnostic tools for localizing the site of uptake in the infection with scintigraphic imaging. Nanoliposomes continue to be very promising carriers for delivery of drugs to inflamed regions of the body. Many articles have been published using imaging that demonstrate the potential of nanoliposomes as carriers of anti-infective and anti-inflammatory drugs [9–12, 30], and scintigraphic imaging continues to be a very useful method to investigate nanoliposomal targeting to infections [19, 29, 43, 99, 139]; however, to date, there have been no clinical products that have specifically taken advantage of the inflammatory targeting of liposomes. It is apparent that the increased uptake of nanoliposomes in inflammation may be an important aspect of the clinically approved liposome-encapsulated antifungal agent amphotericin B [36].

4.6.2.1 Nanoliposomes for imaging and treatment of inflammatory arthritis

Liposome imaging has been used in the preclinical development of anti-inflammatory liposomes containing prednisolone [89]. Long circulating PEG-coated liposomes containing prednisolone were labeled with ^{111}In-oxine. Scintigraphic imaging was used to show that these anti-inflammatory liposomes accumulated in the inflamed joints of small animal models. The increased uptake was clearly visualized on the images and the ^{111}In labeling of the liposomes was used to determine the clearance from the blood and the biodistribution in the tissues. A sevenfold increased liposome activity was detected in rats with inflamed hind paws compared with the hind paws of normal rats. The encapsulated prednisolone was much more effective than free unencapsulated prednisolone. A single dose of liposome encapsulated prednisolone resulted in complete remission of the inflammatory response in 1 week [89].

Scintigraphic imaging has also been recently used to develop liposomes for arthritis that carry superoxide dismutase (SOD) on their surface [41]. Liposomes carrying SOD greatly improved the pharmacokinetic behavior of SOD, allowing this powerful antioxidant to reach the site of inflammation in the joint. The investigators have named this type of liposome an enzymosome.

4.6.2.2 Nanoliposome imaging of external beam radiation induced inflammation

External beam radiation has been investigated by our group as a method to increase nanoliposome uptake in the tumor secondary to radiation induced inflammation. In this study, rats with a subcutaneous prostate tumor on the thigh were treated with 5 Grays of radiation. Twenty-four hours following the external beam radiation directed only to the tumor, the rats were injected with ^{99m}Tc-nanoliposomes, and 24 hours following the nanoliposome injection the rats were imaged. External beam radiation therapy was associated with a threefold increased uptake of activity in the tumor that had received the dose of radiation than in a control rat with a tumor of the same size that did not receive radiation. This increased uptake in the tumor that received radiation was likely due to radiation-induced inflammation from the 5 Gray dose. In Fig. 4.4, planar scintigraphic images show the increased uptake in the tumor that was irradiated compared with the nonspecific uptake in the un-irradiated tumor. These planar images demonstrate the capability of scintigraphic imaging for noninvasively quantifying the percentage of the injected dose that accumulates in a targeted region of the body.

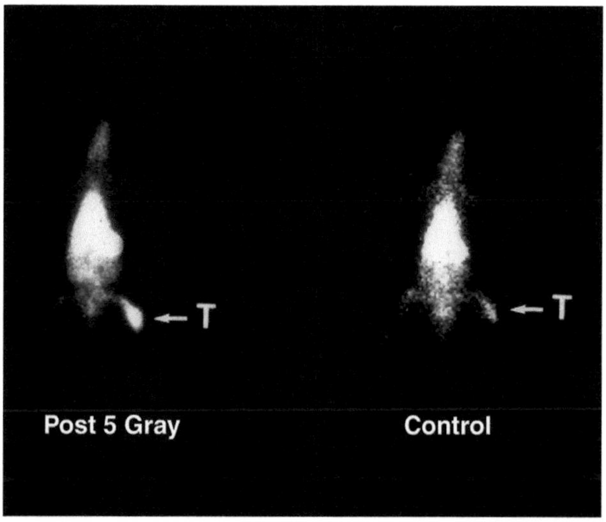

Figure 4.4 Scintigraphic images of rats bearing a MATLy Lu prostate tumor implanted in the thigh and receiving 5 Gray (left panel) or no ionizing radiation directed at the tumor. Images were acquired 20 hours after intravenous injection of ^{99m}Tc-PEG-liposomes. Ionizing radiation was delivered 24 hours before the injection of the ^{99m}Tc-PEG-liposomes. There was prominent increased uptake of ^{99m}Tc-PEG-liposomes in the tumor in the rat receiving 5 Gray of ionizing radiation.

The increased uptake of nanoliposomes following external beam radiation therapy are promising and suggest that combined modality therapy with external beam radiation and nanoliposome chemotherapy may be more effective for the treatment of tumors than administering these therapies sequentially.

Another research group has also performed similar studies that combined external beam radiation and nanoliposomes, but they did not find statistical differences [55]. Further investigations will need to be done to confirm or refute these findings. Because clinically available nanoliposomal doxorubicin can be radiolabeled using ^{99m}Tc-BMEDA, it will be possible to determine if radiation therapy administered to clinical cancer patients can increase liposomal uptake by using noninvasive scintigraphic imaging of the distribution of nanoliposomal doxorubicin administered immediately following radiation therapy.

4.6.3 Nanoliposomes as Carriers for Lymphatic Drug Delivery

Nanoliposomes have received much attention as lymph node drug delivery agents and imaging of these nanoliposomes has proven to be very valuable in the development of nanoliposome formulations that specifically trap in lymph nodes. This interest in the development of new lymph node drug delivery systems stems from the increasing awareness of the importance of lymph nodes in cancer metastasis and prognosis, the increasing understanding of the importance of lymph nodes for vaccine immune stimulation and the fact that the lymph nodes harbor human immunodeficiency virus (HIV) as well as other infectious diseases [31, 59, 96, 106].

The lymphatic system consists of a network of lymphatic vessels and lymph nodes that serve as a secondary vascular system to return fluid that leaks from the blood vessels in the extremities and other organs back to the vasculature [128]. As this fluid moves through the lymphatic system, it generally passes through a chain of lymph nodes that act as filters to remove and detect foreign agents.

The lymphatic system also moves substantial volumes of fluid that leak into the peritoneal cavity and pleural cavity back into the blood circulation. In this process, fluid from the cavities eventually moves into lymphatic vessels where it will pass through a lymph node prior to entering the systemic circulation.

Nanoliposomes are ideal carriers for delivery of diagnostic and therapeutic agents through to the lymphatic system and lymph nodes because following interstitial or intracavitary administration, nanoliposomes are too large to be directly absorbed into the blood stream or the linings of cavities, yet they are small enough to enter the lymphatic vessels and lymph nodes following interstitial or intracavitary administration.

Following subcutaneous injection or other injection directly into interstitial tissue, a certain portion of liposomes are taken up locally and retained for a prolonged time, while another portion of the liposomes are cleared from this local site and move into the lymphatic vessels where they are either trapped in lymph nodes or else move completely through the lymphatic system and return to the blood at the thoracic duct. The smaller the size of the liposomes administered the greater the clearance from the local site and the higher the fraction that moves into the lymphatic vessels [103], so that these nanoliposome formulations are cleared from the local site to a greater extent than larger liposome formulations. Following intracavitary administration in the peritoneal or the pleural cavity, almost all of the nanoliposomes are cleared from the cavity and move into the lymphatic vessels.

With standard liposome formulations, only a small fraction of the nanoliposomes injected subcutaneously are retained in each lymph node encountered (approximately 1% of the injected dose), so that the majority of the liposomes that are cleared from the injection site end up returning to the systemic circulation [96, 97, 103]. The relatively low retention of nanoliposomes in lymph nodes led our group to search for new ways to improve liposome retention in lymph nodes. This research resulted in development of a new method of increasing lymph node retention of subcutaneously injected liposomes [106]. This lymph node targeting method utilizes the high-affinity ligands biotin and avidin. Biotin is a naturally occurring cofactor and avidin is a protein derived from eggs. Biotin is attached to the surface of the nanoliposomes and these biotin-liposomes are either intracavitarily or interstitially injected. Around the same time as the biotin-liposomes are injected, avidin is injected at a nearby location so that the avidin and biotin-liposomes eventually meet within the lymphatics and lymph nodes and then become aggregated and trapped in the draining lymph nodes. Avidin and biotin have an extremely high affinity for each other. Avidin has four receptor sites for biotin associated with each molecule. These four receptor sites permit the binding of multiple biotin molecules, which causes aggregation of nanoliposomes that have biotin on their surface. This aggregation can be described as *in vivo* nanoassembly of artificially produced nanoparticles and this is the first description of this occurrence that the authors are aware of. This *in vivo* nanoassembly has also been recently refered to as "functional vesicle assemblies" [83]. Nanoaggregation can be assumed because of the very high retention in lymph nodes when the avidin is given compared with when the biotin-nanoliposomes are administered without avidin. When avidin is co-administered, ~20% of the nanoliposomes that are cleared from the injection site become trapped in the first lymph node encountered while only ~2% of the nanoliposomes remain trapped in the lymph node without avidin. This difference is clearly shown in Fig. 4.5, where ^{99m}Tc radiolabeled biotin-liposomes were injected into both rabbit hind paws, but only one hind paw received an adjacent injection of avidin, so the other leg of the rabbit served as a control. Scintigraphic imaging in Fig. 4.5 clearly demonstrates the high accumulation of the nanoliposomes in the popliteal lymph node located in the rabbit's hind leg that also received avidin while the control side has only minimal uptake in the popliteal lymph node.

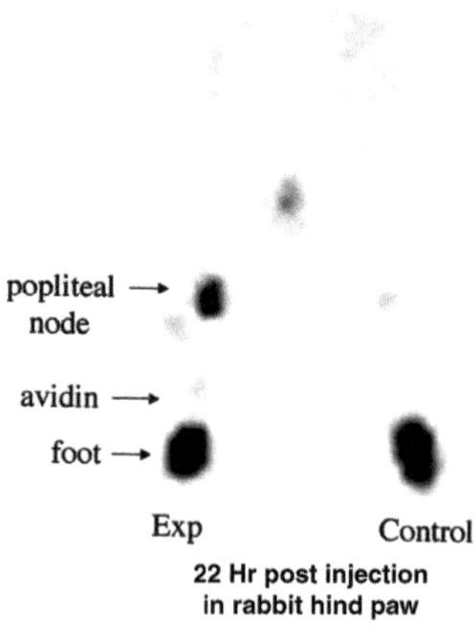

Figure 4.5 Scintigraphic image of legs and lower abdomen of rabbit at 22 hours after subcutaneous injection of ^{99m}Tc-biotin-liposomes in both hind paws. This was followed by subcutaneous injection of avidin in the left hind paw. Increased retention of ^{99m}Tc activity in left popliteal node compared with right popliteal node is easily visualized.

Our group has also shown that when biotin-nanoliposomes encapsulating blue dye are injected into the peritoneal cavity in conjunction with avidin, the nanoliposomes become trapped in either peritoneal cavity or in the lymph vessels and lymph nodes that drain from the peritoneal cavity through the diaphragm and mediastinal lymph nodes [108] (Fig. 4.6). Our group has utilized this peritoneal trapping as a method to treat a rat model of intraperitoneally disseminated ovarian cancer metastasis with rhenium-labeled blue-biotin-nanoliposomes [145]. Direct imaging of ^{186}Re-blue-biotin-nanoliposomes with or without avidin co-administration was possible because of the 10% emission of a gamma photon from rhenium in the 140 keV range.

As can be seen in Fig. 4.6, when avidin was co-administered, the blue-biotin-nanoliposomes were well retained in the peritoneal cavity and mediastinal lymph nodes with minimal activity visualized in the liver and the spleen.

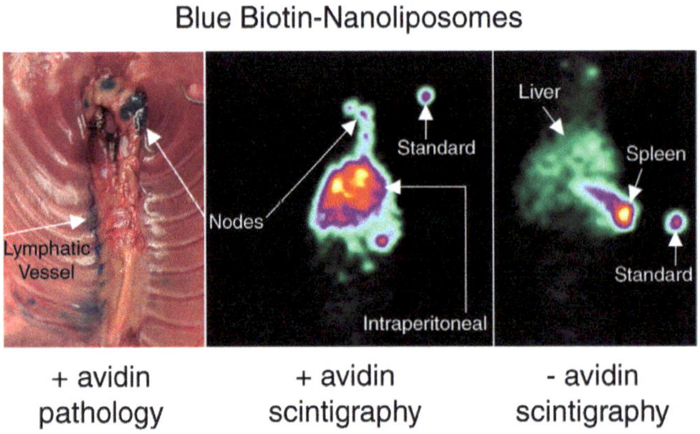

Figure 4.6 *Left panel*: Digital image of blue stained mediastinal nodes and lymphatic channel running along the spine acquired 22 hours after injection of ^{99m}Tc-blue-biotin liposomes in peritoneal cavity of rats with interperitoneal ovarian cancer followed 30 minutes later by avidin. *Middle panel*: Scintigraphic image acquired 22 hours after intraperitoneal injection of ^{99m}Tc-blue-biotin liposomes and avidin in rat with intraperitoneal ovarian cancer. *Right panel*: Scintigraphic image of rat with intraperitoneal ovarian cancer acquired 22 hours after injection of only ^{99m}Tc-blue-biotin-liposomes.

When the avidin was not co-administered with the ^{186}Re-blue-biotin nanoliposomes, the nanoliposomes returned to the systemic circulation and were taken up by the liver and spleen which are well visualized on the image in Fig. 4.6. In the rat model with disseminated ovarian cancer there was an improved cancer therapy effect of the ^{186}Re-blue-biotin-nanoliposomes when the avidin was administered than when it was not. This was demonstrated by significantly decreased PET imaged fluorine-18-fluorodeoxyglucose uptake in the peritoneal region and by a decrease on malignant cells that were detected in the ascitic peritoneal fluid [145]. This trapping in the peritoneal cavity should have significant application for drug delivery and retention in the peritoneal cavity.

4.6.4 Scintigraphic Imaging for the Development of Ligand-Specific Targeted Nanoliposome Agents

Scintigraphic imaging can be particularly valuable in the development of targeted nanoliposomes by surface modification with specific ligands. A variety of specifically targeted nanoliposomes are currently under development [134]. Targeted liposomes carrying a chemotherapeutic agent have been shown by imaging studies to have substantially greater uptake in the tumor and be more effective for treating murine lung carcinomas than control liposomes [32]. Targeting may also be helpful for specific diagnostic imaging applications. Use of imaging for the development of one type of specifically targeted nanoliposome is described in Chapter 5 of this book by Elbayoumi and Torchilin.

4.6.5 Scintigraphic Imaging of Bone Marrow–Targeted Liposomes as an Example of an Imaging Modality That Can Be Translated Across All Species, Small and Large

Scintigraphic imaging can be performed for determination of the location of nanoparticles in almost all living species. This makes scintigraphic imaging particularly valuable for development of therapeutic nanoparticles because findings in small animals have the potential to be directly translated to humans who can also be imaged with scintigraphic imaging. Small animal SPECT camera systems have been specially developed to image mice and rats and many of these systems can perform CT imaging with the mouse or rat in the same position, providing a more detailed anatomic location of the scintigraphic image when it is fused with the CT image (Fig. 4.2).

Scintigraphy is a valuable imaging technique in drug development because a wide variety of species can be imaged with this imaging modality. This capability is proving to be valuable for the testing in a variety of therapeutic agents, including nanoparticles in larger species that may better represent a particular disease or physiologic processes found in humans. Scintigraphic imaging has been performed in rabbits, dogs, cats, primates, pigs and horses in addition to humans [21, 27, 42, 56, 106, 107, 111, 122, 129, 130]. Imaging in multiple species may be particularly important in studies of nanoparticles because studies have shown that there are significant differences in the circulation times and distribution of nanoparticles in larger animals and humans as compared with mice and rats.

The ability to image a wide variety of species noninvasively can provide very useful information to assess potential clinical usefulness of different nanoparticle formulations. As an example, as part of an investigation of a special nanoliposome formulation that was serendipitously found to have high bone marrow uptake in rabbits as part of an investigation of ideal formulations of liposomes for hemoglobin encapsulated blood substitutes, we observed a very different distribution of the same formulation in rats and mice. This unexpected biodistribution might not have been detected without the use of scintigraphic imaging, which clearly demonstrated whole body distribution in rabbits. This particular nanoliposome formulation having a surface modified with an anionic amphiphile; L-glutamic acid, *N*-(3-carboxy-1-oxopropyl)-,1,5,-dihexadecyl ester (SA) (5% of total lipid) had a very high uptake in rabbit bone marrow with over 60% of the injected dose of nanoliposomes being taken up in the bone marrow in rabbits by 24 hours [126]. A high fraction of this bone marrow uptake occurred within 6 hours of intravenous injection. Histological examination of the bone marrow in rabbits revealed that these bone marrow–targeting nanoliposomes were taken up in phagocytic cells of the bone marrow. We hypothesized that these bone marrow–targeted liposomes had a surface that made them appear similar to aging erythrocytes or white blood cells which are heavily removed by marrow. Alternatively, the surface of the nanoliposomes may mimic erythroblast nuclei that are expelled by the erythroblast shortly before maturing erythrocytes enter into the circulation. These erythroblast nuclei are taken up by bone marrow macrophages.

Delivery to bone marrow may be very useful for various drug delivery strategies due to the importance of the bone marrow for erythrocyte, platelet and immune cell production. Because we were interested in further investigation of these bone marrow–targeted liposomes in various experimental therapeutic models, we proceeded to study their distribution in rats and mice as a prelude to interventional drug delivery studies.

To our surprise, these bone marrow–targeted nanoliposomes had much lower uptake in the bone marrow of rats and mice. The quantitatively measured uptake in rats at 6 hours obtained from direct tissue counting was 9.3% of the total injected dose, which although much lower than rabbits and primates, was significantly higher than standard liposome formulations which generally have no more than 1–2% uptake in the bone marrow. Rats were also found to have very high percentage uptake in the spleen with more than 22% of the injected dose of nanoliposomes found in the spleen of rats at 6 hours post injection. This contrasts with only 5% of the injected dose taken up by the spleen of rabbits at 6 hours post injection [126]. Mice also were found to have a relatively low uptake in the bone marrow with 12.5% of the dose in the marrow at 6 hours. With such discrepant results, it was very hard to determine if this formulation would have a high biodistribution in human subjects as that observed in rabbits. It was possible that rabbits had a unique response that would not be applicable to other mammals or to humans.

Fortunately, with scintigraphic imaging, it was possible to perform noninvasive studies of the ^{99m}Tc-bone marrow–targeted nanoliposomes in a rhesus primate model which should be considered a good predictor of their biodistribution in humans. The primate model was very similar to the rabbit model, with an even larger percentage of the injected dose (70%) determined to be in the bone marrow as early as 3 hours post-injection [125]. As depicted in Fig. 4.7, the rat image had a low uptake in the bone marrow while the rabbit and rhesus primate have very obvious high uptake in the bone marrow at 6 and 3 hours post injection, respectively.

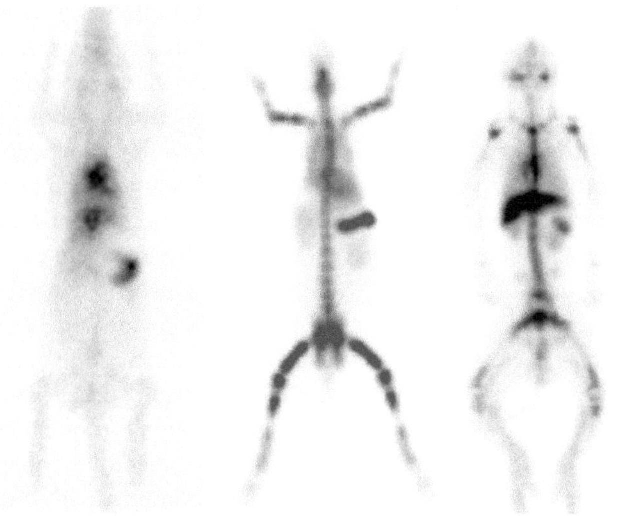

Figure 4.7 Scintigraphic images of rat, rabbit, and rhesus primate injected intravenously with ^{99m}Tc-bone marrow–targeted liposomes.

The difference in distribution of these bone marrow–targeted nanoliposomes between the mice, rat, rabbit, and rhesus primate is not known; however, these observations may be applicable to other nanoparticles and should be taken into account by investigators working with any type of intravenously administered nanoparticles. It may be important to study distributions in other species and it is possible that the rabbit will be a very useful model for use in nanoparticle studies to predict nanoparticle distribution in humans. One possibility to explain the very different inter-species distributions may be related to the relatively larger spleen to body weight ratios found in rats and mice as compared with rabbits and primates. The spleen to body weight ratio of the rat is 3 times greater than that of the primate and 6 times greater than the spleen to body weight ratio of the rabbit. If the nanoliposomes are being taken up in relationship to total cell turnover, it may be that rats have a very high rate of lymphocyte production in their large spleen as compared with other species and this fact is related to their high uptake in the spleen and relatively lower uptake in the bone marrow.

The significant differences in *in vivo* biodistribution found with this particular liposomal nanoparticle formulation between the rat and rabbit/rhesus primate models demonstrates the importance of studying nanoparticle distributions in a variety of animal species, particular larger animals such as rabbits and nonhuman primates. Quantitative imaging studies of the distribution of nanoparticles in larger animals can best be performed with scintigraphic imaging.

4.7 NANOLIPOSOMES AS A THERANOSTIC AGENT

Nanoparticles that can be labeled with one or multiple tracer agents so that they can be imaged *in vivo* are very promising as theranostic agents. A theranostic agent has been defined as an agent that has the ability to affect therapy or the treatment of disease [101]. Theranostics have three principal applications that can result in a personalized treatment that will be more effective for a particular patient. These include (i) identification of subgroups of patients likely to give a positive response to a given therapy, (ii) identification of subgroups of patients at risk of aggravated side effects during treatment with a given therapy, and (iii) an agent that can monitor the response to a given treatment [101]. Noninvasive medical imaging has been suggested as an important area for theranostics [82, 101] and nanoparticles in particular have been described as having an important role in theranostics due to the ability to carry multiple contrast agents that permit multimodal imaging [82]. Nanoparticles such as nanoliposomes labeled with diagnostic imaging agents can be imaged to determine if they will have an effective biodistribution in a particular patient and then the same nanoparticle carrying a therapeutic agent can be administered to the patient. It has also been suggested that the ability to determine whether a nanoparticle will be able to reach its intended target in preclinical and initial clinical testing using diagnostic agents will accelerate the pace of drug development [101].

Nanoliposomes, like other nanoparticles, are good candidates as theranostic agents since they can simultaneously carry both diagnostic and therapeutic agents. Nanoliposomes as diagnostic agents can determine and quantify liposome drug delivery to disease processes in a specific patient prior to administration of same nanoliposome formulation that carries the therapeutic payload. Liposomes that simultaneously carry both diagnostic and therapeutic agents can be imaged to verify and quantify delivery.

An obvious theranostic cancer application with radiolabeled nanoliposomes is provided by the ability to label the same nanoliposome with the most commonly used diagnostic SPECT imaging agent, ^{99m}Tc, and the therapeutic radionuclide ^{186}Re or ^{188}Re. This is because the chemistry of the diagnostic ^{99m}Tc is very close to that of the rhenium radionuclides. With this system, theranostic imaging of the nanoliposomes prior to treatment can permit determination of the precise dose of a nanoliposome carried therapeutic agent to be administered in a particular patient. When the radiotherapeutic nanoliposome is administered, the same distribution can be anticipated because the nanoliposomes will be labeled using the same chemistry and methods. This ^{99m}Tc/^{186}Re theranostic system also fufills the last criteria of an effective theranostic because the precise distribution of the radioactive rhenium nanoliposome can be monitored due to the gamma photon emission of the rhenium isotopes (^{186}Re or ^{188}Re) following therapy to ensure complete treatment of the targeted tumor and determine the exact dose that was actually delivered. In addition, because nanoliposomes can be multi-modal carriers of contrast agents, these same theranostic nanoliposomes could also be labeled with MRI contrast agents to provide much higher noninvasive resolution of nanoliposome distribution than is possible with SPECT imaging at the local level or with optical contrast imaging agents to provide microscopic resolution of the tissue level following organ removal and tissue sectioning.

4.7.1 Nanoliposomes as Carriers of Therapeutic Radionuclides

Nanoliposomes for radiotherapy applications can be radiolabeled with either alpha-particle-emitting radionuclides or beta-particle-emitting radionuclides and there has been significant recent research interest in labeling of nanoliposomes with either of these two types of therapeutic radionuclides [15, 23, 25, 33, 44, 53, 67, 88, 127, 140, 145]. The use of nanoliposomes as carriers of therapeutic radionuclides has many differences from the use of nanoliposomes as carriers of chemotherapeutic agents. These differences with advantages and disadvantages are listed in Table 4.1. One distinct advantage is that therapeutic radionuclide nanoliposomes do not have to have a mechanism for releasing the radionuclide from the nanoliposomes because the therapeutic alpha or beta radio-particle released from the nanoliposomes delivers therapeutic radiation through many cell diameters. This same penetration of therapeutic radioparticles also means that the nanoliposomes do not have to reach all the cells in a tumor to be effective and so the requirements of a homogeneous distribution of nanoliposomes in the tumor are less rigorous than for a chemotherapeutic nanoliposome agent. This radio-particle penetration can treat tumor cells that are not reached by the radionuclide labeled nanoliposomes by a cross-fire mechanism as illustrated in Fig. 4.8.

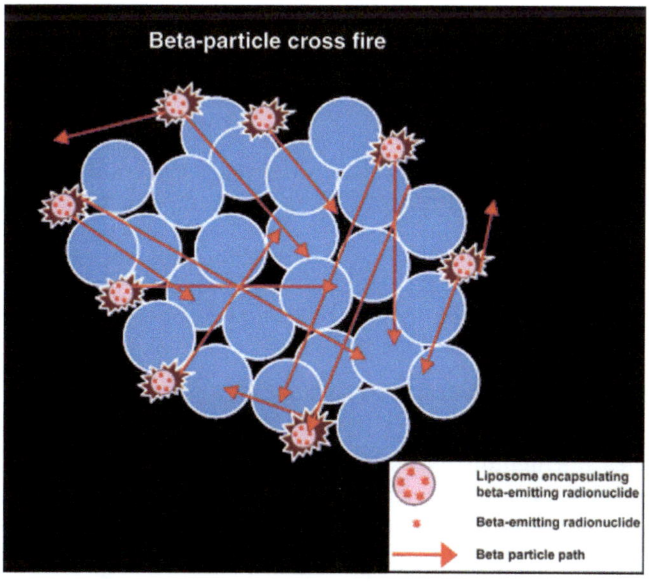

Figure 4.8 Diagram depicting the field effect of Beta-particle emission.

Table 4.1 Comparison of nanoparticle beta-particle-emitting radionuclide therapy (nanobrachytherapy) versus nanoparticle chemotherapy for treatment of cancer

Nanoparticle-beta emitting radionuclide therapy	*Nanoparticle chemotherapy*
No release from the nanoparticle is required	Chemotherapeutic agent must be released from the nanoparticle to be effective
May have improved retention in the tumor if the radionuclide is not released from nanoparticle	Since release of the chemotherapeutic agents from the nanoparticle is required, the chemotherapeutic may rapidly leave the region of the tumor following release
Requirements for homogeneity of distribution are less rigorous due to cross-fire "effect" resulting from 1–5 mm average pathlength of the beta particle	A more homogeneous distribution is required for effective treatment of tumor due to limited extension of encapsulation of chemotherapeutic agent from the nanoparticle
Can be effective when administered intracavitarily or intratumorally	Can also be effective when administered intracavitarily or intratumorally; however, more rigorous homogeneity required
No requirement of therapeutic radionuclide to reach the nucleus or cytoplasm	Chemotherapeutic agent must cross cell membrane and often nuclear membrane to be effective
Imaging of regional localization of radionuclide labeled liposomes possible because many also emit gamma photons permitting determination of *in vivo* biodistribution	Additional procedure required to radiolabel liposomal carrier or encapsulated chemotherapeutic agent for noninvasive quantification of regional dose distribution
Radionuclide therapy always actively delivering radiation dose until decay	Potential to control the exact timing of release of chemotherapeutic agent in the local region by a variety of methods including pH, thermal release, enzyme activation, etc.

4.7.1.1 Alpha particles versus beta particles for use with nanoliposomes

Alpha particles are helium nuclei that are emitted from certain high molecular weight radionuclides generally have molecular weights of over 200 MW. They deposit a high amount of energy along a short path that is very damaging to DNA. Alpha particle carrying nanoparticles are potential theranostic agents because they also emit gamma photons that can be imaged [124]. Beta particles are electrons that are emitted from radionuclides that are unstable and they tend to have lower molecular weights (50–190 MW) than alpha-particle-emitting radionuclides. Alpha particles have a high energy transfer and very short path length through tissue compared with beta particles. The DNA damaging energy alpha particles release along their short path length is 100 to 1000 times greater than the energy released from beta particles [120]. This means that a small number of alpha particle tracks through a cell nucleus can lead to death of a cancer cell, whereas a much greater number of tracks are required for beta particles to cause cell death. Obviously, alpha-particle-emitting radionuclides can be very effective and highly lethal when in very close range to the targeted cell, however, their disadvantage is that they have to reach very close to the targeted cell to be effective due to their very short path length of 0.05–0.1 mm or 5–10 cell diameters. In comparison, beta-particle-emitting radionuclides have average path lengths of 1–4 mm or 40–160 cell diameters. The short path length and high energy release can be viewed as an advantage or a disadvantage. Its advantages include that it is very toxic to nearby cells within the length of its short path length. An obvious disadvantage of alpha particles carried by nanoparticles is that the nanoparticle will need very high homogeneity within the tumor to achieve effective therapy as compared with beta particles. Although there has been much theoretical discussion and publications concerning alpha particle dosimetry related to the possible use of liposomes for carriers of alpha-particle-emitting radionuclides, there has been, to date, almost no investigations with actual alpha particle carrying liposomes. Two radionuclides that have been discussed for actual treatment of nanoliposomes with the alpha-particle-emitting radionuclides are actinium-225 and radium-223 [22, 58, 62, 123, 124]. Recently, our group has developed beta-particle-emitting radionuclides, rhenium-186, and rhenium-188, for labeling nanoliposomes [15, 23, 25, 44, 53, 88, 127, 140, 145]. Our group has investigated ^{186}Re-labeled nanoliposomes using BMEDA radiolabeling method for treatment of head and neck cancer and ovarian cancer [127, 140–141, 145]. Another research group has also used the same BMEDA chemistry to label liposomes with ^{188}Re [23, 24].

4.7.1.2 Advantages of rhenium for nanoliposome radionuclide therapy

One significant advantage of both ^{186}Re and ^{188}Re therapeutic radionuclides compared with many other therapeutic radionuclides is that they emit a low ratio of photons in the ideal range of scintigraphic imaging. For every 10 and 6.5 beta emissions, ^{186}Re and ^{188}Re radionuclides emit a gamma photon in the ideal energy range near that of ^{99m}Tc (140 keV). These ratios of beta photons to gamma photon are ideal for use as a theranostic agent, permitting accurate imaging and quantification of the distribution of the therapeutic agent with either planar or SPECT imaging. This ability to accurately image the distribution of ^{186}Re-liposomes provides a significant advantage as compared with other therapeutic radionuclides such as yttrium-90 (^{90}Y) and phosphorus-32 (^{32}P), which are pure beta-emitting radionuclides without gamma photons for imaging. A second significant advantage of ^{188}Re and ^{186}Re compared with many other therapeutic radionuclides such as ^{90}Y and ^{32}P is that they have almost no affinity for bone. Bone uptake occurs when the radioisotope eventually becomes separated from its chelator during metabolism of its carrier in the body. High bone uptake of therapeutic radionuclides leading to the suppression of hematopoietic cells in bone marrow is generally the dose limiting factor with current antibody-based radionuclide therapies. In contrast, rhenium radionuclides are mainly cleared through the liver and kidneys in a similar manner as ^{99m}Tc without significant bone marrow deposition.

Previous theoretical dosimetry studies have addressed the potential use of radiotherapeutic liposomes for treatment of tumors via intravenous injection [33, 67, 68]. In a recent experimental study with long circulating pegylated liposomes radiolabeled with ^{188}Re using the BMEDA method, ^{188}Re liposomes had much higher uptake in the tumor as compared with ^{188}Re-BMEDA alone. ^{188}Re-liposomes were found to have a 7.1-fold higher tumor–to-muscle ratio as compared with intravenously administered unencapsulated ^{188}Re-BMEDA in a C26 murine colon carcinoma solid tumor animal model [23]. It is also a great advantage to have the availability of using either of two isotopes of rhenium, ^{186}Re or ^{188}Re. This is because these two rhenium isotopes have different half-lives and different average tissue path lengths which can permit selection of the best radionuclide for a particular application. ^{186}Re has a 90 hours half-life and a mean tissue pathlength of its emitted beta particle of 2 mm. ^{188}Re has a shorter 17 hours half-life and a mean tissue pathlength of its emitted beta particle of 4 mm. Both radionuclides are readily available and ^{188}Re can be obtained from a generator, similar to ^{99m}Tc, that can be used for 3 months. The availability of a generator has the potential to make ^{188}Re widely and inexpensively available for nanoliposome cancer therapy applications [66, 78].

4.7.2 Intravenous Tumor Therapy with ^{186}Re-liposomal Doxorubicin

Intravenously administered ^{186}Re-nanoliposomes have potential for treatment of solid tumors. Theoretical studies have shown that intravenously administered radiotherapeutic nanoliposomes may deliver a larger dose to a solid tumor with less dose delivered to red bone marrow as compared with monoclonal antibodies labeled with therapeutic radionuclides [33]. Multimodal combination therapy with 2 or more therapies used at the same time may also prove more effective than treatment of cancer with single agents. Intravenously administered nanoliposomal radionuclide therapy and combination nanoliposomal chemo/radionuclide therapy have been investigated by our group for the treatment of a head and neck cancer [127]. This combination therapy approach is possible due to the ability to label nanoliposomes encapsulating chemotherapeutic agents with ^{186}Re using the BMEDA radiolabeling method. Our group has investigated the use of commercially available liposomal doxorubicin labeled with ^{186}Re [127]. As shown in Fig 4.9, combinational therapy nanoliposome doxorubicin/^{186}Re was more effective in treating a solid head and neck tumor model than nanoliposomal doxorubicin alone. These studies show the potential of this approach for clinical treatment of solid tumor using intravenously administered radiotherapeutic/chemotherapeutic nanoliposomes.

4.7.3 Intratumoral Delivery of ^{186}Re-Nanoliposomes

Nanoliposomes are proving to have very promising potential for direct interstitial therapy of disease processes. The nanoliposome carrier helps the encapsulated agent disperse through the local tissue while facilitating the retention of the agent in the local region. Direct intratumoral drug delivery of chemotherapeutic agents, radionuclides, and genes has been preliminarily investigated for the treatment of cancer [85, 116, 131, 132, 142]. Our previous studies and the studies by several other groups have shown excellent promise for the use of nanoliposomes as a carrier for image-guided treatment of solid tumors via direct intratumoral administration [17, 52, 92, 93, 114, 117]. This approach appears to be one of the more promising applications of nanotechnology for the treatment of cancer.

4.7.3.1 CED administration of nanoliposomes

The area of greatest progress and largest number of studies has been performed investigating local infusion into the brain for the treatment of various brain pathologies, primarily brain tumors. Investigators in this area have developed a technique termed convection enhanced delivery (CED) for infusion of the brain as a method to achieve the largest amount of dispersion from the needle tip throughout the brain. CED is a method in which the therapeutic agents are slowly injected through an infusion pump to provide a constant pressure force that facilitates the dispersion of the agent through the tissue. To reach a broad dispersion of the therapeutic agents after administration, it has been recognized that CED is a preferred method of direct intratumoral drug delivery [80, 113, 138]. Nanoparticles, in particular nanoliposomes, have been shown to have the most ideal characteristics for infusion by CED. Studies by Bankiewicz's group have also shown that liposomal topotecan when infused by CED in the brain had sustained days-long high concentration retention in brain tumor, while free topotecan was quickly cleared in 2–3 hours after administration [116]. In previous intratumoral micro-distribution and retention studies, liposomal therapeutic agents have been investigated in detail [17, 114, 141, 145]. Our group has used small animal SPECT imaging to monitor the intratumoral distribution of radiolabeled nanoliposomes in a 1.5 cm diameter head and neck cancer tumor in nude rats. In this study, imaging clearly demonstrated the advantages of nanoliposomes administered directly into a solid tumor by CED as compared with the administration of the same radionuclide as a free drug not associated with a liposome.

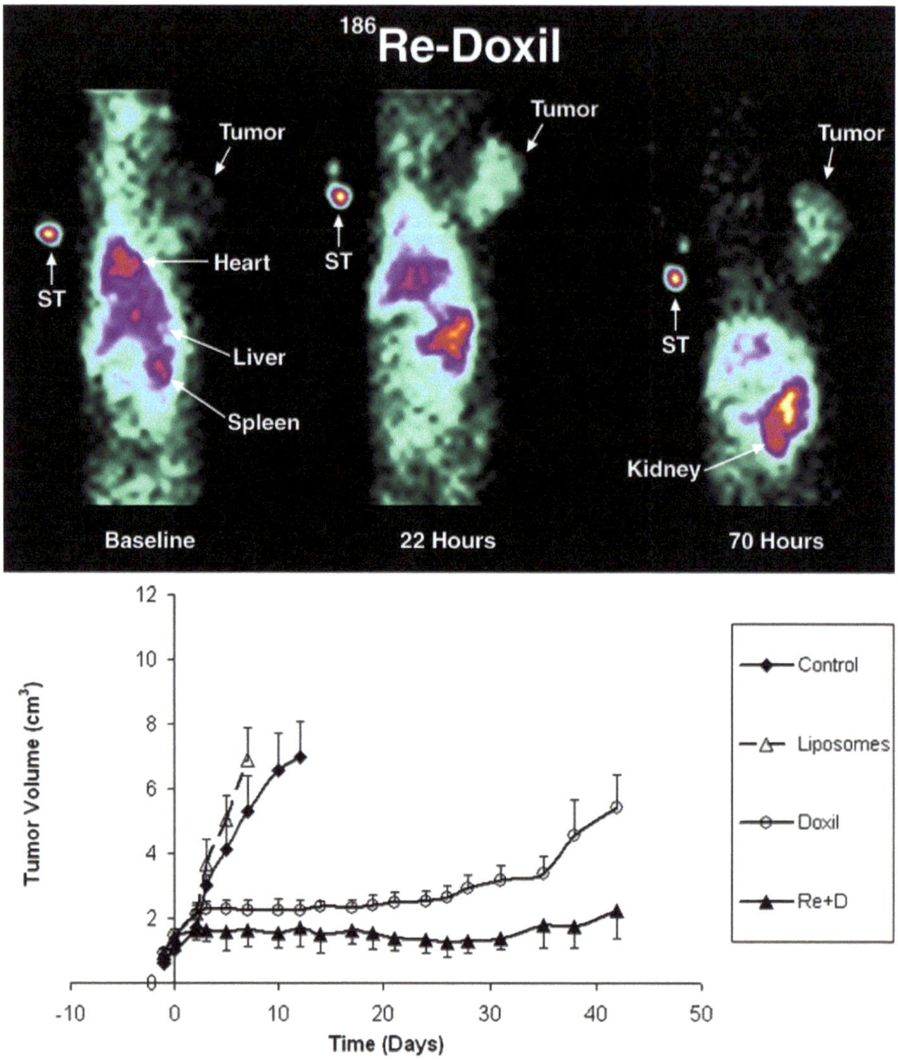

Figure 4.9 *Top panel*: Scintigraphic images of head and neck tumor bearing rats at 0, 22 and 70 hours after intravenous injection of ^{186}Re-liposomal doxorubicin. *Bottom panel*: Tumor growth following intravenous injection of ^{186}Re-Liposomal doxorubicin was controlled about 7 days longer than Doxil therapy alone.

As shown in Fig. 4.10, the radiolabeled nanoliposomes spread through the tumor and are relatively well retained 4 hours after administration whereas the radiolabeled BMEDA chelator spread less and had less tumor retention. The ability of the nanoliposomes to spread through the tumor tissue is likely related to their nanoparticle size which prevents direct uptake into the capillaries or localized binding to tissue near the site of the injection.

The *in vivo* distribution of liposomes can also be obtained with MR imaging [52, 92–93, 114, 116], if liposomes are surface-tagged or encapsulated with MR contrast agents, such as Gd-DTPA. Real-time MR imaging during CED of liposomes or liposomes encapsulating therapeutic agents into animal brain or brain cancer xenografts has been previously investigated. These studies have shown that intratumoral distribution volume of nanoliposome drug carriers has a linear relationship with infusion volume, while the lipid concentration does not affect this distribution [69]. In another study, nanoliposomes were shown to disperse and spread well through a primate brain when administered by CED with no evidence of toxicity[70].

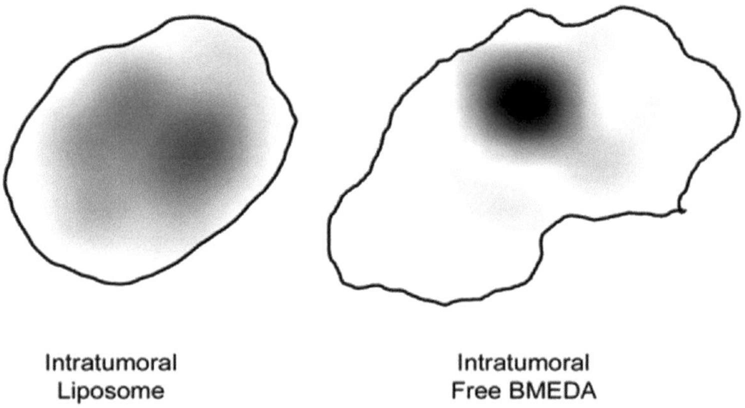

Figure 4.10 Pinhole collimator SPECT images of excised head and neck xenograft tumors acquired 20 hours after direct intratumoral injection of ^{99m}Tc-Liposomes (*left*) or ^{99m}Tc-BMEDA (*right*). Black lines trace outline of tumor.

When considering the practical challenges in micro-distribution homogeneity in a solid tumor and the required mechanisms of tumor cell killing, therapeutic radionuclides appear to have an advantages when delivered by direct intratumoral administration [85], since the path range of energetic particles generated following radionuclide decay may be able to compensate for the intratumoral micro-distribution inhomogeneity [146]. This is particularly true with beta-emitting radionuclides that have average pathlength ranging from 1–5 mm.

4.7.3.2 CED delivery of ^{186}Re-nanoliposomes

A direct method of radiolabeling liposomes developed by the authors' group has been used [14–16]. By using this convenient after-loading labeling method, both diagnostic radionuclide, ^{99m}Tc, and therapeutic beta-emission radionuclides, ^{186}Re and ^{188}Re, can be efficiently and stably encapsulated into liposomes. The use of animal imaging scanner equipped with 1 mm pinhole collimators was used to obtain intratumoral distribution of liposomes at 1 mm spatial resolution [17, 141].

In a study by our group, ^{186}Re-nanoliposomes were infused by CED into solid tumor of 1.5 cm in diameter. A SPECT image showing the distribution of the rhenium-liposomes through the tumor is shown in Fig 4.11. Although homogeneity within the tumor is less than ideal, it can clearly be seen that the ^{186}Re-liposomes have spread through the tumor.

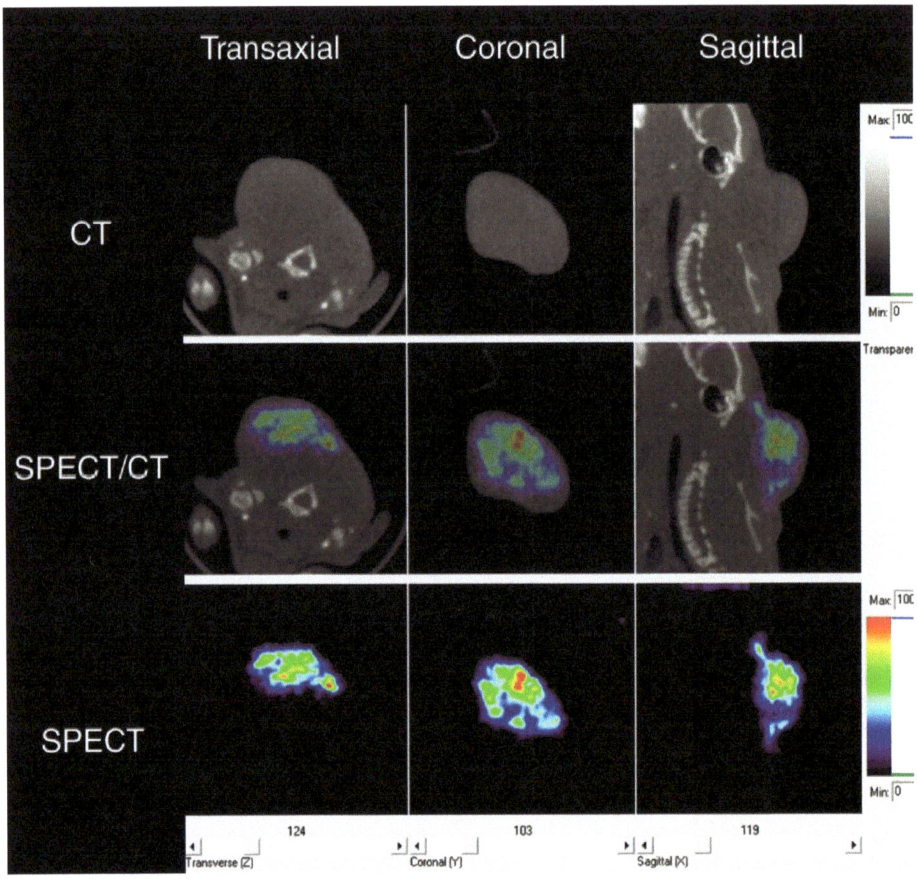

Figure 4.11 Micro CT, fused SPECT/CT and pinhole collimator SPECT images in transaxial coronal and sagittal views of head and neck tumor bearing rat acquired at 20 hours after intratumoral CED injection of ^{186}Re-nanoliposomes.

The retention of the rhenium liposomes was compared to the same dose of ^{186}Re in either the form of free ^{186}Re-perrhenate or free ^{186}Re-BMEDA. After convection-enhanced delivery of radiolabeled liposomes, a broad dispersion of the ^{186}Re-liposomes in head and neck squamous cell carcinoma xenografts was observed. Approximately 50 % of nanoliposomes were initially retained in the tumor followed by a stable, prolonged intratumoral retention, whereas smaller molecules of ^{186}Re-perrhenate and ^{186}Re-BMEDA had a poor intratumoral distribution homogeneity and a more rapid clearance from tumor. A graph of tumor clearance is shown in Fig. 4.12.

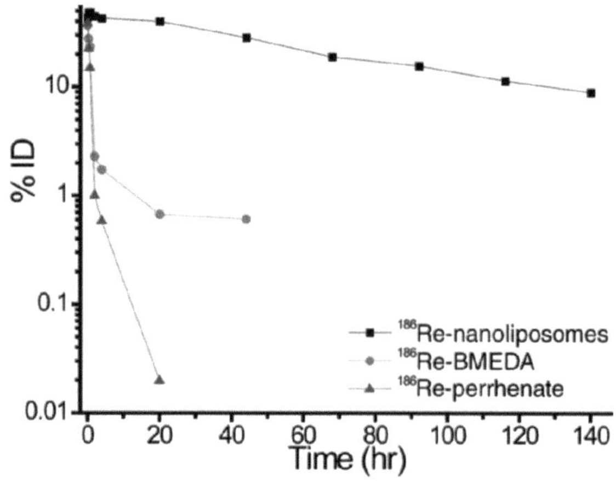

Figure 4.12 Average percent of injected ^{186}Re radioactivity present in the head and neck tumors over time following CED intratumoral injection of ^{186}Re-nanoliposomes, ^{186}Re-perrhenate, or ^{186}Re-BMEDA. Retention was significantly improved by encapsulation in nanoliposomes.

Only the ^{186}Re-nanoliposomes were effective in treating the tumor as can be seen in Fig 4.13 because more ^{186}Re-nanoliposomes were more homogeneously distributed and retained in the tumors. This type of "nanobrachytherapy" can be much less toxic to normal tissue that is adjacent to the local region. The large difference in therapeutic effectiveness clearly demonstrates the significant advantage of the nanoliposome carrier for the treatment of solid tumors by CED. When carefully administered by image guidance with sophisticated imaging technology in order to paint the tumor, rhenium-nanoliposome CED local therapy has potential to supplant other external beam radiation therapies.

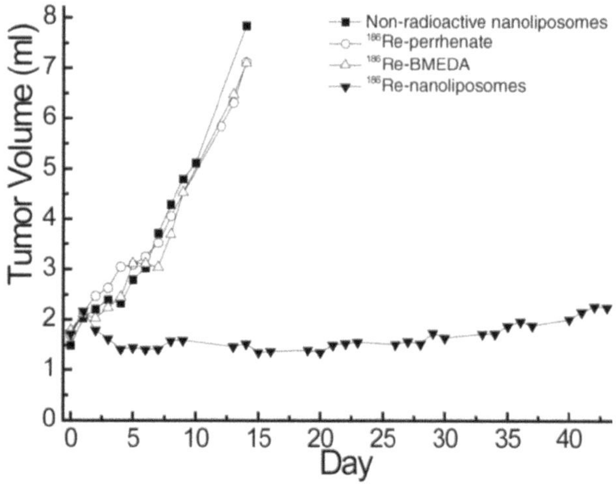

Figure 4.13 Average tumor growth after CED intratumoral injection of ^{186}Re-nanoliposomes versus control groups of ^{186}Re-perrhenate, ^{186}Re-BMEDA, or nonradioactive nanoliposomes. Encapsulation of ^{186}Re inside nanoliposomes greatly improved tumor control.

4.8 SUMMARY AND CONCLUDING REMARKS

The performance of nanoliposome scintigraphic imaging is an excellent tool for development of novel nanoliposome formulations. Theranostic nanoliposomes have great flexibility as not only multi-modality diagnostic imaging agents but also multi-therapeutic agent carriers. Scintigraphic imaging can facilitate the development of useful diagnostic and therapeutic agents across all sizes of species and into humans. Theranostic scintigraphic imaging of liposomal nanoparticles offers the promise of greatly improving dosimetry in individual patients and the noninvasive monitoring of therapeutic delivery so that the amount of a therapeutic agent delivered to a disease target can be easily correlated with the response of the disease target to the therapy. This theranostic ability promises to improve disease treatment and accelerate clinical translation of new, better-understood therapeutic agents.

References

1. Ahmed M., and Goldberg S. N. (2004) Combination radiofrequency thermal ablation and adjuvant IV liposomal doxorubicin increases tissue coagulation and intratumoural drug accumulation. *Int J Hyperthermia* **20**, 781–802.
2. Ahmed M., Lukyanov A. N., Torchilin V., Tournier H., Schneider A. N., and Goldberg S. N. (2005) Combined radiofrequency ablation and adjuvant liposomal chemotherapy: effect of chemotherapeutic agent, nanoparticle size, and circulation time. *J Vasc Interv Radiol* **16**, 1365–1371.
3. Allen T. M., Hansen C., Martin F., Redemann C., and Yau-Young A. (1991) Liposomes containing synthetic lipid derivative of poly(ethylene) show prolonged circulation half-lives *in vivo*. *Biochimica et Biophysica Acta* **1066**, 29–36.
4. Allen T. M., Sapra P., and Moase E. (2002) Use of the post-insertion method for the formation of ligand-coupled liposomes. *Cell Mol Biol Lett* **7**, 889–894.
5. Awasthi V., Goins B., McManus L., Klipper R., and Phillipsa W. T. (2003) [99mTc] liposomes for localizing experimental colitis in a rabbit model. *Nucl Med Biol* **30**, 159–168.
6. Awasthi V., Yee S. H., Jerabek P., Goins B., and Phillips W. T. (2007) Cerebral oxygen delivery by liposome-encapsulated hemoglobin: a positron-emission tomographic evaluation in a rat model of hemorrhagic shock. *J Appl Physiol* **103**, 28–38.
7. Awasthi V. D., Garcia D., Goins B. A., and Phillips W. T. (2003) Circulation and biodistribution profiles of long-circulating PEG-liposomes of various sizes in rabbits. *Int J Pharm* **253**, 121–132.
8. Awasthi V. D., Goins B., Klipper R., and Phillips W. T. (1998) Dual radiolabeled liposomes: Biodistribution studies and localization of focal sites of infection in rats. *Nuclear Medicine and Biology* **25**, 155–160.
9. Baker-Woudenberg I. A. J. M., Lokerse A. F., ten Kate M. T., Mouton J. W., Woodle M. C., and Storm G. (1993) Liposomes with prolonged blood circulation and selective localization

in *Klebsiella pneumoniae*-infected lung tissue. *Journal of Infectious Diseases* **168**, 164–171.

10. Baker-Woudenberg I. A. J. M., Lokerse A. F., ten Kate M. T., and Storm G. (1992) Enhanced localization of liposomes with prolonged blood circulation time in infected lung tissue. *Biochim Biophys Acta* **1138**, 318–326.

11. Bakker-Woudenberg I. A., Schiffelers R. M., Storm G., Becker M. J., and Guo L. (2005) Long-circulating sterically stabilized liposomes in the treatment of infections. *Methods Enzymol* **391**, 228–260.

12. Bakker-Woudenberg I. A. J. M., ten Kate M. T., Stearne-Cullen L. E. T., and Woodle M. C. (1995) Efficacy of gentamicin or ceftazidime entrapped in liposomes with prolonged blood circulation and enhanced localization in *Klebsiella pneumoniae*-infected lung tissue. *Jornal of Infectious Diseases* **171**, 938–947.

13. Bangham A. D. (1993) Liposomes: the Babraham connection. *Chem Phys Lipids* **64**, 275–285.

14. Bao A., Goins B., Klipper R., Negrete G., Mahindaratne M., and Phillips W. T. (2003) A novel liposome radiolabeling method using 99mTc-"SNS/S" complexes: *in vitro* and *in vivo* evaluation. *J Pharm Sci* **92**, 1893–1904.

15. Bao A., Goins B., Klipper R., Negrete G., and Phillips W. T. (2003) 186Re-liposome labeling using 186Re-SNS/S complexes: *in vitro* stability, imaging, and biodistribution in rats. *J Nucl Med* **44**, 1992–1999.

16. Bao A., Goins B., Klipper R., Negrete G., and Phillips W. T. (2004) Direct 99mTc labeling of pegylated liposomal doxorubicin (Doxil) for pharmacokinetic and non-invasive imaging studies. *J Pharmacol Exp Ther* **308**, 419–425.

17. Bao A., Phillips W. T., Goins B., Zheng X., Sabour S., Natarajan M., Ross Woolley F., Zavaleta C., and Otto R. A. (2006) Potential use of drug carried-liposomes for cancer therapy via direct intratumoral injection. *Int J Pharm* **316**, 162–169.

18. Barratt G., and Legrand P. (2005) Comparison of the efficacy and pharmacology of formulations of amphotericin B used in treatment of leishmaniasis. *Curr Opin Infect Dis* **18**, 527–530.

19. Becker M. J., Dams E. T., de Marie S., Oyen W. J., Boerman O. C., Fens M. H., Verbrugh H. A., and Bakker-Woudenberg I. A. (2002) Scintigraphic imaging using 99mTc-labeled PEG liposomes allows early detection of experimental invasive pulmonary aspergillosis in neutropenic rats. *Nucl Med Biol* **29**, 177–184.

20. Boerman O. C., Storm G., Oyen W. J. G., van Bloois L., van der Meer J. W. M., Claessens R. A. M. J., Crommelin D. J. A., and Corstens F. H. M. (1995) Sterically stabilized liposomes labeled with indium-111 to image focal infection. *J Nucl Med* **36**, 1639–1644.

21. Boothe H. W., Boothe D. M., Komkov A., and Hightower D. (1992) Use of hepatobiliary scintigraphy in the diagnosis of extrahepatic biliary obstruction in dogs and cats: 25 cases (1982–1989). *J Am Vet Med Assoc* **201**, 134–141.

22. Chang M. Y., Seideman J., and Sofou S. (2008) Enhanced loading efficiency and retention of 225Ac in rigid liposomes for potential targeted therapy of micrometastases. *Bioconjug Chem* **19**, 1274–1282.

23. Chang Y. J., Chang C. H., Chang T. J., Yu C. Y., Chen L. C., Jan M. L., Luo T. Y., Lee T. W., and Ting G. (2007) Biodistribution, pharmacokinetics and microSPECT/CT imaging of 188Re-bMEDA-liposome in a C26 murine colon carcinoma solid tumor animal model. *Anticancer Res* **27**, 2217–2225.

24. Chen L. C., Chang C. H., Yu C. Y., Chang Y. J., Hsu W. C., Ho C. L., Yeh C. H., Luo T. Y., Lee T. W., and Ting G. (2007) Biodistribution, pharmacokinetics and imaging of (188)Re-BMEDA-labeled pegylated liposomes after intraperitoneal injection in a C26 colon carcinoma ascites mouse model. *Nucl Med Biol* **34**, 415–423.

25. Chen L. C., Chang C. H., Yu C. Y., Chang Y. J., Wu Y. H., Lee W. C., Yeh C. H., Lee T. W., and Ting G. (2008) Pharmacokinetics, micro-SPECT/CT imaging and therapeutic efficacy of (188)Re-DXR-liposome in C26 colon carcinoma ascites mice model. *Nucl Med Biol* **35**, 883–893.

26. Chowdhury F. U., and Scarsbrook A. F. (2008) The role of hybrid SPECT-CT in oncology: current and emerging clinical applications. *Clin Radiol* **63**, 241–251.

27. Dabareiner R. M., and Cole R. C. (2009) Fractures of the tuber coxa of the ilium in horses: 29 cases (1996–2007). *J Am Vet Med Assoc* **234**, 1303–1307.

28. Dagar S., Krishnadas A., Rubinstein I., Blend M. J., and Onyuksel H. (2003) VIP grafted sterically stabilized liposomes for targeted imaging of breast cancer: *in vivo* studies. *J Control Release* **91**, 123–133.

29. Dams E. T. M., Oyen W. J. G., Boerman O. C., Storm G., Laverman P., van der Meer J. M. W., and Corstens F. H. M. (1997) Detecting focal infection in granulocytopenic rats: A comparative study of three Tc-99m-labeled radioactive agents. *J. Nucl. Med.* **38**, 270P.

30. de Steenwinkel J. E., van Vianen W., Ten Kate M. T., Verbrugh H. A., van Agtmael M. A., Schiffelers R. M., and Bakker-Woudenberg I. A. (2007) Targeted drug delivery to enhance efficacy and shorten treatment duration in disseminated Mycobacterium avium infection in mice. *J Antimicrob Chemother* **60**, 1064–1073.

31. Duzgunes N., Simoes S., Slepushkin V., Pretzer E., Rossi J. J., De Clercq E., Antao V. P., Collins M. L., and de Lima M. C. (2001) Enhanced inhibition of HIV-1 replication in macrophages by antisense oligonucleotides, ribozymes and acyclic nucleoside phosphonate analogs delivered in pH-sensitive liposomes. *Nucleosides Nucleotides Nucleic Acids* **20**, 515–523.

32. Elbayoumi T. A., and Torchilin V. P. (2008) Tumor-Specific Anti-Nucleosome Antibody Improves Therapeutic Efficacy of Doxorubicin-Loaded Long-Circulating Liposomes against Primary and Metastatic Tumor in Mice. *Mol Pharm*

33. Emfietzoglou D., Kostarelos K., and Sgouros G. (2001) An analytic dosimetry study for the use of radionuclide-liposome conjugates in internal radiotherapy. *J Nucl Med* **42**, 499–504.

34. Erdogan S., Ozer A. Y., Ercan M. T., and Hincal A. A. (2000) Scintigraphic imaging of infections with 99m-Tc-labelled glutathione liposomes. *J Microencapsul* **17**, 459–465.

35. Erdogan S., Roby A., and Torchilin V. P. (2006) Enhanced tumor visualization by gamma-scintigraphy with 111In-labeled polychelating-polymer-containing immunoliposomes. *Mol Pharm* **3**, 525–530.

36. Fujii G. (1996) Liposomal amphotericin B (AmBisome): Realization of the drug delivery concept. In: Rosoff M. (ed.) *Vesicles*. Marcel Dekker, New York, pp. 491–520.

37. Gabizon A., Huberty J., Straubinger R. M., Price D. C., Papahadjopoulos D. (1988–89) An improved method for *in vivo* tracing and imaging of liposomes using a gallium 67-deferoxamine complex. *J liposome Res* **1**, 123–135.

38. Gabizon A., Catane R., Uziely B., Kaufman B., Safra T., Cohen R., Martin F., Huang A., and Barenholz Y. (1994) Prolonged circulation time and enhanced accumulation in malignant exudates of doxorubicin encapsulated in polyethylene-glycol coated liposomes. *Cancer Research* **54**, 987–992.

39. Gabizon A., Chisin R., Amselem S., Druckmann S., Cohen R., Goren D., Fromer I., Peretz T., Sulkes A., and Barenholz Y. (1991) Pharmacokinetic and imaging studies in patients receiving a formulation of liposome-associated adriamycin. *Br J Cancer* **64**, 1125–1132.

40. Gabizon A., Shmeeda H., and Barenholz Y. (2003) Pharmacokinetics of pegylated liposomal Doxorubicin: review of animal and human studies. *Clin Pharmacokinet* **42**, 419–436.

41. Gaspar M. M., Boerman O. C., Laverman P., Corvo M. L., Storm G., and Cruz M. E. (2007) Enzymosomes with surface-exposed superoxide dismutase: *in vivo* behaviour and therapeutic activity in a model of adjuvant arthritis. *J Control Release* **117**, 186–195.

42. Gillen A., Dyson S., and Murray R. (2009) Nuclear scintigraphic assessment of the thoracolumbar synovial intervertebral articulations. *Equine Vet J* **41**, 534–540.

43. Goins B, Klipper R, Rudolph AS, Cliff RO, Blumhardt R, and Phillips W. T. (1993) Biodistribution and imaging studies of technetium-99m labeled liposomes in rats with focal infection. *J Nucl Med* **34**, 2160–2168.

44. Goins B., Bao A., and Phillips W. T. Techniques for loading technetium-99m and rhenium-186/188 radionuclides into pre-formed liposomes for diagnostic imaging and radionuclide therapy. *Methods Mol Biol* **606**, 469–491.

45. Goins B., Klipper R., Rudolph A. S., Cliff R. O., Blumhardt R., and Phillips W. T. (1993) Biodistribution and imaging studies of technetium-99m-labeled liposomes in rats with focal infection. *Journal of Nuclear Medicine* **34**, 2160–2168.

46. Goins B. A. (2008) Radiolabeled Lipid Nanoparticles for Cancer Diagnosis and Treatment. In: Kumar M. N. V. R. (ed.) *Handbook of Particulate Drug Devliery*. American Scientific Publishers, pp. 65–82.

47. Goins B. A., and Phillips W. T. (2003) Radiolabelled Liposomes for Imaging and Biodistribution Studies. In: Torchilin V. P. and Weissig V. (eds.) *Liposomes: A Practical Approach*. Oxford University Press, Oxford, pp. 319–336.

48. Goins B. A., and Phillips W. T. (2007) Methods for tracking radiolabeled after injection in the body. In: Gregoriadis G. (ed.) *Liposome Technololgy*, vol 3. Informa Healthcare, New York, pp. 191–209

49. Goldberg S. N., Kamel I. R., Kruskal J. B., Reynolds K., Monsky W. L., Stuart K. E., Ahmed M., and Raptopoulos V. (2002) Radiofrequency ablation of hepatic tumors: increased tumor destruction with adjuvant liposomal doxorubicin therapy. *AJR Am J Roentgenol* **179**, 93–101.

50. Gregoriadis G., and Ryman B. E. (1971) Liposomes as carriers of enzymes or drugs: a new approach to the treatment of storage diseases. *Biochem J* **124**, 58P.

51. Gupta B., and Torchilin V. P. (2007) Monoclonal antibody 2C5-modified doxorubicin-loaded liposomes with significantly enhanced therapeutic activity against intracranial human brain U-87 MG tumor xenografts in nude mice. *Cancer Immunol Immunother* **56**, 1215–1223.

52. Gupta H., and Weissleder R. (1996) Targeted contrast agents in MR imaging. *Magn Reson Imaging Clin N Am* **4**, 171–184.

53. Hafeli U., Tiefenauer L. X., Schbiger P. A., and Weder H. G. (1991) A lipophilic complex with 186Re/188Re incorporated in liposomes suitable for radiotherapy. *Int J Rad Appl Instrum B* **18**, 449–454.

54. Harrington K. J., Mohammadtaghi S., Uster P. S., Glass D., Peters A. M., Vile R. G., and Stewart J. S. W. (2001) Effective targeting of solid tumors in patients with locally advanced cancers by radiolabeled pegylated liposomes. *Clin. Cancer Res.* **7**, 243–254.

55. Harrington K. J., Rowlinson-Busza G., Syrigos K. N., Uster P. S., Vile R. G., Peters A. M., and Stewart J. S. (2001) The effect of irradiation on the biodistribution of radiolabeled pegylated liposomes. *Int J Radiat Oncol Biol Phys* **50**, 809–820.

56. Harvey A. M., Hibbert A., Barrett E. L., Day M. J., Quiggin A. V., Brannan R. M., and Caney S. M. (2009) Scintigraphic findings in 120 hyperthyroid cats. *J Feline Med Surg* **11**, 96–106.

57. Head H., Dodd G. D., 3rd, Bao A., Garcia-Rojas X., Goins B., Soundararajan A., McManus L., Prihoda T., and Phillips W. T. (2010) Combination radiofrequency ablation and intravenous radiolabeled liposomal doxorubicin: imaging and qunatification of increased drug delivery to tumors. *Radiology* **255**, 405–414.

58. Henriksen G., Schoultz B. W., Michaelsen T. E., Bruland O. S., and Larsen R. H. (2004) Sterically stabilized liposomes as a carrier for alpha-emitting radium and actinium radionuclides. *Nucl Med Biol* **31**, 441–449.

59. Hirnle P. (1997) Liposomes for drug targeting in the lymphatic system. *Hybridoma* **16**, 127–132.

60. Holmberg E., Maruyama K., Litzinger D. C., Wright S., Davis M., Kabalka G. W., Kennel S. J., and Huang L. (1989) Highly efficent immunoliposomes prepared with a method which is compatible with various lipid compositions. *Biochem Biophys Res Commun* **165**, 1272–1278.

61. Hwang K. J., Merriam J. E., Beaumier P. L., and Luk K. S. (1982) Encapsulation, with high efficiency, of radioactive metal ions in liposomes. *Biochim Biophys Acta* **716**, 101–109.

62. Jonasdottir T. J., Fisher D. R., Borrebaek J., Bruland O. S., and Larsen R. H. (2006) First *in vivo* evaluation of liposome-encapsulated 223Ra as a potential alpha-particle-emitting cancer therapeutic agent. *Anticancer Res* **26**, 2841–2848.

63. Kikuchi T., Saito R., Sugiyama S., Yamashita Y., Kumabe T., Krauze M., Bankiewicz K., and Tominaga T. (2008) Convection-enhanced delivery of polyethylene glycol-coated liposomal doxorubicin: characterization and efficacy in rat intracranial glioma models. *J Neurosurg* **109**, 867–873.

64. Kleiter M. M., Yu D., Mohammadian L. A., Niehaus N., Spasojevic I., Sanders L., Viglianti B. L., Yarmolenko P. S., Hauck M., Petry N. A., Wong T. Z., Dewhirst M. W., and Thrall D. E. (2006) A tracer dose of technetium-99m-labeled liposomes can estimate the effect of hyperthermia on intratumoral doxil extravasation. *Clin Cancer Res* **12**, 6800–6807.

65. Klibanov A., Maruyama K., Torchilin V. P., and Huang L. (1990) Amphipathic polyethylene glycols effectively prolong the circulation time of liposomes. *FEBS Letters* **268**, 235–237.

66. Knapp F. F., Jr. (1998) Rhenium-188--a generator-derived radioisotope for cancer therapy. *Cancer Biother Radiopharm* **13**, 337–349.

67. Kostarelos K., and Emfietzoglou D. (2000) Tissue dosimetry of liposome-radionuclide complexes for internal radiotherapy: toward liposome-targeted therapeutic radiopharmaceuticals. *Anticancer Res* **20**, 3339–3345.

68. Kostarelos K., Emfietzoglou D., Papakostas A., Yang W. H., Ballangrud A., and Sgouros G. (2004) Binding and interstitial penetration of liposomes within avascular tumor spheroids. *Int J Cancer* **112**, 713–721.

69. Krauze M. T., Forsayeth J., Yin D., and Bankiewicz K. S. (2009) Convection-enhanced delivery of liposomes to primate brain. *Methods Enzymol* **465**, 349–362.

70. Krauze M. T., Vandenberg S. R., Yamashita Y., Saito R., Forsayeth J., Noble C., Park J., and Bankiewicz K. S. (2008) Safety of real-time convection-enhanced delivery of liposomes to primate brain: a long-term retrospective. *Exp Neurol* **210**, 638–644.

71. Lau W. Y., and Lai E. C. (2009) The current role of radiofrequency ablation in the management of hepatocellular carcinoma: a systematic review. *Ann Surg* **249**, 20–25.

72. Laverman P., Boerman O. C., Oyen W. J., Dams E. T., Storm G., and Corstens F. H. (1999) Liposomes for scintigraphic detection of infection and inflammation. *Adv Drug Deliv Rev* **37**, 225–235.

73. Laverman P., Boerman O. C., and Storm G. (2003) Radiolabeling of liposomes for scintigraphic imaging. *Methods Enzymol* **373**, 234–248.

74. Laverman P., Dams E. T., Storm G., Hafmans T. G., Croes H. J., Oyen W. J., Corstens F. H., and Boerman O. C. (2001) Microscopic localization of PEG-liposomes in a rat model of focal infection. *J Control Release* **75**, 347–355.

75. Laverman P., Dams E. T. M., Oyen W. J. G., Storm G., Koenders E. B., Prevost R., Corstens F. H. M., and Storm G. (1998) A novel method to label liposomes with Tc-99m via the hydrazino nicotinyl derivative. *J Nucl Med* **39**, 7P.

76. Laverman P., Storm G., Phillips W. T., Bao A., and Goins B. (2007) Radiolabeling of liposome for scintigraphic imaging. In: Gregoriadis G. (ed.), *Liposome Technology*, vol 3. Informa Healthcare, New York, pp. 169–185.

77. Lee C. M., Choi Y., Huh E. J., Lee K. Y., Song H. C., Sun M. J., Jeong H. J., Cho C. S., and Bom H. S. (2005) Polyethylene glycol (PEG) modified 99mTc-HMPAO-liposome for improving blood circulation and biodistribution: the effect of the extent of PEGylation. *Cancer Biother Radiopharm* **20**, 620–628.

78. Lee J. S., Park U. J., Son K. J., and Han H. S. (2009) Development of a high performance (188)

W/(188)Re generator by using a synthetic alumina. *Appl Radiat Isot* **67**, 1162–1166.

79. Liang H., Yang Y., Yang K., Wu Y., Boone J. M., and Cherry S. R. (2007) A microPET/CT system for *in vivo* small animal imaging. *Phys Med Biol* **52**, 3881–3894.

80. Linninger A. A., Somayaji M. R., Mekarski M., and Zhang L. (2008) Prediction of convection-enhanced drug delivery to the human brain. *J Theor Biol* **250**, 125–138.

81. Love W. G., Amos N., Williams B. D., and Kellaway I. W. (1989) Effect of liposome surface charge on the stability of technetium (^{99m}Tc) radiolabelled liposomes. *Annals of the Rheumatic Diseases* **6**, 105–113.

82. Lucignani G. (2009) Nanoparticles for concurrent multimodality imaging and therapy: the dawn of new theragnostic synergies. *Eur J Nucl Med Mol Imaging* **36**, 869–874.

83. Mart R. J., Liem K. P., and Webb S. J. (2009) Creating functional vesicle assemblies from vesicles and nanoparticles. *Pharm Res* **26**, 1701–1710.

84. Mauk M. R., and Gamble R. C. (1979) Preparation of lipid vesicles containing high levels of entrapped radioactive cations. *Analytical Biochemistry* **94**, 302–307.

85. McCready V. R., and Cornes P. (2001) The potential of intratumoural unsealed radioactive source therapy. *Eur J Nucl Med* **28**, 567–569.

86. McGhana J. P., and Dodd G. D., III (2001) Radiofrequency ablation of the liver: current status. *AJR Am J Roentgenol* **176**, 3–16.

87. Medina L. A., Calixto S. M., Klipper R., Phillips W. T., and Goins B. (2004) Avidin/biotin-liposome system injected in the pleural space for drug delivery to mediastinal lymph nodes. *J Pharm Sci* **93**, 2595–2608.

88. Medina L. A., Goins B., Rodriguez-Villafuerte M., Bao A., Martinez-Davalos A., Awasthi V., Galvan O. O., Santoyo C., Phillips W. T., and Brandan M. E. (2007) Spatial dose distributions in solid tumors from 186Re transported by liposomes using HS radiochromic media. *Eur J Nucl Med Mol Imaging* **34**, 1039–1049.

89. Metselaar J. M., Wauben M. H., Wagenaar-Hilbers J. P., Boerman O. C., and Storm G. (2003) Complete remission of experimental arthritis by joint targeting of glucocorticoids with long-circulating liposomes. *Arthritis Rheum* **48**, 2059–2066.

90. Morgan JR, Williams KE, Davies RL, Leach K, Thomson M, and LAP W. (1981) Localisation of experimental staphylococcal abscesses by ^{99m}Tc-technetium-labelled liposomes. *Journal of Medical Microbiology* **14**, 213–217.

91. Mougin-Degraef M., Jestin E., Bruel D., Remaud-Le Saec P., Morandeau L., Faivre-Chauvet A., and Barbet J. (2006) High-activity radio-iodine labeling of conventional and stealth liposomes. *J Liposome Res* **16**, 91–102.

92. Mulder W. J., Strijkers G. J., van Tilborg G. A., Griffioen A. W., and Nicolay K. (2006) Lipid-based nanoparticles for contrast-enhanced MRI and molecular imaging. *NMR Biomed* **19**, 142–164.

93. Mulder W. J., van der Schaft D. W., Hautvast P. A., Strijkers G. J., Koning G. A., Storm G., Mayo K. H., Griffioen A. W., and Nicolay K. (2007) Early *in vivo* assessment of angiostatic therapy efficacy by molecular MRI. *FASEB J* **21**, 378–383.

94. O'Connor M. K., and Kemp B. J. (2006) Single-photon emission computed tomography/computed tomography: basic instrumentation and innovations. *Semin Nucl Med* **36**, 258–266.

95. Ogihara I., Kojima S., and Jay M. (1986) Differential uptake of gallium-67-labeled liposomes between tumors and inflammatory lesions in rats. *J Nucl Med* **27**, 1300–1307.

96. Oussoren C., and Storm G. (2001) Liposomes to target the lymphatics by subcutaneous administration. *Adv Drug Deliv Rev* **50**, 143–156.

97. Oussoren C., Zuidema J., Crommelin D. J., and Storm G. (1997) Lymphatic uptake and biodistribution of liposomes after subcutaneous injection. II. Influence of liposomal size, lipid compostion and lipid dose. *Biochim Biophys Acta* **1328**, 261–272.

98. Oyen W. J., Boerman O. C., Storm G., van Bloois L., Koenders E. B., Claessens R. A., Perenboom R. M., Crommelin D. J., van der Meer J. W., and Corstens F. H. (1996) Detecting infection and inflammation with technetium-99m-labeled Stealth liposomes. *J Nucl Med* **37**, 1392–1397.

99. Oyen W. J. G., Boerman O. C., Storm G., Bloois L. V., Koenders E. B., Crommelin D. J. A., Meer J. W. M. V. D., and Corstens F. H. M. (1996) Labelled Stealth liposomes in experimental infection: An alternative to leukocyte scintigraphy? *Nuclear Medicine Communications* **17**, 742–748.

100. Paoli E. E., Kruse D. E., Seo J. W., Zhang H., Kheirolomoom A., Watson K. D., Chiu P., Stahlberg H., and Ferrara K. W. (2009) An optical and microPET assessment of thermally-sensitive liposome biodistribution in the Met-1 tumor model: Importance of formulation. *J Control Release* **143**, 13–22.

101. Pene F., Courtine E., Cariou A., and Mira J. P. (2009) Toward theragnostics. *Crit Care Med* **37**, S50–58.

102. Phillips WT, Rudolph AS, Goins B, Timmons JH, Klipper R, and Blumhardt R. (1992) A simple method for producing a technetium-99m-labeled liposome which is stable *in vivo*. *Nucl Med Biol* **19**, 539–547.

103. Phillips W. T., Andrews T., Liu H., Klipper R., Landry A. J., Blumhardt R., and Goins B. (2001) Evaluation of [(99m)Tc] liposomes as lymphoscintigraphic agents: comparison with [(99m)Tc] sulfur colloid and [(99m)Tc] human serum albumin. *Nucl Med Biol* **28**, 435–444.

104. Phillips W. T., and Goins B. (2002) Assessment of liposome delivery using scintigraphic imaging. *J Liposome Res* **12**, 71–80.

105. Phillips W. T., Goins B. A., and Bao A. (2009) Radioactive liposomes. *Wiley Interdiscip Rev Nanomed Nanobiotechnol* **1**, 69–83.

106. Phillips W. T., Klipper R., and Goins B. (2000) Novel method of greatly enhanced delivery of liposomes to lymph nodes. *J Pharmacol Exp Ther* **295**, 309–313.

107. Phillips W. T., Klipper R. W., Awasthi V. D., Rudolph A. S., Cliff R., Kwasiborski V., and Goins B. A. (1999) Polyethylene glycol-modified liposome-encapsulated hemoglobin: a long circulating red cell substitute. *J Pharmacol Exp Ther* **288**, 665–670.

108. Phillips W. T., Medina L. A., Klipper R., and Goins B. (2002) A novel approach for the

increased delivery of pharmaceutical agents to peritoneum and associated lymph nodes. *J Pharmacol Exp Ther* **303**, 11–16.

109. Phillips W. T., Rudolph A. S., Goins B., Timmons J. H., Klipper R., and Blumhardt R. (1992) A simple method for producing a technetium-99m-labeled liposome which is stable *in vivo*. *Nucl Med Biol* **19**, 539–547.

110. Phillips W. T., Rudolph A. S., Goins B., Timmons J. H., Klipper R., and Blumhardt R. (1992) A simple method for producing a technetium-99m-labeled liposome which is stable *in vivo*. *Int J Rad Appl Instrum B* **19**, 539–547.

111. Pittet J. C., Pepin M., Normier G., Binz H., and Le Pape A. (1995) Lymphoscintigraphy via the targeting of macrophages with 99mTc-J001X poly-galactoside in a model of pyogranulomas developed in sheep lymph nodes. *Nucl Med Biol* **22**, 355–365.

112. Proffitt R., Williams L., Presant C., Tin G., Uliana J., Gamble R., and Baldeschwieler J. (1983) Tumor-imaging potential of liposomes loaded with In-111-NTA: biodistribution in mice. *J Nucl Med* **24**, 45–51.

113. Raghavan R., Brady M. L., Rodriguez-Ponce M. I., Hartlep A., Pedain C., and Sampson J. H. (2006) Convection-enhanced delivery of therapeutics for brain disease, and its optimization. *Neurosurg Focus* **20**, E12.

114. Saito R., Bringas J. R., McKnight T. R., Wendland M. F., Mamot C., Drummond D. C., Kirpotin D. B., Park J. W., Berger M. S., and Bankiewicz K. S. (2004) Distribution of liposomes into brain and rat brain tumor models by convection-enhanced delivery monitored with magnetic resonance imaging. *Cancer Res* **64**, 2572–2579.

115. Saito R., Krauze M. T., Bringas J. R., Noble C., McKnight T. R., Jackson P., Wendland M. F., Mamot C., Drummond D. C., Kirpotin D. B., Hong K., Berger M. S., Park J. W., and Bankiewicz K. S. (2005) Gadolinium-loaded liposomes allow for real-time magnetic resonance imaging of convection-enhanced delivery in the primate brain. *Exp Neurol* **196**, 381–389.

116. Saito R., Krauze M. T., Noble C. O., Drummond D. C., Kirpotin D. B., Berger M. S., Park J. W., and Bankiewicz K. S. (2006) Convection-enhanced delivery of Ls-TPT enables an effective, continuous, low-dose chemotherapy against malignant glioma xenograft model. *Neuro Oncol* **8**, 205–214.

117. Saito R., Krauze M. T., Noble C. O., Tamas M., Drummond D. C., Kirpotin D. B., Berger M. S., Park J. W., and Bankiewicz K. S. (2006) Tissue affinity of the infusate affects the distribution volume during convection-enhanced delivery into rodent brains: implications for local drug delivery. *J Neurosci Methods* **154**, 225–232.

118. Sapra P., Tyagi P., and Allen T. M. (2005) Ligand-targeted liposomes for cancer treatment. *Curr Drug Deliv* **2**, 369–381.

119. Seo J. W., Zhang H., Kukis D. L., Meares C. F., and Ferrara K. W. (2008) A novel method to label preformed liposomes with 64Cu for positron emission tomography (PET) imaging. *Bioconjug Chem* **19**, 2577–2584.

120. Sgouros G. (2008) Alpha-particles for targeted therapy. *Adv Drug Deliv Rev* **60**, 1402–1406.

121. Shiina S. (2009) Image-guided percutaneous ablation therapies for hepatocellular

carcinoma. *J Gastroenterol* **44** (Suppl 19), 122–131.

122. Shim K. M., Kim S. E., Moon C., Kim S. H., Bae C. S., Han H. J., and Kang S. S. (2009) A detailed examination of pulmonary uptake of (99m)Tc-Tin colloid in healthy mature miniature pigs. *In Vivo* **23**, 551–554.

123. Sofou S., Kappel B. J., Jaggi J. S., McDevitt M. R., Scheinberg D. A., and Sgouros G. (2007) Enhanced retention of the alpha-particle-emitting daughters of Actinium-225 by liposome carriers. *Bioconjug Chem* **18**, 2061–2067.

124. Sofou S., Thomas J. L., Lin H. Y., McDevitt M. R., Scheinberg D. A., and Sgouros G. (2004) Engineered liposomes for potential alpha-particle therapy of metastatic cancer. *J Nucl Med* **45**, 253–260.

125. Sou K., Goins B., Leland M. M., Tsuchida E., and Phillips W. T. Bone marrow-targeted liposomal carriers: a feasibility study in nonhuman primates. *Nanomed* **5**, 41–49.

126. Sou K., Goins B., Takeoka S., Tsuchida E., and Phillips W. T. (2007) Selective uptake of surface-modified phospholipid vesicles by bone marrow macrophages *in vivo*. *Biomaterials* **28**, 2655–2666.

127. Soundararajan A., Bao A., Phillips W. T., Perez R., III, and Goins B. A. (2009) [(186) Re]Liposomal doxorubicin (Doxil): *in vitro* stability, pharmacokinetics, imaging and biodistribution in a head and neck squamous cell carcinoma xenograft model. *Nucl Med Biol* **36**, 515–524.

128. Swartz M. A. (2001) The physiology of the lymphatic system. *Adv Drug Deliv Rev* **50**, 3–20.

129. Sweers L., Kirberger R. M., Leisewitz A. L., Dormehl I. C., Killian E., and Naude F. (2008) The scintigraphic evaluation of the pulmonary perfusion pattern of dogs hospitalised with babesiosis. *J S Afr Vet Assoc* **79**, 76–83.

130. Swinebroad E. L., Dabareiner R. M., Swor T. M., Carter G. K., Watkins J. P., Walker M., Schmitz D. G., and Honnas C. M. (2003) Osteomyelitis secondary to trauma involving the proximal end of the radius in horses: five cases (1987–2001). *J Am Vet Med Assoc* **223**, 486–491.

131. Takakura Y., Nishikawa M., Yamashita F., and Hashida M. (2001) Development of gene drug delivery systems based on pharmacokinetic studies. *Eur J Pharm Sci* **13**, 71–76.

132. Thakur M. L., Coss R., Howell R., Vassileva-Belnikolovska D., Liu J., Rao S. P., Spana G., Wachsberger P., and Leeper D. L. (2003) Role of lipid-soluble complexes in targeted tumor therapy. *J Nucl Med* **44**, 1293–1300.

133. Tilcock C., Yap M., Szucs M., and Utkhede D. (1994) PEG-coated lipid vesicles with encapsulated technetium-99m as blood pool agents for nuclear medicine. *Nucl Med Biol* **21**, 165–170.

134. Torchilin V. P. (2007) Targeted pharmaceutical nanocarriers for cancer therapy and imaging. *Aaps J* **9**, E128–147.

135. Torchilin V. P., and Papisov M. I. (1994) Why do polyethylene glycol-coated liposomes circulate so long? *Journal of Liposome Research* **4**, 725–739.

136. Townsend D. W., Carney J. P., Yap J. T., and Hall N. C. (2004) PET/CT today and tomorrow.

J Nucl Med **45** (Suppl 1), 4S–14S.

137. Urakami T., Akai S., Katayama Y., Harada N., Tsukada H., and Oku N. (2007) Novel amphiphilic probes for [18F]-radiolabeling preformed liposomes and determination of liposomal trafficking by positron emission tomography. *J Med Chem* **50**, 6454–6457.

138. Vogelbaum M. A. (2007) Convection enhanced delivery for treating brain tumors and selected neurological disorders: symposium review. *J Neurooncol* **83**, 97–109.

139. Vyas S. P., Khatri K., and Goyal A. K. (2009) Functionalized nanocarrier(s) to image and target fungi infected immune cells. *Med Mycol* **47** (Suppl 1), S362–368.

140. Wang S. X., Bao A., Herrera S. J., Phillips W. T., Goins B., Santoyo C., Miller F. R., and Otto R. A. (2008) Intraoperative 186Re-liposome radionuclide therapy in a head and neck squamous cell carcinoma xenograft positive surgical margin model. *Clin Cancer Res* **14**, 3975–3983.

141. Wang S. X., Bao A., Phillips W. T., Goins B., Herrera S. J., Santoyo C., Miller F. R., and Otto R. A. (2009) Intraoperative therapy with liposomal drug delivery: retention and distribution in human head and neck squamous cell carcinoma xenograft model. *Int J Pharm* **373**, 156–164.

142. Wang Y., and Yuan F. (2006) Delivery of viral vectors to tumor cells: extracellular transport, systemic distribution, and strategies for improvement. *Ann Biomed Eng* **34**, 114–127.

143. Zaidi H. (2006) Recent developments and future trends in nuclear medicine instrumentation. *Z Med Phys* **16**, 5–17.

144. Zaidi H., Montandon M. L., and Alavi A. The clinical role of fusion imaging using PET, CT, and MR imaging. *Magn Reson Imaging Clin N Am* **18**, 133–149.

145. Zavaleta C. L., Goins B. A., Bao A., McManus L. M., McMahan C. A., and Phillips W. T. (2008) Imaging of 186Re-liposome therapy in ovarian cancer xenograft model of peritoneal carcinomatosis. *J Drug Target* **16**, 626–637.

146. Zweit J. (1996) Radionuclides and carrier molecules for therapy. *Phys Med Biol* **41**, 1905–1914.

Chapter 5

ANTIBODY-TARGETED LIPOSOMES AND MICELLES FOR IMAGING APPLICATIONS

Tamer Elbayoumi [a] and Vladimir Torchilin[b]

[a] *Department of Pharmaceutical Sciences Midwestern University, Glendale, AZ 85308, USA*
telbayoumi@midwestern.edu

[b] *Department of Pharmaceutical Sciences and Center for Pharmaceutical Biotechnology and Nanomedicine, Northeastern University, Boston, MA 02115, USA*
v.torchilin@neu.edu

5.1 INTRODUCTION

In addition to the recent technical advances in software and hardware in the various diagnostic imaging modalities, the biggest impact in the field is currently seen through the development of tunable and targetable diagnostic nanocarrier systems. In this respect, several nanoscale carriers (nanoparticles, liposomes, water-soluble polymers, micelles, and dendrimers) have been developed for improved delivery of a multitude of diagnostic and therapeutic agents. These carriers can selectively target various pathological sites and carry large payloads, thereby improving disease detection and therapy effectiveness. For almost three decades now, liposomes and more recently micelles have been identified as promising vehicles to deliver a wide range of encapsulated and/or membrane-incorporated therapeutic or diagnostic agents into the required areas [25, 35, 41]. In particular, these nanocarriers have been utilized for carrying the diagnostic moieties used with all imaging modalities: gamma-scintigraphy, magnetic resonance imaging (MRI), computed tomography (CT) imaging, and sonography [55, 57]. More specifically, immunoliposomes and immunomicelles, directed to clinically relevant cell-surface molecules with antibodies, antibody fragments, or peptides, have been widely recognized for their potential clinical application for site-specific diagnostic evaluation along with the delivery of therapeutic agents.

Nanoimaging
Edited by Beth Goins and William Phillips

www.panstanford.com

Non-enhanced imaging techniques are useful only when relatively large tissue areas are involved in the pathological process. In addition, though attenuations (i.e., the ability of a tissue to absorb a certain signal, such as X-rays, sound waves, radiation, or radiofrequencies) of different tissues differ; however, in the majority of cases this difference is not sufficient for clear discrimination between normal and pathological areas. To solve this problem and to achieve a sufficient attenuation, contrast agents are used able to absorb certain types of signal (irradiation) much stronger than surrounding tissues. The main task in this case is to accumulate a sufficient quantity of a contrast agent in the area of interest and to keep its presence in normal tissues and organs on a minimum level.

No matter which imaging modality is used, precise medical diagnostic imaging requires that the sufficient intensity of a corresponding signal from an area of interest be achieved in order to differentiate the target area from surrounding tissues. Currently used medical imaging modalities include (i) gamma scintigraphy (based on the application of gamma-emitting radioactive materials; either single-photon emission computed tomography [SPECT] or dual-photon annilation image detection, positron emission tomography [PET]); (ii) magnetic resonance (MR, phenomenon involving the transition between different energy levels of atomic nuclei under the action of radiofrequency signal); (iii) computed tomography (CT, the modality utilizing ionizing radiation—X-rays—with the aid of computers to acquire cross-images of the body and three-dimensional images of areas of interest); (iv) ultrasonography (US, the modality utilizes the irradiation with ultrasound and is based on the different passage rate of ultrasound through different tissues); (v) near-infrared (NIR, relies on differentiating long-wavelength fluorescent emissions of diagnostic fluorophores from normal tissue auto- fluorescence via whole-body CCD cameras with specific high cut-off filters). See also Table 5.1 for diagnostic moieties (reporter groups) used in different imaging modalities.

Table 5.1 Diagnostic Moiety Required for Appropriate Tissue Attenuation in Different Imaging Modalitites

Modality	*Reporter group*	*Required tissue concentration*
Gamma-scintigraphy	Radionuclide, e.g. ^{111}In, ^{99m}Tc, ^{67}Ga	10^{-10} M
PET imaging	Radionuclide (with short half-life), e.g. ^{11}C, ^{18}F, ^{15}O	10^{-8} M
MR Imaging	Paramagnetic metal, e.g. Gd, Mn, and iron oxide	10^{-4} M
CT Imaging	Heavy elements, e.g. I, Br, Ba	10^{-2} M
Near IR imaging	Fluorescent probe / Quantum dots	10^{-2} M
Ultrasonography	Gas (air, argon, nitrogen)	

The combination of newer imaging techniques providing high sensitivity and spatial resolution and use of nanoscale devices to deliver diagnostic agents with high target specificity would indeed enable more accurate detection and staging, leading to more accurate therapy planning of the disease state.

5.2 LIPOSOMES AND MICELLES AS CARRIERS OF CONTRAST AGENTS FOR VARIOUS IMAGING MODALITIES

Among particulate drug carriers, micro/nano-reservoir-type systems, such as liposomes (mainly, for water-soluble drugs) and micelles (mainly, for water-insoluble drugs) are the most extensively studied and have certain advantages over other delivery systems and possess the most suitable characteristics for encapsulation of many drugs, genes, and diagnostic (imaging) agents [64, 65]. Simply, the possibility to easily control composition, size, and *in vivo* stability of a nanoreservoir is a major advantage. More importantly, these nanovesicular systems retain remarkable loading capacity of various imaging and contrast agents, of either hydrophilic or lipophilic nature, or both [16].

Liposomes are artificial phospholipid vesicles of size varying from 50 to 1000 nm and even more, which can be loaded with a variety of water-soluble drugs (into their inner aqueous compartment) and water-insoluble drugs (into the hydrophobic compartment of the phospholipid bilayer), and are considered as promising drug carriers for well over two decades [16]. The use of targeted liposomes, i.e., liposomes selectively accumulating inside the affected organ or tissue, may increase the efficacy of the liposome incorporated drug and decrease the loss of liposomes and their contents in reticuloendothelial system [RES] [16]. Several successful protocols have been employed to actively and passively target liposomal nanocarierrs to various pathological tissues and into target cells in particular, via various targeting ligands and homing moieties [33]. In a similar fashion, micelles, including polymeric micelles, represent another promising type of pharmaceutical carrier. Micelles are amphiphilic compound-formed colloidal particles with hydrophobic core and hydrophilic corona with size range between 5 and 100 nm.[31] An important property of micelles is their ability to increase the solubility and bioavailability of poorly soluble pharmaceuticals [63]. The use of certain special amphiphilic molecules as micelle building blocks can extended blood half-life of the micellar carrier, in combination with different targeting ligands, strategically incorporated into the micelle structure. There are several key micelle properties (size, critical micelle concentration or CMC, and loading capacity of the hydrophobic core of the micelle) important for the preparation of micelle-incorporated pharmaceuticals and contrast agents [61]. Usually, these parameters are as follows: the size of a pharmaceutical micelle is between 10 and 80 nm, an optimal CMC value is in a low micromolar region, and the loading efficiency toward a hydrophobic drug between 5 and 25 wt.%. In case of targeted micelles, a local release of a free drug from micelles in the target organ should lead to the increased efficacy of the drug, while the stability of the micelles en route to the target organ or tissue should contribute to drug solubility and toxicity reduction due to less interaction with nontargeted organs [63, 65].

5.3 FORMULATION OF LIPOSOMAL AND MICELLAR CARRIERS FOR IMAGING APPLICATIONS

Pursuing different *in vivo* delivery purposes, one can easily change the size, charge and surface properties of these carriers simply by adding new ingredients to the lipid or amphiphilic compound mixture before liposome or micelles preparation and/or by variation of preparation methods [62]. As follows from Fig. 5.1, while micelles may be loaded with a contrast agent only into the micelle core in the process of micelle assembly, liposomes may incorporate contrast agents in both internal water compartment and membrane. Gamma-scintigraphy and MR imaging both require the sufficient quantity of radionuclide or paramagnetic metal to be associated with liposome.

Independently on the supposed imaging modality, there exist several general approaches to liposome/micelle loading with a contrast agent (reporter group, label). The contrast agent can be: (i) added and incorporated into the aqueous interior of liposome or into the liposome membrane during the manufacturing process to liposomes; (ii) adsorbed onto the surface of preformed liposomes; (iii) incorporated into the lipid bilayer of preformed liposomes; (iv) loaded into preformed liposomes using membrane-incorporated transporters or ion channels. Two very general approaches are the most often used and efficient to prepare liposomes/micelles for gamma- and MR-imaging. First, metal is chelated into a soluble chelate (such as diethylene triamine pentaacetic acid [DTPA]) and then entrapped into the water interior of a liposome [50]. Alternatively, DTPA or a similar chelating compound may be chemically derivatized by the incorporation of a hydrophobic group, which can anchor the chelating moiety on the liposome or micelle surface during or after liposome preparation [29]. Different chelators and different hydrophobic anchors were tried for the preparation of indium-111 (^{111}In), technetium -99m (^{99m}Tc), manganese (Mn-), and gadolinium (Gd) Gd-liposomes [1, 47]. Several detailed reviews describing the various methods and chelators for radiolabeling liposomes were recently published [15, 36, 43] .

Taking into account that most clinically relevant radioisotopes have rather short half-life (under 5 days), the last step of the preparation of contrast liposomes for gamma-imaging takes place directly before the application moment. With this in mind, liposomes have to be prepared which can be sufficiently loaded with a contrast label by applying simple and fast labeling protocol. Stability issues and remedies for gamma-imaging with liposomes will be discussed in another chapter in this book by Phillips *et al.*

In the case of MR imaging, for a better MR signal, all reporter atoms should be freely exposed for interaction with water. This requirement makes metal encapsulation into the liposome less attractive than metal coupled with chelators exposed into the water space. In addition, low-molecular-weight water-soluble paramagnetic probes may leak from liposomes upon contact with body fluids, which destabilizes most liposomal membranes. Moreover, it has been shown that when too high concentrations of Gd-DTPA are encapsulated inside liposomes in an attempt to achieve better enhancement, the relaxivity of the compound might be even lower than for non-encapsulated Gd-DTPA complex, probably because of decreased residence lifetime of water molecules inside vesicles [50]. Membranotropic chelating agents, such as DTPA-stearylamine or DTPA-phosphatidyl ethanolamine [30], consist of the polar head containing a chelated paramagnetic atom, and a lipid moiety which anchors the metal-chelate complex in the liposome membrane [1]. This approach has been shown to be far more superior in terms of the relaxivity of the final preparation when compared with liposome-encapsulated paramagnetic ions due to the decrease in the rotational correlation times of the paramagnetic moiety rigidly connected with relatively large particle. Liposomes with membrane-bound paramagnetic ion also demonstrate a reduced risk of leakage in the body [47]. Membranotropic chelates are suitable for micelle incorporation (they anchor in the hydrophobic micelle core), and may also serve to load micelles with heavy radiometals (see Fig. 5.1) [58].

Since it is rather difficult to enhance the signal intensity from a given reporter metal in liposomes and micelles, one may attempt to increase the quantity of carrier-associated reporter metal (such as Gd or ^{111}In). The approach developed to solve this issue was by using polychelating polymers additionally modified with a hydrophobic residue to assure their firm incorporation into the liposome membrane or micelle core. With this in mind, we developed PAP, polychelating amphiphilic polymers, a family of co-polymers containing a hydrophilic fragment with multiple chelating groups and relatively short but highly hydrophobic phospholipid fragment on one end of a polymeric chain suitable for incorporation into liposomes and micelles [60, 70]. This polylysine (PL)-based polychelator easily incorporates into the liposomal membrane or micelle core in the process of liposome or micelle preparation, and sharply increases the number of chelated heavy metal atoms attached to a single lipid anchor (Fig. 5.1). This allowed us to increase the number of bound reporter metal atoms per vesicle and to decrease the dosage of an administered lipid without compromising the image signal intensity. Upon incorporation into the bilayer, the phospholipid anchor grafted with a chelating polymer forms a "coat" of chelating groups around the liposomal membrane or within the outer shell of a micelle. In case of MR, metal atoms chelated into these groups are directly exposed to the water environment that enhances the relaxivity of the paramagnetic ions and leads to the corresponding enhancement of the vesicle contrast properties. In addition, liposomes may be modified with PEG to achieve an improved stability and longevity in the body [60].

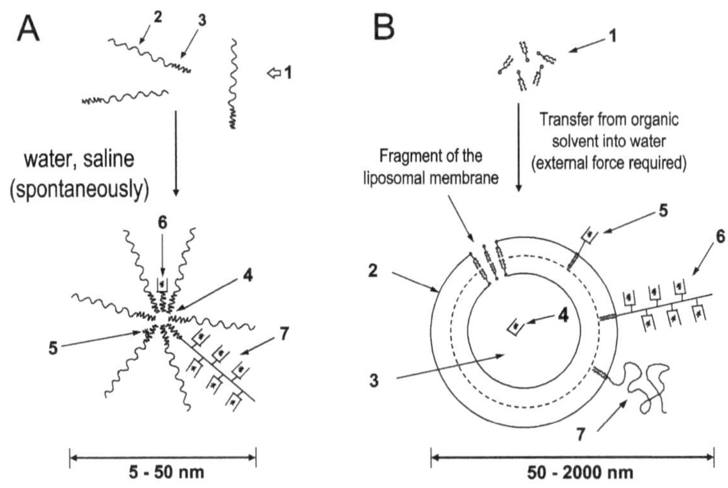

Figure 5.1 Schematic structures of micelle and liposome, their formation and loading with a contrast agent. (A) Micelle is formed spontaneously in aqueous media from amphiphilic compound (1) the molecule of which consists of distinct hydrophilic (2) and hydrophobic (3) moieties. Hydrophobic moieties form the micelle core (4). Contrast agent (asterisk; gamma- or MR-active metal-loaded chelating group, or heavy element, such as iodine or bromine) can be directly coupled to the hydrophobic moiety within the micelle core (5), or incorporated into the micelle as an individual monomeric (6) or polymeric (7) amphiphilic unit. (B) Liposome can be artificially prepared from individual phospholipid molecules (1), and consists of bilayered membrane (2) and internal aqueous compartment (3). Contrast agent (asterisk) can be entrapped into inner water space of liposome as a soluble entity (4), or incorporated into the liposome membrane as a part of monomeric (5) or polymeric (6) amphiphilic unit (similar to that in case of micelle). Additionally, liposome can be sterically protected by amphiphilic derivative of PEG or PEG-like polymer (7) [58].

Micelles formed by self-assembled amphiphilic polymers (such as PEG-phosphatidyl ethanolamine, PEG-PE) can also be loaded with amphiphilic PL-based chelates carrying diagnostically important metal ions such as ^{111}In and Gd [69].

5.4 TARGETED LIPOSOMES AND MICELLES FOR DIAGNOSTIC IMAGING APPLICATIONS

Further, these nanocarriers can be surface modified by a variety of different moieties to impart them with certain properties and functionalities. A small required quantity of cell-targeting components confers the ability to specifically recognize and bind target tissues or cells via the surface-attached specific ligands (monoclonal antibodies as well as their Fab fragments and some other moieties, e.g. folate or transferrin, are used for this purpose) [4, 10, 17, 46]. Manipulation of the exterior wall/membrane components allows the ability to respond to local stimuli characteristic of the pathological site by, for example, releasing an entrapped drug or specifically acting on cellular membranes under the abnormal pH or temperature in disease sites (this property could be provided by surface-attached pH- or temperature- sensitive components); and the surface-attachment of cell-penetration molecules confers the ability to penetrate inside cells bypassing lysosomal degradation for efficient targeting of intracellular components [32, 53].

To increase the quantity of liposomes/micelles accumulating in the required areas, the use of targeted liposomes or micelles (with a surface-attached specific ligand) was suggested [7, 42, 54]. Liposomes (micelles) with a specific affinity for an affected organ or tissue were believed to increase the efficacy of loaded pharmaceutical agents, including the imaging ones. Immunoglobulins, primarily of the IgG class, are the most promising and widely used targeting moieties for various drugs and drug carriers including liposomes and (in this case we are talking about immunoliposomes and immuomicelles). Numerous methods for antibody coupling to liposomes are reviewed in [59]. Usually, as much as 50 to 200 antibody molecules should be bound per single 100–250 nm liposome [59]. Despite evident success in the development of antibody-to-liposome coupling technique and improvements in the targeting efficacy, the majority of immunoliposomes still ended up in the liver, which was usually a consequence of insufficient time for the interaction between the target and targeted liposome. This is especially true in cases when a target of choice has a diminished blood supply (ischemic or necrotic areas), and only a small number of liposomes pass through the target with the blood during the time period when liposomes are present in the circulation. The same insufficient targeting occurs if the concentration of the target antigen is very low, and the number of productive collisions between target antigens and immunoliposomes is too small. So, it was a natural development to combine the properties of targeted and long-circulating liposomes in one preparation [54, 59]. Clearly, such combination can significantly enhance the specificity and intensity of radiologic signal localization in the target tissue or pathology.

5.4.1 Antibody-Targeted Liposomes and Micelles for Tumor Imaging

Since long-circulating liposomes are now widely used in biomedical *in vitro* and *in vivo* studies and have even found their way into clinical practice [23–24, 49], the use of ligand-targeted PEGylated liposomes as carriers for contrast agents represents now an important area of research.

As reported recently by our group, the EPR-dependent passive tumor-targeting of long-circulating liposomes for the delivery of imaging agents could be further improved by adding a specific targeting moiety to the liposomal composition [14]. This was achieved via the attachment to the liposomes of the anti-cancer monoclonal antibody 2C5 (mAb 2C5) with nuclesome (NS)-restricted specificity that can recognize the surface of various tumor but not normal cells [9, 28]. Attached antibody allowed us to specifically target pharmaceutical carriers to tumor cells *in vitro* and *in vivo* [18, 38, 67]. These mAb 2C5-modified PEG-liposomes demonstrated 3-to-

8-fold increases in the *in vitro* specific cell binding with various cancer cell lines of diverse origins compared with nonmodified PEG-liposomes. In addition, using ^{111}In-labeled liposomes, the biodistribution data showed almost doubled tumor-specific accumulation of the mAb 2C5-modified liposomes compared with nonspecific controls. As the same time, the gamma-imaging, starting 6 hours and up to 12 hours post-injection, of tumor-bearing mice clearly demonstrate the far superior tumor-accumulation of the mAb 2C5-modified liposomes compared with both nontargeted controls, in both 4T1 and LLC tumor models (S.C. tumor sizes were between 4–8mm in diameter). Moreover, the fast and distinct tumor visualization acquired with the 2C5-liposomes at 6 hours compared with 12–24 hours with PEG-liposomes, suggests that the EPR effect responsible for tumor accumulation of nanocarriers through the leaky vasculature is much less significant for the accumulation of 2C5-liposomes, which proceeds via a more specific mechanism. These results clearly demonstrate an evident potential of specific antibody (including mAb 2C5)-modified long-circulating liposomes as targeted vehicles capable of specifically delivering diagnostic moieties to a variety of tumors decreasing the image acquisition time. It also may allow for the noninvasive quantification and characterization of tumor delivery and therapeutic efficacy of various nano-particulate formulations of anti-cancer agents [17, 18].

This same concept was applied in case of monoclonal antibody 2C5-targeted micelles, to monitor their temporal accumulation in murine tumor model of Lewis lung carcinoma, LLC [45]. Although this effort was primarily directed as a tool for evaluation of tumor-targeting capability of these nanoemulsifier particles to deliver photodynamic therapy drug, the accelerated preferential distribution of these immunomicelles emphasized their potential for enhanced detection of highly perfused localilized tumors.

Recently, another group has investigated a similar approach to study the biodistribution of immunoliposomes, labeled with ^{99m}Tc in tumor xenografted mice [34]. The authors used polyethyleneglycol-coated (PEGylated) liposome surface-grafted with F (ab') (2) fragment of mAb GAH to target human gastric cancer, MKN45, in murine model. Employing two labeling strategies, either using ^{99m}Tc-labeled antibody on liposome or immunoliposome encapsulating ^{99m}Tc, they demonstrated tumor targeting with both labeled formulations. Despite the uncertain tracer stability, their finding showed some promise, in case of the radiolabel-encapsulating immunoliposomes, for improved tumor uptake and retention, for both radioimaging and radiotherapy applications.

Using the MR imaging modality, a few recent reports investigated 100 nm cationic immunoliposomes, encapsulating Gd-complex (gadopentetate dimeglumine), as MRI contrast agent [22]. The anti-transferrin receptor single-chain antibody fragment (TfRscFv) attached to the liposome surface would increase the sensitivity of detection of lung metastases *in vitro* and *in vivo*. In addition, these immunotargeted liposomes loaded with the conventional MRI agent, Magnevist, allowed for an increased resolution and image intensity in an *in vivo* orthotopic mouse model of pancreatic cancer [44].

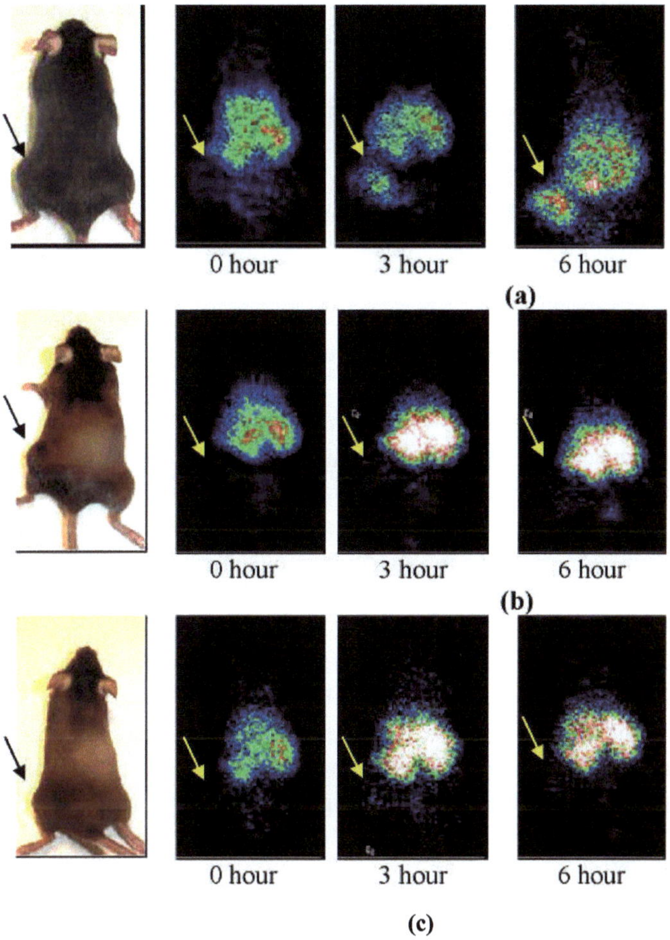

Figure 5.2 Whole body imaging of LLC tumor bearing mice at different time points after the injection of ^{111}In-labeled PAP-containing PEGylated liposomes: (a) 2C5-modified, (b) control IgG-modified, and (c) unmodified. Arrows indicate tumor locations [52].

It is important to remember that tumor targeting of liposomal diagnostic moieties should not be limited to ligands known to be tumor-specific, or are overexpressed on tumor cells, but it encompasses ligands specific for tumor microenvironment as a whole. For instance, MR imaging of tumor angiogenesis using sterically stabilized Gd-DTPA liposomes targeted to CD105 endothelial ligands (CD105-Gd-SLs), to depict tumor angiogenesis via the expression of CD105 in tumor-bearing rats [71]. Axial T1/WI MRI images collected at baseline up to 120 min post-intravenous injection in glioma-bearing rats indicated the enhancement area was restricted for group CD105-Gd-SLs, while diffused for controls. In addition, the signal intensity (SI) in case of CD105-Gd-SLs persistently increased over 120 min, compared with Gd-DTPA and nontargeted Gd-DTPA-liposomes. Immunohistochemistry and immunofluorescence both confirmed the specificity of these immuno-liposomal contrasts towards new microvessel formations, indicating the possibility to detect early tumor angiogenesis on MR images.

A natural progression of these tumor-specific immuno-targeted diagnostic liposomes and micelles, several reports exploited the PAP-exaggerated contrast loading strategy in combination with immuno-targeted vesicles. This combination approach was used for MRI signal amplification of the multitude of Gd-moeities loaded onto mAb 2C5-modified liposomes. Following the encouraging *in vitro* data on various cancer cell lines [20], Gd-PAP-containing 200 nm immuno-liposomes were tested for whole body MRI. The T1 inversion recovery maps demonstrated a significant increase in tumor-associated R1 in animals injected with antibody-modified liposomes 4 hours post-injection, followed by a gradual decrease consistent with clearance of the agent, from the tumor region, in comparison to the antibody-free Gd-PAP liposomes [19]. Analogous work, to illustrate the wide applicability of this PAP-immuno-liposomal platform for tumor visualization, using the different imaging modalities, was performed by the same group. It was demonstrated that the tumor accumulation of mAb 2C5 modified liposomes, heavily loaded with ^{111}In via the liposome-incorporated polylysine-based (PLL-based) polychelating amphiphilic polymer (PAP), was significantly (3-to-5-fold) higher than in the neighboring muscle tissue at all times after administration (6, 24, and 48 hour) in mice bearing murine Lewis lung carcinoma (LLC) and human HT-29 tumors (Fig. 5.2). The whole body direct gamma-imaging of LLC tumor-bearing mice at different times has demonstrated the superior *in vivo* tumor accumulation of the targeted mAb 2C5-modified PAP-containing PEGylated liposomes compared with nontargeted liposomal control formulations, consequently resulting in better and faster detection of the small tumors (about 8 mm) [21].

An interesting–inverse imaging strategy, employing the immuno-liposomal platform, should be mentioned, despite the limited success achieved through this approach. The principle is based on using liposomally entrapped second antibody to improve tumor imaging with radiolabelled first anti-tumor antibody [6]. In this scenario, patients with gastrointestinal carcinomas received intravenous ^{131}I-labelled goat anti-CEA or mouse monoclonal 17–1A as first anti-tumour antibodies (specific for carcino-embryonic antigen) given systematically for tumor imaging. Then, a second antibody, entrapped in liposomes, directed against the first anti-tumor antibody was utilized to accelerate reticuloendothelial (RES) clearance of first antibody, thus improving discrimination between tumor and normal tissues. Tumor imaging was enhanced among 15 patients with gastrointestinal cancer and tumor was correctly identified at five sites, where it was not seen by a background subtraction [5]. The complexities of this approach, along with the varied stability of the liposomal second antibody, obscured the proposed diagnostic improvements.

5.4.2 Immuno-Targeted Liposomes and Micelles for Imaging Within the Cardiovascular System

Monoclonal antibodies provide an effective method for noninvasive detection and visualization of different cardiac disorders, acute myocardial infarction being among them. Considering the accelerated incidence rate of this disease, it's rapid and accurate diagnosis and staging is of primary importance. One strategy towards enhancing antibody-mediated infarct visualization via liposomal attachment is discussed here [54]. It is based on the fact that following ischemic myocardial cell death, exposed cardiac myosin from disrupted cell membranes, is recognized as the target antigen, characteristic of infarcted myocardium. Hence, attempts have been made to use anti-myosin antibody, mAb 2G4, to actively target radiolabeled liposomes to infarcted myocardium [66]. Membrane-labeled PEGylated liposomes with ^{111}In were coupled to anti-myosin antibody and injected in rabbit model of acute experimental myocardial infarct. Both gamma-scintigraphy and histological staining indicated that the areas of myocardium with increased radioactivity always coincided with infarct-positive staining, which illustrated the possibility of this strategy for application in diagnostic detection of cardiac infarcts [68].

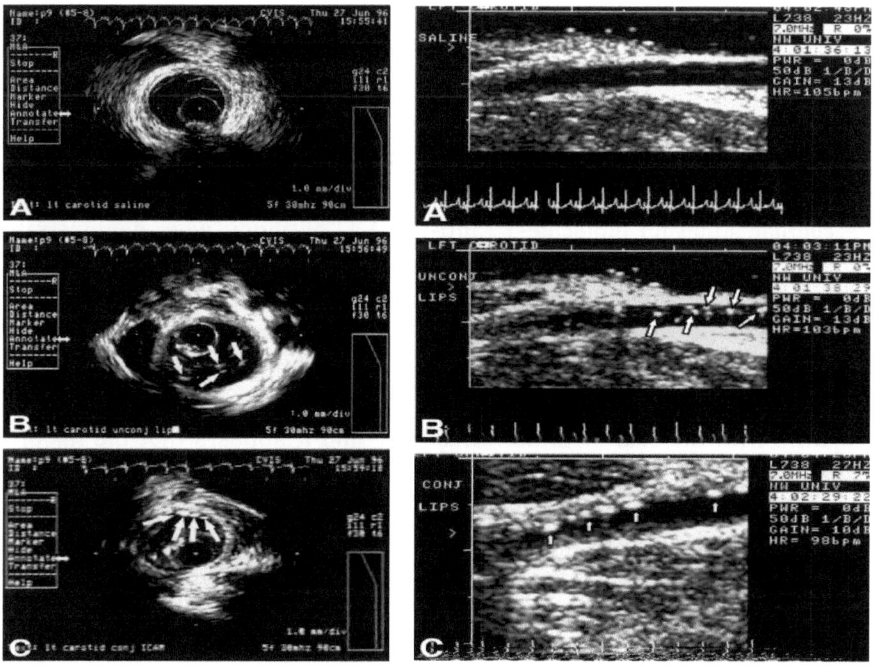

Figure 5.3 Intravascular (*left*) and transvascular (*right*) ultrasound images of the atherosclerotic left carotid artery of a Yucatan miniswine. (A) After injection of saline. (B) After injection of unconjugated liposomes (arrows point to liposomes within the lumen). (C) After injection of anti–ICAM-1 labeled liposomes (arrows point to liposomes attached to the atherosclerotic plaque) [11].

It is very hard to consider drug targeting to heart and lungs separately from the rest of the vascular system. Heart and lung diseases are closely related to those of blood vessels. Thus, for example, a disease such as myocardial infarction is usually caused by thrombosis or atherosclerosis of the coronary blood vessel; pulmonary thromboembolism is the result of thrombotic occlusion of the pulmonary artery; etc. hence, it is important to understand the same regularities underlay the prognosis and therapy of thrombi and atheromas in any particular compartment of the circulation. So, many approaches considered within the following vascular section are of a general character, and might be applied to the diagnosis of similar pathologies located in the pulmonary or cardiac compartment as well [56].

Immunoliposomes were shown to be suitable carriers for targeting blood vessel injuries, using different imaging modalities. The acoustically-reflective –liposome platform has been repeatedly explored in this area, with applications not only limited to enhanced diagnosis of localized thrombi and lesions, but also to drive site-specific release of therapeutic drugs. Earlier efforts used echogenic liposomes covalently linked to fibrin(ogen)-specific antibodies and their Fab' fragments for the ultrasonic imaging of atherosclerotic plaques and subsequently for targeted delivery of thrombolytic proteins [52]. One of the approaches in this area is based on the concept of acoustically reflective liposomes that can be targeted for site-specific acoustic enhancement [2, 26]. Liposome compositions containing maleimido-activated phospholipids were conjugated with anti-fibrinogen antibodies, and shown to acquire the ability to attach to fibrin-coated surfaces and thrombi in cell culture and flow models [12–13]. In addition, these anti-fibrinogen acoustically-reflective liposomes were shown to attach to fibrous atheroma and thrombi in a Yucatan miniswine model of induced atherosclerosis, whereas liposomes conjugated to anti-intercellular adhesion molecule-1 (anti-ICAM-1) antibody, were demonstrated to target early stage atherosclerotic plaques (Fig. 5.3) [11]. This concept was further developed into targeted delivery of echogenic liposomes with thrombolytic loading [52], where tPA-loaded echogenic liposomes demonstrated an effective clot lysis similar to the treatment with free tPA [52]. Additionally, the ultrasound exposure of tPA-loaded echogenic liposomes resulted in two-fold enhanced thrombolysis compared with the case with no ultrasound, and it appears that the majority of the ultrasound effect was related to the increase in the localized drug release from the tPA–liposome complex. In a recent study, these tPA-loaded echogenic liposomes demonstrated twice the affinity for fibrin than free tPA, and the evaluation of the clots demonstrated an enhanced highlighting of clots treated with tPA-echogenic liposomes compared with controls [51].

An interesting new report exploited similar echogenic liposomal platform to specifically recruit exogenous stem cells to the atherosclerotic plaque tissue, which may repopulate vascular cells and help repair the arterial tissue. Bifunctional echogenic immunoliposomes (BF-ELIP) were covalently coupled to two antibodies; the first one specific for CD34 antigens on the surface of stem cells and the second directed against the intercellular adhesion molecule-1 (ICAM-1) antigens on the inflammatory endothelium covering atheroma [27]. It was demonstrated that ICAM-1-expressing endothelium of the aorta of swine fed high-cholesterol diets, which was preloaded with BF-ELIP, significantly increased the adherence of adult bone marrow CD34+ stem cells. Subsequent penetration of the CD34+ stem cells was detected, in particular in the aortic segments treated with 1 MHz low-amplitude continuous wave ultrasound. Fluorescence and scanning electron microscopy confirmed the presence of BF-ELIP-bound CD34+ cells in the intimal compartment of the atheromatous arterial wall, indicating ultrasound treatment-mediated increase in endothelial cell progenitors migrating into the intima. The data suggest, under ultrasound enhancement, that BF-ELIP-bound CD34+ stem cells appear to selectively bind to the ICAM-1 expressing endothelium of atherosclerotic lesions.

The marked sensitivity and anatomical resolution of the MR imaging modality always favors the clinical use of this technique. Thus, it has attracted significant efforts to show the feasibility of the magnetic resonance imaging of atherosclerotic lesions in mice using immuno-nanovesicles. Several targets were identified, specific to the pathological progression of atheroma and atherosclerotic lesion formation, including oxidized low-density lipoprotein in lipid pool, as well as scavenger receptor (MSR) characteristic of macrophage/foam cell depositions. Immuno-micellar formulations containing gadolinium were coupled to murine (MDA2 and E06) or human (IK17) antibodies that bind unique oxidation-specific epitopes. In MR images of apolipoprotein E-deficient (apoE(-/-)) mice, performed at 9.4 Tesla, MDA2 and IK17, micelles showed maximal arterial wall uptake at 72 hours and E06 micelles at 96 hours, manifested by 125% to 231% enhancement in magnetic resonance signal compared with adjacent muscle, indicative of specific targeting and providing excellent image quality of oxidation-rich atherosclerotic lesions [8].

5.4.3 Antibody-Targeted Vesicles for Molecular Imaging

Although anatomic relationships remain important, increased understanding of cell structure and function is rapidly moving us toward diagnosis and treatment at the cellular level. This intrinsic advantage of molecular imaging is being rapidly developed through the use of targeted imaging agents along with nanovesicles, a broad platform being developed for manipulating and probing of biological functions on the cellular and sub-cellular level. Compared with traditional molecular-based contrast agents, or therapeutic drugs, this marker-specific nanoimaging paradigm enables a highly integrated platform, which incorporates multiple functions, such as cell-targeting, ultrasensitive imaging and sometimes even therapy in one system. Early experiences in these parallel fields show great promise, with widespread clinical applications expected in the near future.

Towards a liposomal system for contrast-enhanced magnetic resonance imaging of molecular targets, PEGylated paramagnetic and fluorescent immunoliposomes were designed to enable the parallel detection of the induced expression of molecular markers on endothelial cells, using MR imaging and fluorescence microscopy [40].

As a proof of principle, *in vitro* model of E-selectin-expressing human umbilical vein endothelial cells (HUVEC) was proposed. After co-incubation with PEGylated paramagnetic fluorescently-labeled liposomes, carrying anti-E-selectin monoclonal antibody as a targeting ligand, both MRI and fluorescence microscopy revealed the specific association of the liposomal MR contrast agent with stimulated HUVEC.

The MRI-based visualization of the expression of specific molecules expressed on vascular endothelium during the disease was also shown in the intact brain of mice with experimental autoimmune encephalitis (EAE) [48]. Antibody-conjugated paramagnetic liposomes were targeted to intercellular adhesion molecule-1 (ICAM-1), an endothelial leukocyte receptor upregulated on the cerebral microvasculature during EAE. High-resolution MRI of mouse brains *ex-vivo* demonstrated that these ICAM-immunotargeted paramagnetic liposome binding conferred significant enhancement of signal intensity as compared with control images.

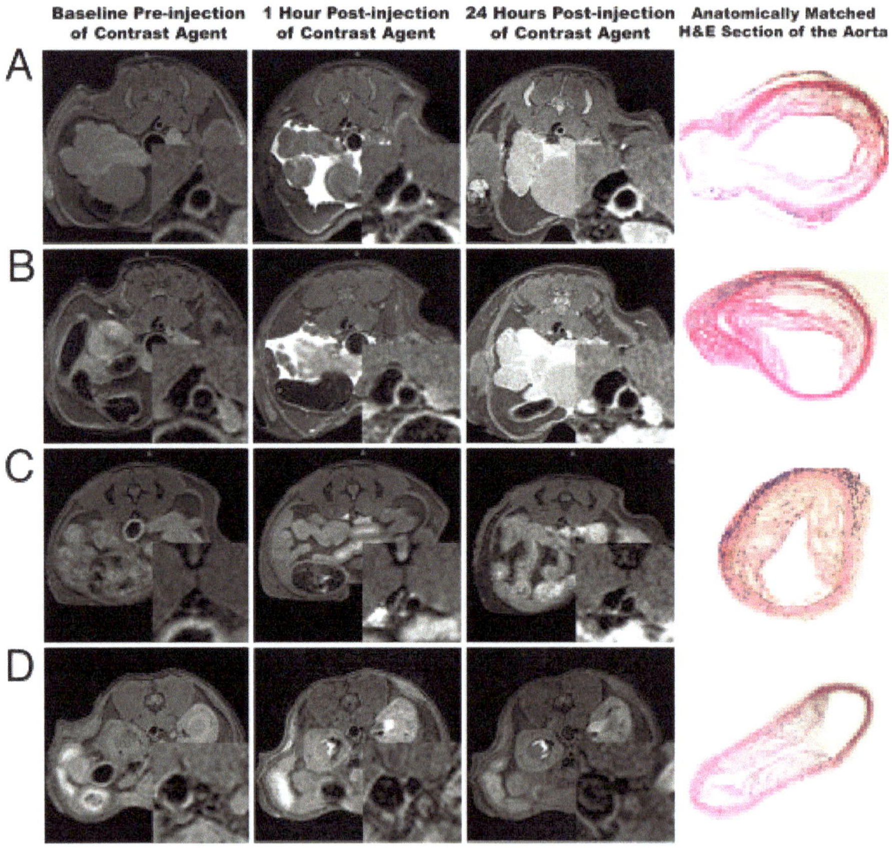

Figure 5.4 *In vivo* MRI images obtained at baseline and post-injection of macrophage-targeted immunomicelles (A and B), untargeted micelles (C), and Gd-DTPA (D) in ApoE / mice. The MRI insets are enlargements of the aortas. A–D right show H&E sections of the aorta at the identical anatomic level as the MRI images from the same animal [3].

As noted earlier, MRI-based diagnosis of atherosclerotic lesion molecular markers, via immunomicelles gained significant momentum in recent years. Molecular imaging of macrophages in atherosclerotic plaques using bimodal PEG-micelles was investigated via the utilization of PEGylated, fluorescent, and paramagnetic micelles conjugated with macrophage scavenger receptor (MSR)-specific antibodies [37]. T_1-weighted high-resolution MR imaging of atherosclerotic apolipoprotein E knockout (apoE-KO) mice demonstrated pronounced signal enhancement (up to 200%) was observed for apolipoprotein E knockout (apoE-KO) mice that were injected with MSR-targeted micelles, while the aortic vessel wall of mice injected with nontargeted micelles showed little signal enhancement [39]. Confirmed via fluorescence imaging, this approach was extremely valuable in detecting and assessing macrophages *in vivo* to evaluate atherosclerosis noninvasively using molecular MRI, providing excellent enhancement of atherosclerotic plaques (Fig. 5.4). As the MRI-enhancement observed is proportional to the macrophage content of the atherosclerotic vessel areas imaged, MSR-specific immunomicelles may aid in the detection of high macrophage content associated with plaques vulnerable to rupture [3].

5.5 CONCLUDING REMARKS

Perhaps the key benefit of an immunotargeted liposome- or micelle-based diagnostic is the ability to "image" (i.e., measure) functional characteristic of a tissue or other region of interest. The development of the effective nanocarrier mechanism does not only mean the execution of delivery, but also the positive confirmation of the correct and specific delivery of the payload. Based on the biochemical understanding of the characteristics of the microparticulate carrier and how it interacts via its targeting ligands with the biological milieu, it is possible to design antibody-targeted diagnostics that make these functional measurements possible. The advantages of antibody-guided imaging with liposomes (and, more recently, with micelles) have been already demonstrated in many experimental and clinical studies, in most, providing the evidence of the successful visualization of numerous organs, tissues, and disease sites in all imaging modalities. More importantly, as the ability to track and image the fate of any immuno-vesicle starting from the systemic down to the sub-cellular level becomes available, more research is expected to properly implement these antibody-targeted diagnostic vesicles for imaging, visualization, tracking and monitoring cellular and even intra-cellular alterations, subsequent to or indicative of therapeutic delivery. Still, we have to learn more about possible toxicity of metal-loaded liposomes and micelles owing to the intracellular liberation of the loaded contrast and any potential interaction/ interference with the antibody part, to properly assess their safe corresponding clinical doses in each specific imaging application.

References

1. Ahkong Q. F., and Tilcock C. (1992) Attachment of 99mTc to lipid vesicles containing the lipophilic chelate dipalmitoylphosphatidylethanolamine-DTTA. *Int J Rad Appl Instrum B* **19**, 831–840.

2. Alkan-Onyuksel H., Demos S. M., Lanza G. M., Vonesh M. J., Klegerman M. E., Kane B. J., Kuszak J., and McPherson D. D. (1996) Development of inherently echogenic liposomes as an ultrasonic contrast agent. *J Pharm Sci* **85**, 486–490.

3. Amirbekian V., Lipinski M. J., Briley-Saebo K. C., Amirbekian S., Aguinaldo J. G., Weinreb D. B., Vucic E., Frias J. C., Hyafil F., Mani V., Fisher E. A., and Fayad Z. A. (2007) Detecting and assessing macrophages *in vivo* to evaluate atherosclerosis noninvasively using molecular MRI. *Proc Natl Acad Sci U S A* **104**, 961–966.

4. Baum P., Muller D., Ruger R., and Kontermann R. E. (2007) Single-chain Fv immunoliposomes for the targeting of fibroblast activation protein-expressing tumor stromal cells. *J Drug Target* **15**, 399–406.

5. Begent H. J., Chester K. A., Bagshawe K. D., Keep P. A., Searle F., Boden J., Barratt G. M., Green A. J., Riggs S. J., and Woodrow D. F. (1989) Second antibody for improvement of antibody imaging: liposome-entrapped and free preparations in animal and human studies. *Clin Exp Immunol* **78**, 307–313.

6. Begent R. H., Keep P. A., Green A. J., Searle F., Bagshawe K. D., Jewkes R. F., Jones B. E., Barratt G. M., and Ryman B. E. (1982) Liposomally entrapped second antibody improves tumour imaging with radiolabelled (first) antitumour antibody. *Lancet* **2**, 739–742.

7. Blume G., Cevc G., Crommelin M. D., Bakker-Woudenberg I. A., Kluft C., and Storm G. (1993) Specific targeting with poly(ethylene glycol)-modified liposomes: coupling of homing devices to the ends of the polymeric chains combines effective target binding with long circulation times. *Biochim Biophys Acta* **1149**, 180–184.

8. Briley-Saebo K. C., Shaw P. X., Mulder W. J., Choi S. H., Vucic E., Aguinaldo J. G., Witztum J. L., Fuster V., Tsimikas S., and Fayad Z. A. (2008) Targeted molecular probes for imaging atherosclerotic lesions with magnetic resonance using antibodies that recognize oxidation-specific epitopes. *Circulation* **117**, 3206–3215.

9. Chakilam A. R., Pabba, S., Mongayt, D., Iakoubov, L.Z., and Torchilin, V.P. (2004) A single monoclonal antinuclear autoantibody with nucleosome-restricted specificity inhibits the growth of diverse human tumors in nude mice. *Cancer Therapy* **2**, 353–364.

10. Constantinides P. P., Chaubal M. V., and Shorr R. (2008) Advances in lipid nanodispersions for parenteral drug delivery and targeting. *Adv Drug Deliv Rev* **60**, 757–767.

11. Demos S. M., Alkan-Onyuksel H., Kane B. J., Ramani K., Nagaraj A., Greene R., Klegerman M., and McPherson D. D. (1999) *In vivo* targeting of acoustically reflective liposomes for intravascular and transvascular ultrasonic enhancement. *J Am Coll Cardiol* **33**, 867–875.

12. Demos S. M., Dagar S., Klegerman M., Nagaraj A., McPherson D. D., and Onyuksel H. (1998) *In vitro* targeting of acoustically reflective immunoliposomes to fibrin under various flow conditions. *J Drug Target* **5**, 507–518.

13. Demos S. M., Onyuksel H., Gilbert J., Roth S. I., Kane B., Jungblut P., Pinto J. V., McPherson D. D., and Klegerman M. E. (1997) *In vitro* targeting of antibody-conjugated echogenic liposomes for site-specific ultrasonic image enhancement. *J Pharm Sci* **86**, 167–171.

14. Elbayoumi T. A., and Torchilin V. P. (2006) Enhanced accumulation of long-circulating liposomes modified with the nucleosome-specific monoclonal antibody 2C5 in various tumours in mice: gamma-imaging studies. *Eur J Nucl Med Mol Imaging*

15. Elbayoumi T. A., and Torchilin V. P. (2007) Use of Radiolabeled Liposomes for Tumor Imaging. In: Bulte JWM, Modo M (eds) Nanoparticles in Biomedical Imaging, vol 102. Springer New York, pp 211–236

16. Elbayoumi T. A., and Torchilin V. P. (2009) Current trends in liposome research, in *Liposomes*, vol. 1 (ed. Weissig V.), Springer & Humana Press Totowa, NJ.

17. Elbayoumi T. A., and Torchilin V. P. (2009) Tumor-specific anti-nucleosome antibody improves therapeutic efficacy of doxorubicin-loaded long-circulating liposomes against primary and metastatic tumor in mice. *Mol Pharm* **6**, 246–254.

18. Elbayoumi T. A., and Torchilin V. P. (March 15, 2009) Tumor-targeted nanomedicines: enhanced antitumor efficacy *in vivo* of doxorubicin-loaded, long-circulating liposomes modified with cancer-specific monoclonal antibody. *Clin Cancer Res*, 1973–1980.

19. Erdogan S., Medarova Z. O., Roby A., Moore A., and Torchilin V. P. (2008) Enhanced tumor MR imaging with gadolinium-loaded polychelating polymer-containing tumor-targeted liposomes. *J Magn Reson Imaging* **27**, 574–580.

20. Erdogan S., Roby A., Sawant R., Hurley J., and Torchilin V. P. (2006) Gadolinium-loaded polychelating polymer-containing cancer cell-specific immunoliposomes. *J Liposome Res* **16**, 45–55.

21. Erdogan S., Roby A., and Torchilin V. P. (2006) Enhanced tumor visualization by gamma-scintigraphy with 111In-labeled polychelating-polymer-containing immunoliposomes. *Mol Pharm* **3**, 525–530.

22. Freedman M., Chang E. H., Zhou Q., and Pirollo K. F. (2009) Nanodelivery of MRI contrast agent enhances sensitivity of detection of lung cancer metastases. *Acad Radiol* **16**, 627–637.

23. Gabizon A., Shmeeda H., and Barenholz Y. (2003) Pharmacokinetics of pegylated liposomal Doxorubicin: review of animal and human studies. *Clin Pharmacokinet* **42**, 419–436.

24. Gabizon A. A., Lyass O., Berry G. J., and Wildgust M. (2004) Cardiac safety of pegylated liposomal doxorubicin (Doxil/Caelyx) demonstrated by endomyocardial biopsy in patients with advanced malignancies. *Cancer Invest* **22**, 663–669.

25. Gregoriadis G., Wills E. J., Swain C. P., and Tavill A. S. (1974) Drug-carrier potential of liposomes in cancer chemotherapy. *Lancet* **1**, 1313–1316.

26. Hamilton A., Huang S. L., Warnick D., Stein A., Rabbat M., Madhav T., Kane B., Nagaraj A., Klegerman M., MacDonald R., and McPherson D. (2002) Left ventricular thrombus enhancement after intravenous injection of echogenic immunoliposomes: studies in a new experimental model. *Circulation* **105**, 2772–2778.

27. Herbst S. M., Klegerman M. E., Kim H., Shelat H., Wassler M., Moody M. R., Yang C. M., Ge X., Zou Y., Kopechek J. A., Clubb F. J., Kraemer D. C., Huang S., Holland C. K., McPherson D. D., and Geng Y. J. (2009) Delivery of stem cells to porcine arterial wall with echogenic liposomes conjugated to antibodies against CD34 and intercellular adhesion molecule-1. *Mol Pharm* **7**, 3–11.

28. Iakoubov L. Z., and Torchilin V. P. (1997) A novel class of antitumor antibodies: nucleosome-restricted antinuclear autoantibodies (ANA) from healthy aged nonautoimmune mice. *Oncol Res* **9**, 439–446.

29. Kabalka G. W., Davis M. A., Holmberg E., Maruyama K., and Huang L. (1991) Gadolinium-labeled liposomes containing amphiphilic Gd-DTPA derivatives of varying chain length: targeted MRI contrast enhancement agents for the liver. *Magn Reson Imaging* **9**, 373–377.

30. Kabalka G. W., Davis M. A., Moss T. H., Buonocore E., Hubner K., Holmberg E., Maruyama K., and Huang L. (1991) Gadolinium-labeled liposomes containing various amphiphilic Gd-DTPA derivatives: targeted MRI contrast enhancement agents for the liver. *Magn Reson Med* **19**, 406–415.

31. Kakizawa Y., and Kataoka K. (2002) Block copolymer micelles for delivery of gene and related compounds. *Adv Drug Deliv Rev* **54**, 203–222.

32. Khalil I. A., Kogure K., Futaki S., and Harashima H. (2008) Octaarginine-modified liposomes: enhanced cellular uptake and controlled intracellular trafficking. *Int J Pharm* **354**, 39–48.

33. Kirpotin D. B., Drummond D. C., Shao Y., Shalaby M. R., Hong K., Nielsen U. B., Marks J. D., Benz C. C., and Park J. W. (2006) Antibody targeting of long-circulating lipidic nanoparticles does not increase tumor localization but does increase internalization in animal models. *Cancer Res* **66**, 6732–6740.

34. Kitamura N., Shigematsu N., Nakahara T., Kanoh M., Hashimoto J., Kunieda E., and Kubo A. (2009) Biodistribution of immunoliposome labeled with Tc-99m in tumor xenografted mice. *Ann Nucl Med* **23**, 149–153.

35. Lasic D. D., and Papahadjopoulos D. (eds.) (1998) Medical applications of liposomes. Elsevier New York

36. Laverman P., Boerman O. C., and Storm G. (2003) Radiolabeling of liposomes for scintigraphic imaging. *Methods Enzymol* **373**, 234–248.

37. Lipinski M. J., Amirbekian V., Frias J. C., Aguinaldo J. G., Mani V., Briley-Saebo K. C., Fuster V., Fallon J. T., Fisher E. A., and Fayad Z. A. (2006) MRI to detect atherosclerosis with gadolinium-containing immunomicelles targeting the macrophage scavenger receptor. *Magn Reson Med* **56**, 601–610.

38. Lukyanov A. N., Elbayoumi T. A., Chakilam A. R., and Torchilin V. P. (2004) Tumor-targeted liposomes: doxorubicin-loaded long-circulating liposomes modified with anti-cancer antibody. *J Control Release* **100**, 135–144.

39. Mulder W. J., Strijkers G. J., Briley-Saboe K. C., Frias J. C., Aguinaldo J. G., Vucic E., Amirbekian V., Tang C., Chin P. T., Nicolay K., and Fayad Z. A. (2007) Molecular imaging of macrophages in atherosclerotic plaques using bimodal PEG-micelles. *Magn Reson Med*

58, 1164–1170.

40. Mulder W. J., Strijkers G. J., Griffioen A. W., van Bloois L., Molema G., Storm G., Koning G. A., and Nicolay K. (2004) A liposomal system for contrast-enhanced magnetic resonance imaging of molecular targets. *Bioconjug Chem* **15**, 799–806.
41. Papahadjopoulos D. (1988) Liposome formation and properties: an evolutionary profile. *Biochem Soc Trans* **16**, 910–912.
42. Phillips T. W., and Goins B. (1995) Targeted delivery of imaging agents by liposomes, in *Handbook of Targeted Delivery of Imaging Agents* (ed. Torchilin V. P.), CRS Press, Boca Raton.
43. Phillips W. T., Goins B., and Bao A. (2009) Radioactive liposomes. *Wiley Interdisciplin Rev Nanomed Nanobiotechnol* **1**, 69–83.
44. Pirollo K. F., Dagata J., Wang P., Freedman M., Vladar A., Fricke S., Ileva L., Zhou Q., and Chang E. H. (2006) A tumor-targeted nanodelivery system to improve early MRI detection of cancer. *Mol Imaging* **5**, 41–52.
45. Roby A., Erdogan S., and Torchilin V. P. (2007) Enhanced *in vivo* antitumor efficacy of poorly soluble PDT agent, meso-tetraphenylporphine, in PEG-PE-based tumor-targeted immunomicelles. *Cancer Biol Ther* **6**, 1136–1142.
46. Sadava D., Phillips T., Lin C., and Kane S. E. (2002) Transferrin overcomes drug resistance to artemisinin in human small-cell lung carcinoma cells. *Cancer Lett* **179**, 151–156.
47. Schwendener R. A., Wuthrich R., Duewell S., Wehrli E., and von Schulthess G. K. (1990) A pharmacokinetic and MRI study of unilamellar gadolinium-, manganese-, and iron-DTPA-stearate liposomes as organ-specific contrast agents. *Invest Radiol* **25**, 922–932.
48. Sipkins D. A., Gijbels K., Tropper F. D., Bednarski M., Li K. C., and Steinman L. (2000) ICAM-1 expression in autoimmune encephalitis visualized using magnetic resonance imaging. *J Neuroimmunol* **104**, 1–9.
49. Swenson C. E., Bolcsak L. E., Batist G., Guthrie T. H., Jr., Tkaczuk K. H., Boxenbaum H., Welles L., Chow S. C., Bhamra R., and Chaikin P. (2003) Pharmacokinetics of doxorubicin administered i.v. as Myocet (TLC D-99; liposome-encapsulated doxorubicin citrate) compared with conventional doxorubicin when given in combination with cyclophosphamide in patients with metastatic breast cancer. *Anticancer Drugs* **14**, 239–246.
50. Tilcock C., Unger E., Cullis P., and MacDougall P. (1989) Liposomal Gd-DTPA: preparation and characterization of relaxivity. *Radiology* **171**, 77–80.
51. Tiukinhoy-Laing S. D., Buchanan K., Parikh D., Huang S., MacDonald R. C., McPherson D. D., and Klegerman M. E. (2007) Fibrin targeting of tissue plasminogen activator-loaded echogenic liposomes. *J Drug Target* **15**, 109–114.
52. Tiukinhoy-Laing S. D., Huang S., Klegerman M., Holland C. K., and McPherson D. D. (2007) Ultrasound-facilitated thrombolysis using tissue-plasminogen activator-loaded echogenic liposomes. *Thromb Res* **119**, 777–784.
53. Torchilin V. (2009) Multifunctional and stimuli-sensitive pharmaceutical nanocarriers. *Eur J Pharm Biopharm* **71**, 431–444.

54. Torchilin V. P. (1994) Immunoliposomes and PEGylated immunoliposomes: possible use for targeted delivery of imaging agents. *Immunomethods* **4**, 244–258.

55. Torchilin V. P. (ed.) (1995) *Handbook of Targeted Delivery of Imaging Agents*, CRS Press, Boca Raton.

56. Torchilin V. P. (1995) Targeting of drugs and drug carriers within the cardiovascular system. *Adv Drug Deliv Rev* **17**, 75–102.

57. Torchilin V. P. (1996) Liposomes as delivery agents for medical imaging. *Mol Med Today* **2**, 242–249.

58. Torchilin V. P. (1997) Pharmacokinetic considerations in the development of labeled liposomes and micelles for diagnostic imaging. *Q J Nucl Med* **41**, 141–153.

59. Torchilin V. P. (2000) Drug Targeting. *Eur J Pharm Sci* **11**(Suppl 2), S81–91.

60. Torchilin V. P. (2000) Polymeric contrast agents for medical imaging. *Curr Pharm Biotechnol* **1**, 183–215.

61. Torchilin V. P. (2005) Lipid-core micelles for targeted drug delivery. *Curr Drug Deliv* **2**, 319–327.

62. Torchilin V. P. (2006) Multifunctional nanocarriers. *Adv Drug Deliv Rev* **58**, 1532–1555.

63. Torchilin V. P. (2007) Micellar nanocarriers: pharmaceutical perspectives. *Pharm Res* **24**, 1–16.

64. Torchilin V. P. (2007) Nanocarriers. *Pharm Res* **24**, 2333–2334.

65. Torchilin V. P. (2007) Targeted pharmaceutical nanocarriers for cancer therapy and imaging. *Aaps J* **9**, E128–147.

66. Torchilin V. P., Klibanov A. L., Huang L., O'Donnell S., Nossiff N. D., and Khaw B. A. (1992) Targeted accumulation of polyethylene glycol-coated immunoliposomes in infarcted rabbit myocardium. *Faseb J* **6**, 2716–2719.Torchilin V. P., Lukyanov A. N., Gao Z., and Papahadjopoulos-Sternberg B. (2003) Immunomicelles: Targeted pharmaceutical carriers for poorly soluble drugs. *Proc Natl Acad Sci U S A* **100**, 6039–6044.

67. Torchilin V. P., Narula J., Halpern E., and Khaw B. A. (1996) Poly(ethylene glycol)-coated anti-cardiac myosin immunoliposomes: factors influencing targeted accumulation in the infarcted myocardium. *Biochim Biophys Acta* **1279**, 75–83.

68. Trubetskoy V. S., Frank-Kamenetsky M. D., Whiteman K. R., Wolf G. L., and Torchilin V. P. (1996) Stable polymeric micelles: lymphangiographic contrast media for gamma scintigraphy and magnetic resonance imaging. *Acad Radiol* **3**, 232–238.

69. Trubetskoy V. S., and Torchilin V. P. (1994) New approaches in the chemical design of Gd-containing liposomes for use in magnetic resonance imaging of lymph nodes. *J Liposome Res* **4**, 961–980.

70. Zhang D., Feng X. Y., Henning T. D., Wen L., Lu W. Y., Pan H., Wu X., and Zou L. G. (2009) MR imaging of tumor angiogenesis using sterically stabilized Gd-DTPA liposomes targeted to CD105. *Eur J Radiol* **70**, 180–189.

Chapter 6

QUANTUM DOT-BASED MULTIMODALITY IMAGING AGENTS

Weibo Cai[a,b] and Xiaoyuan Chen[c]

[a] *Departments of Radiology and Medical Physics, School of Medicine and Public Health, University of Wisconsin – Madison, Madison, Wisconsin, USA*

[b] *University of Wisconsin Paul P. Carbone Comprehensive Cancer Center, Madison, Wisconsin, USA*

wcai@uwhealth.org

[c] *Laboratory of Molecular Imaging and Nanomedicine, National Institute of Biomedical Imaging and Bioengineering, National Institutes of Health, Bethesda, Maryland, USA*

shawn.chen@nih.gov

6.1 INTRODUCTION

Nanotechnology, an interdisciplinary research field involving chemistry, engineering, biology, material science, and medicine, has enormous potential for early detection, accurate diagnosis, and personalized treatment of diseases. With the size of many orders of magnitude smaller than human cells, nanoparticles can offer unprecedented interactions with biomolecules both on the surface of and inside the cells, which may revolutionize disease diagnosis and treatment. Over the last decade, numerous nanotechnology centers have been established worldwide, and it is expected that nanotechnology will mature into a clinically useful field in the near future [55, 116]. To date, the most well-studied nanoparticles include liposomes [37, 42], quantum dots (QDs) [16, 80], paramagnetic nanoparticles [117], nanowires [100], carbon nanotubes [62], gold nanoparticles [15], nanoshells [45], and many others [11, 38].

Organic fluorophores and dyes have been widely used to label cells and tissues for both *in vitro* and *in vivo* imaging [84]. However, due to the inherent photophysical properties such as low photobleaching thresholds, broad absorption/emission spectra, and small Stokes shifts, their use is quite limited and they are not ideal agents for multiplexed, long-term, and/or real-time imaging. On the other hand, QDs are

Nanoimaging
Edited by Beth Goins and William Phillips

www.panstanford.com

inorganic fluorescent semiconductor nanoparticles with superior optical properties than organic fluorophores [35, 36]. They have unique size- and composition-dependent optical and electrical properties due to quantum confinement; hence the name "quantum dots" [20, 81]. QDs have many desirable properties for biological imaging, such as high quantum yields, high molar extinction coefficients (1–2 orders of magnitude higher than organic dyes), strong resistance to photobleaching and chemical degradation, continuous absorption spectra spanning UV to near-infrared (NIR; 700–900 nm), long fluorescence lifetimes (>10 nanoseconds), narrow emission spectra (typically 20–30 nm full width at half maximum), and large effective Stokes shifts [3, 27, 28, 67, 78–80, 90].

Since the first demonstration of the biomedical potential of QDs in 1998 [7, 21], QD-based research has increased exponentially in recent years. In the last decade, QDs have become powerful tools in fields such as molecular biology, cell biology, molecular imaging, and medical diagnostics. In this chapter, we will first briefly describe the broad applications of QDs *in vitro* and *in vivo*. Then we will discuss in detail QD-based multimodality imaging agents.

6.2 GENERAL APPLICATIONS OF QDs

Numerous applications have been discovered for QDs because of their unique photophysical properties [2, 16, 67, 78, 80]. QDs can replace organic dyes in virtually any system and outperform them in the majority of cases. The major advantage of QDs is their strong resistance to photobleaching over long periods of time (minutes to hours), allowing acquisition of images with good contrast and signal intensity. In addition, most QDs are much brighter than organic dyes due to the combination of higher extinction coefficients ($0.5–5 \times 10^6\ M^{-1}cm^{-1}$) and higher quantum yields [78, 80]. The various applications of QDs are briefly summarized below.

6.2.1 Cell Labeling and Tracking

In recent years, QDs have made the most progress and drawn the greatest interest in the area of cellular labeling. Numerous cellular components/proteins and membrane proteins/receptors have been labeled and visualized with functionalized QDs (in live or fixed cells) [29, 48, 50, 68, 78, 80, 126]. High sensitivity combined with virtually an unlimited number of well-separated colors, all excitable by a single light source, makes QDs ideal probes for multiplexed cellular imaging (Fig. 6.1A) [63, 78]. As a result of their high photostability, QDs can be effectively tracked over an extended period of time in cells [29, 32, 50], embryos [32, 104], as well as small animal models [69, 120]. All these cell tracking studies would not have been possible to perform with conventional organic dyes.

6.2.2 Fluorescent *in situ* Hybridization (FISH)

FISH allows for visualization and mapping of cellular genetic material in order to quantify gene copy numbers within tumor cells that have abnormal gene amplification [91]. DNA or oligonucleotides have been conjugated to QDs, and studies have shown that these conjugates retain the ability to form complementary sequences of Watson-Crick base pairs [32, 64, 71, 83]. The significantly brighter and more photostable fluorescence signal of QDs over organic dyes can allow more stable and quantitative uses of FISH for research and clinical applications [19], as well as direct multicolor imaging of multiple subnuclear genetic sequences [125].

6.2.3 Resonance Energy Transfer

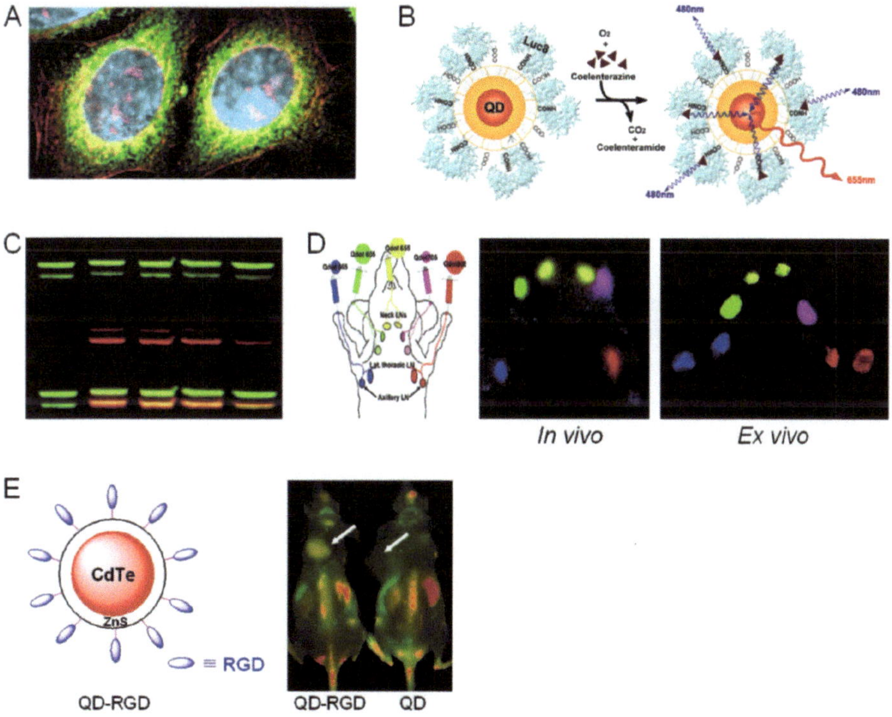

Figure 6.1 The various applications of QDs. (A) A pseudo-colored fluorescence image depicting five-color QD staining of fixed human epithelial cells. The nucleus, Ki-67 protein, mitochondria, microtubules, and actin filaments were each labeled with a QD of different emission wavelength. (B) Self-illuminating QDs based on bioluminescence resonance energy transfer. (C) Western blots of two proteins (top & middle) using two QD–antibody conjugates. An overlay of the two images is shown in the bottom. (D) Simultaneous *in vivo* and *ex vivo* imaging of the primary draining lymph nodes in a mouse injected with five different QDs intracutaneously. (E) RGD peptide–conjugated QD successfully targets the tumor (arrows) *in vivo*. Adapted from Refs. [18, 60, 78, 94, 110].

Fluorescence resonance energy transfer (FRET) is a process in which energy is transferred from an excited donor to an acceptor via a resonant, near-field dipole-dipole interaction [113]. QDs were first reported for FRET applications in 1996 [52–53]. Since then, numerous studies have demonstrated the use of QD-based FRET in biological systems where QDs can be either energy donors or acceptors [75–77, 99, 123]. The two major advantages of using QDs as FRET donors over organic fluorophores is the size-tuned emission of QDs, which can be chosen based on the excitation spectrum of a specific dye, and the ability to attach several dye molecules per QD to improve the FRET efficiency. However, the disadvantages of QDs are the heterogeneity in size, blinking [46], and their limited use as FRET acceptors [24].

Bioluminescence resonance energy transfer (BRET) with QD as an acceptor has been reported [110]. This approach does not require external light excitation for the generation of QD signal, which can give greatly enhanced signal-to-background ratio (Fig. 6.1B). One of the major goals BRET will have to achieve before it can be widely used for *in vivo* imaging is targeting specificity. Besides imaging applications, BRET has also been explored for protease detection, which uses a protease substrate between the donor and acceptor as the sensor [127, 130].

6.2.4 Other *in vitro/ex vivo* Applications of QDs

In addition to the above-mentioned studies, QDs have also been investigated for many other purposes. For example, QD-based probes have been used for co-immunoprecipitation and Western blot analysis, allowing for simpler and faster image acquisition and quantification than traditional techniques (Fig. 6.1C) [70, 94]. QD-based optical barcodes can detect single nucleotide polymorphisms where the DNA sequences differ only at a single nucleotide [44, 129]. QDs have also been reported for tracking RNA interference [22], targeting surface proteins in living cells [47], and detecting bacteria [34].

Over the last decade, QD-based probes have found numerous applications where fluorescent dyes and proteins were previously the only tools available. QDs have allowed for complicated and difficult multiplexed cellular imaging which was impossible for fluorescent dyes and proteins. Overall, QD-based probes have almost completely outperformed traditional organic dyes in *in vitro/ex vivo* applications.

6.2.5 Nontargeted Imaging *in vivo*

By taking advantage of the poor extravasation capability of large-sized QDs, vasculature imaging in living mice has been reported with water-soluble QDs [66, 85]. Using multiphoton microscopy, QDs can differentiate tumor vessels from the perivascular cells and matrix better than traditional fluorescently labeled dextran vessel markers [112].

Lymph node mapping with QDs has been extensively studied in mice [57,58, 134], rats [96], and pigs [57, 97]. In one intriguing study, simultaneous visualization of five separate lymphatic draining sites was demonstrated which may have implications for predicting the route of cancer metastasis into the lymph nodes (Fig. 6.1D) [60]. Sentinel lymph node mapping with QDs may eventually provide highly sensitive, real-time image-guided dissection in cancer patients.

QDs have also been explored for neural imaging. Integrative optical imaging was employed to show that water-soluble QDs diffuse within the extracellular space of adult rat neocortex *in vivo* [118]. Intravenously injected QDs were shown to be taken up by macrophages and localize to experimental glioma in a rat model [86]. Further increases in the injected QD dose by three fold resulted in detectable fluorescence signal in the rat brain [49]. Although it is suggested that these techniques have the potential to be translated into the clinic for image-guided brain tumor biopsies and resections, the enormous amount of QDs needed to generate a detectable signal in the tumor is a major concern in terms of both the toxicity and cost.

6.2.6 Targeted Imaging *in vivo*

In order to make QDs more useful for *in vivo* imaging and other biomedical applications, QDs need to be effectively, specifically, and reliably directed to a specific organ or disease site without alteration. *In vivo* targeting and imaging is very challenging due to the relatively large overall size (typically about 20 nm in diameter) and short circulation half-lives of the QD conjugates. To date, there have been only a handful of successful reports in the literature.

The first report to demonstrate *in vivo* targeting of QD conjugates used peptides as the targeting ligands [1]. Since the QD used in this study emits in the visible range which is not optimal for *in vivo* imaging, *ex vivo* histological analysis was carried out to show that QDs were specifically directed to the tumor vasculature and other organ targets by the peptides. Subsequently, several antibodies have been conjugated to QDs for targeted imaging *in vivo* [41, 114, 131].

We reported the first *in vivo* targeted imaging of tumor vasculature using peptide-conjugated QDs [13, 18]. Integrin $\alpha_v\beta_3$, overexpressed on both tumor vasculature and tumor cells [12, 17], is a prime target for *in vivo* targeted imaging using QDs as extravasation is not required to observe tumor signal. Arginine-glycine-aspartic acid (RGD; potent integrin $\alpha_v\beta_3$ antagonist [9, 14]) containing peptides were conjugated to NIR QDs, and the QD-RGD conjugate exhibited high-affinity integrin $\alpha_v\beta_3$ specific binding in cell culture and *ex vivo*. *In vivo* NIR fluorescence (NIRF) imaging of the tumor vasculature was successfully demonstrated in integrin $\alpha_v\beta_3$-positive tumor-bearing mice (Fig. 6.1E). Subsequently, real-time intravital imaging of these QD-RGD conjugates binding to luminal endothelium in mouse tumor neovasculature was also reported [109]. This study provided direct evidence that these QD conjugates do not extravasate, confirming that targeting tumor vasculature rather than targeting tumor cells is the most feasible way to achieve *in vivo* targeted imaging with QDs.

6.3 MOLECULAR IMAGING AND DUAL-MODALITY IMAGING AGENTS

Molecular imaging, "the visualization, characterization and measurement of biological processes at the molecular and cellular levels in humans and other living systems" [73], has flourished over the last decade. In general, molecular imaging modalities include molecular magnetic resonance imaging (MRI), magnetic resonance spectroscopy (MRS), optical bioluminescence, optical fluorescence, targeted ultrasound, single-photon emission computed tomography (SPECT), and positron emission tomography (PET) [74]. Continued development and wider availability of scanners dedicated to small animal imaging studies, which can provide a similar *in vivo* imaging capability in mice, primates, and humans, can enable smooth transfer of knowledge and molecular measurements between species thereby facilitating clinical translation.

Every imaging modality has its advantages and disadvantages [74]. For example, it is difficult to accurately quantify fluorescence signal in living subjects based on fluorescence imaging alone, particularly in deep tissue, although it is the mostly widely available/used imaging technique; MRI has high resolution and good soft tissue contrast, yet it suffers from very low sensitivity; radionuclide-based imaging techniques (SPECT and PET) are very sensitive and highly quantitative, yet they suffer from relatively poor spatial resolution. Combination of optical imaging and other modalities can offer synergistic advantages over either modality alone.

Multimodality imaging using a small molecule-based agent is very challenging, and sometimes impossible, due to the limited number of functional groups for further modification and the potential interference with its receptor binding affinity. On the other hand, nanoparticles (e.g., QDs) have large surface areas where multiple functional moieties can be attached for multimodality molecular imaging [11]. The remaining sections of this chapter will focus on QD-based dual-modality agents for optical/MR, optical/SPECT, and optical/PET imaging.

6.4 QD-BASED DUAL-MODALITY AGENTS FOR OPTICAL/MR IMAGING

MRI is a noninvasive diagnostic technique based on the interaction of protons (or certain other nuclei) with each other and with surrounding molecules in a tissue of interest [98]. Different tissues have different relaxation times, which can result in endogenous MR contrast. The endogenous tissue contrast can be enhanced by administration of exogenous contrast agents [101, 133], such as gadolinium (Gd^{3+}) chelates for enhancing the T_1 contrast [30] and iron oxide (IO) nanoparticles for increasing the T_2 contrast [117]. Dual-modality agents that can be detected by both MR and optical imaging have been extensively studied, and many of them are based on QDs.

One approach to construct dual-modality optical/MR agents is to combine QD with iron or IO nanoparticles. Polymer-coated IO cores that were overcoated with cadmium selenium (CdSe)–zinc silicon (ZS) QD shells, further functionalized with antibodies, have been used to magnetically capture breast cancer cells and view them with fluorescence imaging [121]. A nanocomposite system was reported that consists of an IO nanoparticle and a CdSe QD, grown onto the preformed IO core to yield either a heterodimer or a homogeneous dispersion of QD around the core [108]. After being coated with a silica shell, these nanoparticles can be readily used for bioconjugation through the amino groups on the surface. Magnetic QDs composed of CdS–iron platinum (FePt) with sizes of around 7 nm were prepared in a one-pot synthesis (Fig. 6.2A) [43]. This synthetic process may also allow the production of large quantities of various types of other heterostructures at the nanoscale level.

Stabilization of single QD micelles by a silica shell using "amphiphilic" and "hydrophilic" silica precursors has been reported [5]. The hydrophobic layer between the QD and the silica shell can be manipulated for incorporating hydrophobic paramagnetic substances for dual-modality optical/MR imaging [4]. Recently, a novel strategy to coat silica particles with a dense monolayer of paramagnetic and PEGylated lipids was reported (Fig. 6.2B) [61]. The silica nanoparticle carrying a QD in the center was conjugated with multiple RGD peptides, and receptor-specific uptake by endothelial cells was demonstrated *in vitro* with both fluorescence microscopy and MRI.

A series of core/shell $CdSe/Zn_{1-x}Mn_xS$ nanoparticles have been synthesized for optical/MR imaging (Fig. 6.2C) [122]. The manganese (Mn^{2+}) content in the nanoparticles varies with the thickness of the shell and the amount of Mn^{2+} introduced to the reaction. It was found that the quantum yield and Mn^{2+} concentration in these nanoparticles were sufficient for generating contrast in both modalities at a relatively low concentration. In another report, ultra-small CdS:Mn/ZnS core-shell QDs were also prepared [105]. Recently, Gd^{3+}-functionalized NIR QDs were synthesized for optical/MR imaging [51]. To test the utility of such Gd^{3+}-functionalized QDs as dual-modality agents for *in vivo* imaging, NIRF and MR imaging of a phantom filled with the contrast agent embedded into a mouse abdomen were successfully achieved.

In one study, chitosan nanoparticles encapsulating CdSe/ZnS QDs and Gd-DTPA (where DTPA denotes diethylene triamine pentaacetic acid) was found to exhibit good aqueous stability and biocompatibility with cultured cells [115]. A biotinylated construct consisting of a lysine wedge with eight Gd-DTPA complexes attached to the periphery, to increase its perceptibility in MRI by increasing the load of Gd-DTPA, was synthesized [103]". The resulting molecule, along with biotinylated annexin A5, was conjugated to QD–streptavidin (Fig. 6.2D). The resulting construct was then used to analyze biological samples and vascular structures with MRI at the anatomical level, and with two-photon laser scanning microscopy at the cellular level.

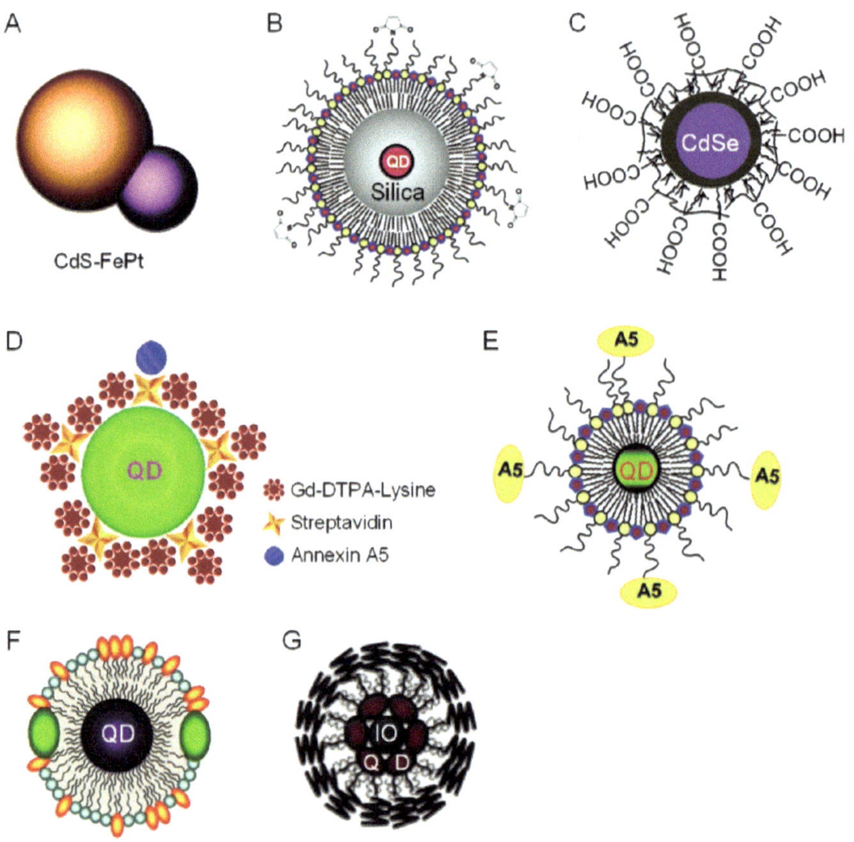

Figure 6.2 The various approaches to construct QD-based dual-modality agents for optical/ MR imaging. (A) Magnetic QDs composed of the CdS-FePt heterostructure. (B) Paramagnetic lipid coated silica nanoparticles with a QD core. (C) A water-soluble core/shell CdSe/$Zn_{1-x}Mn_xS$ nanoparticle. (D) A QD linked to annexin A5 and multiple copies of Gd-DTPA complexes through biotin-streptavidin interaction. (E) A QD with a paramagnetic lipid coating, to which multiple annexin A5 proteins are conjugated. (F) High-density lipoprotein-like nanoparticles with a QD core. (G) A micellar hybrid nanoparticle that encapsulate IO nanoparticles and QDs within a single PEGylated phospholipid micelle. Adapted from Refs. [25, 43, 61, 95, 103, 119, 122].

Using a similar approach, cyclic Asn-Gly-Arg (cNGR) peptide-labeled paramagnetic QDs (cNGR-pQDs) were prepared for the noninvasive assessment of tumor angiogenesis with molecular MRI [93]. cNGR has previously been shown to bind to CD13, an aminopeptidase that is highly overexpressed on angiogenic tumor endothelium [26]. Intravenous injection of cNGR-pQDs in tumor-bearing mice resulted in increased contrast in the tumor rim but not in the center of the tumor or muscle tissue. The MRI results were validated using *ex vivo* two-photon laser scanning microscopy, which showed that cNGR-pQDs were primarily located on the surface of tumor endothelial cells and to a lesser extent in the vessel lumen. In contrast, unlabeled pQDs were not or only sparsely detected with either MRI or optical imaging.

QDs encapsulated in a paramagnetic micelle, with multiple annexin A5 molecules covalently coupled to the nanoparticle surface, have been constructed for targeting apoptosis (Fig. 6.2E) [119]. Target specificity of the annexin A5-conjugated nanoparticles was demonstrated with both fluorescence microscopy and MRI. In another study, the same nanoparticle was functionalized with RGD peptides and tested in cultured endothelial cells [88]. Recently, such RGD peptide conjugated nanoparticles were investigated *in vivo* [87]. Upon intravenous injection into tumor-bearing mice, intravital microscopy allowed the detection of superficial angiogenic endothelium at the cellular resolution, and MRI was used to visualize angiogenesis at the anatomical resolution throughout the tumor. These QD-containing paramagnetic micelles have also been conjugated with antibodies (specific for the macrophage scavenger receptor) for imaging macrophages in atherosclerotic plaques, where T_1-weighted MRI was performed *in vivo* and optical imaging was carried out *ex vivo* on the excised aorta [89]. This study demonstrated that macrophages in mice can be detected by molecular MRI and optical methods upon the administration of such a dual-modality imaging agent.

QD-containing, high-density lipoprotein (HDL)-like nanoparticles were prepared for optical/MR imaging applications (Fig. 6.2F) [25]. The characteristics of the nanoparticle, as well as its *in vitro* and *in vivo* behavior, were reported to closely mimic the naturally occurring HDL. Further, *in vivo* applicability of these HDL-like nanoparticles for atherosclerosis imaging was demonstrated with various imaging modalities, using analogous HDL-like nanoparticles with different core component as the contrast agents (e.g., IO nanoparticle for MRI and gold [Au] nanoparticle for computed tomography [CT], respectively).

For the majority of the above-mentioned QD-based optical/MR imaging agents, *in vivo* targeting has not been achieved, which certainly deserves much further exploration. In some studies, the feasibility of dual-modality imaging *in vivo* was demonstrated, but the two modalities were not equally effective. The less sensitive modality was usually used for *ex vivo* validation of the *in vivo* results obtained from the more sensitive imaging modality. Although only single-modality noninvasive imaging was achieved with these dual-modality agents, which does not take full advantage of the nanoparticle-based approach, the capability of detecting the agent with another imaging modality did provide a convenient way for *ex vivo* validation, which is significantly more reliable and advantageous than the corresponding single-modality agents.

Moving beyond imaging applications only, micellar hybrid nanoparticles that contain IO nanoparticles, QDs, and the anticancer drug doxorubicin were investigated in one study (Fig. 6.2G) [95]. The strong interaction between the PEGylated phospholipids and the hydrophobic chains attached to the QDs and IO nanoparticles resulted in high dispersity and stability of the micellar hybrid nanoparticles, making them suitable for *in vitro* and *in vivo* applications. It was suggested that simultaneous dual-modality imaging and therapy with such a hybrid system may enable more effective early detection and treatment of various types of cancer.

Molecular MRI is still in its infancy, and receptor-specific dual-modality optical/MR imaging is rare in the literature [111]. It is likely that the feasible targets reachable by such optical/MR agents will be mostly vasculature related, since the overall size of these nanoparticles with surface polymer coating and targeting ligands is quite large (>20 nm in diameter). For targeted imaging of cancer, newly developed QDs with smaller diameter (preferably <10 nm) and long circulation half-life (at least a few hours) may allow for extravasation from the leaky tumor vasculature to a certain extent. More importantly, conjugating these nanoparticles with small molecules or peptides should give better targeting efficiency than antibody-conjugated nanoparticles, because of the much larger number of targeting ligands per nanoparticle and significantly smaller overall size. Since the major disadvantage of MRI is its inherent low sensitivity, future development of novel contrast agents capable of targeting the cells in addition to the vasculature may dramatically increase the MR signal and facilitate the biomedical applications of dual-modality optical/MR imaging agents. From the technical perspective, the constantly evolving data acquisition/reconstruction techniques and novel pulse sequences can also improve the sensitivity and temporal resolution of MRI [82, 132].

6.5 QD-BASED DUAL-MODALITY AGENTS FOR OPTICAL/SPECT IMAGING

Radionuclide-based imaging (PET and SPECT) has much higher sensitivity than MRI and much better tissue penetration than optical imaging and ultrasound [74]. The most important advantage of radionuclide-based imaging over other imaging modalities is the ability to quantitatively measure (at virtually any depth) the radionuclide concentration in various organs over time [39], which can provide invaluable information about the pharmacokinetics and full-body distribution of the imaging agent. To date, QD-based dual-modality optical/SPECT imaging has not been well studied.

Radioactive Cd tellurium-125m (^{125m}Te)/ZnS QDs were targeted to the mouse lung using an antibody to mouse lung endothelium as the targeting moiety [124]. Targeting efficacy of the conjugate *in vivo* was evaluated with biodistribution studies and SPECT imaging. It was found that the specific antibody-conjugated Cd^{125m}Te/ZnS QDs primarily accumulated in the mouse lungs, while QDs conjugated with a control antibody mainly localized in the liver and spleen. Although these data provided, for the first time, a quantitative measurement of the *in vivo* targeting efficacy of antibody-conjugated QDs, no optical imaging was performed in this study.

Subsequently, the competition between vascular targeting and interaction of the radioactive Cd^{125m}Te/ZnS QDs with the reticuloendothelial system (RES) was investigated [56]. Again, the QDs were derivatized with the antibody, which binds to murine thrombomodulin expressed in the lumen of lung blood vessels, and tested for *in vivo* targeting with SPECT/CT imaging, tissue autoradiography, and biodistribution studies (Fig. 6.3A,B). Antigen-specific uptake of the targeted QDs was demonstrated by the uniform distribution of radioactivity in the mouse lungs and significantly less accumulation in the liver and spleen. Since QDs can be engulfed by phagocytic cells and removed from the circulation, biodistribution studies were also carried out in mice that had been depleted of phagocytic cells by the use of clodronate-loaded liposomes. As expected, the accumulation of QDs in the liver was reduced fivefold in these mice. Taken together, these two reports indicated that *in vivo* targeting of QDs is a competition between the effectiveness of the targeting ligand and the natural tendency for the RES uptake of the QDs. Temporary inhibition of the RES may enhance the targeting efficacy of QDs.

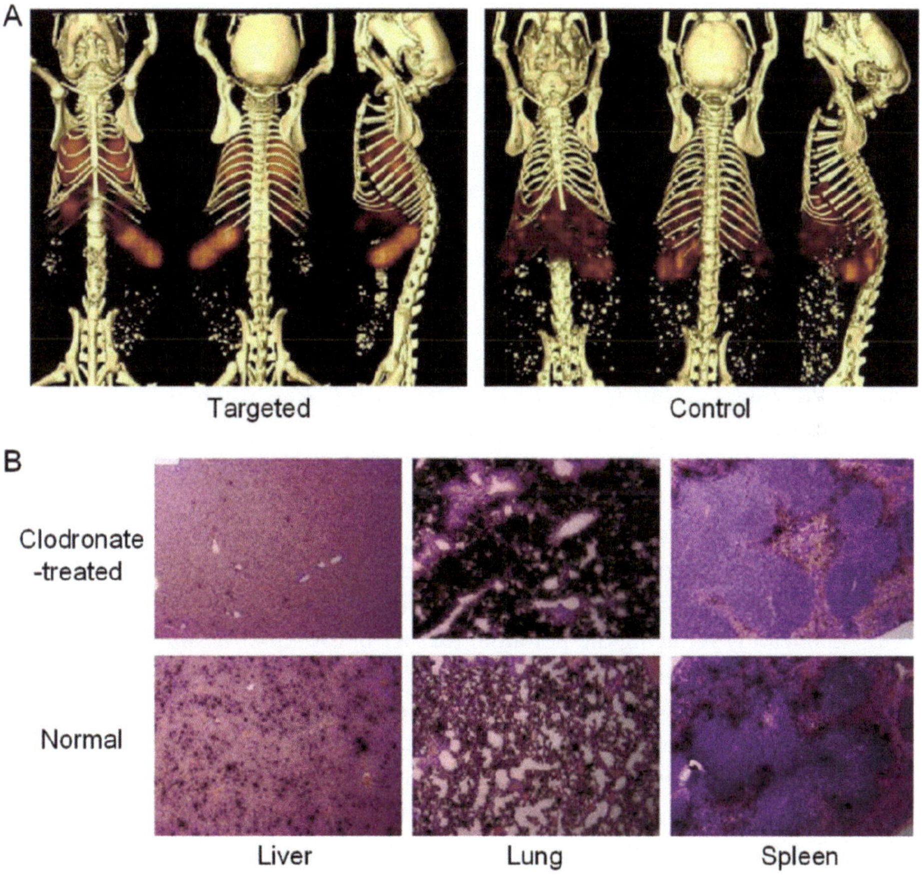

Figure 6.3 The fate of $Cd^{125m}Te/ZnS$ QDs *in vivo*. (A) SPECT/CT images of mice injected with the radioactive QDs one day earlier. Three views with different orientations are shown, and the red color represents the SPECT data. (B) Autoradiographic images of tissues harvested from animals (clodronate-treated or normal mice) injected with the targeted QDs one day earlier. Adapted from Ref. [56].

The main focus of these two reports was on the biodistribution and *in vivo* targeting efficacy of the $Cd^{125m}Te/ZnS$ QDs instead of optical/SPECT imaging with the dual-modality agent. For SPECT imaging, different isotopes that emit different energy gamma rays can be differentiated on the basis of energy [6]. Since QDs are ideal agents for multiplexed fluorescence imaging [60], combination of the multiplexing capabilities of both SPECT and QDs, which can enable the interrogation of a number of biological events simultaneously, should be explored in the future.

6.6 QD-BASED DUAL-MODALITY AGENTS FOR OPTICAL/PET IMAGING

Due to the difficulty in quantifying fluorescence signal *in vivo* and many other technical challenges that remain to be solved, *in vivo* imaging of QDs are so far mostly qualitative or semiquantitative. PET, an imaging modality that is highly quantitative, has been routinely used in the clinic for staging and evaluating many types of cancer [40]. Development of a dual-modality agent containing both an NIR QD and a PET isotope will allow for a sensitive, accurate assessment of the pharmacokinetics and tumor targeting efficacy of NIR QDs by PET, which may greatly facilitate future translation of QDs into clinical applications.We first reported a QD-based dual-modality probe for optical/PET imaging (Fig. 6.4A–C) [8]. In this study, cyclic RGD peptides and 1, 4, 7, 10-tetraazacyclododecane-*N,N′,N″,N‴*-tetraacetic acid (DOTA) chelators were conjugated to an NIR QD for integrin $\alpha_v\beta_3$-targeted dual-modality optical/PET imaging after copper-64 (^{64}Cu) labeling. Using this optical/PET probe, we quantitatively evaluated the tumor targeting efficacy of the probe and found that the majority of the dual-modality agent in the tumor was within the tumor vasculature. Histological examination confirmed that ^{64}Cu-DOTA-QD-RGD targets primarily the tumor vasculature via specific RGD-integrin $\alpha_v\beta_3$ interaction with little extravasation. Further, this optical/PET probe can confer sufficient tumor contrast detectable by PET at much lower concentration than that required for *in vivo* NIRF imaging [18], thus significantly reducing the potential toxicity of cadmium-based QDs, which may greatly facilitate their future biomedical applications [31, 59].

Vascular endothelial growth factor (VEGF)/VEGF receptor (VEGFR, expressed primarily on tumor vasculature but not on tumor cells) signaling pathway plays a pivotal role in regulating both physiological and tumor angiogenesis [10, 12]. Since the radiolabeled QDs primarily targeted the tumor vasculature rather than tumor cells in the previous study, we also investigated VEGFR targeting of these QDs by both optical and PET imaging [23]. The naturally occurring VEGF was used as the targeting ligand and DOTA-QD-VEGF exhibited VEGFR-specific binding in cell-based studies. In tumor-bearing mice, both NIRF and PET imaging showed VEGFR-specific delivery of DOTA-QD-VEGF to the tumor. In both studies (targeting integrin $\alpha_v\beta_3$ or VEGFR), good correlation was observed between the results measured by *ex vivo*

NIRF and PET imaging, which indicated that these QD-based conjugates are quite stable *in vivo* [8, 23].

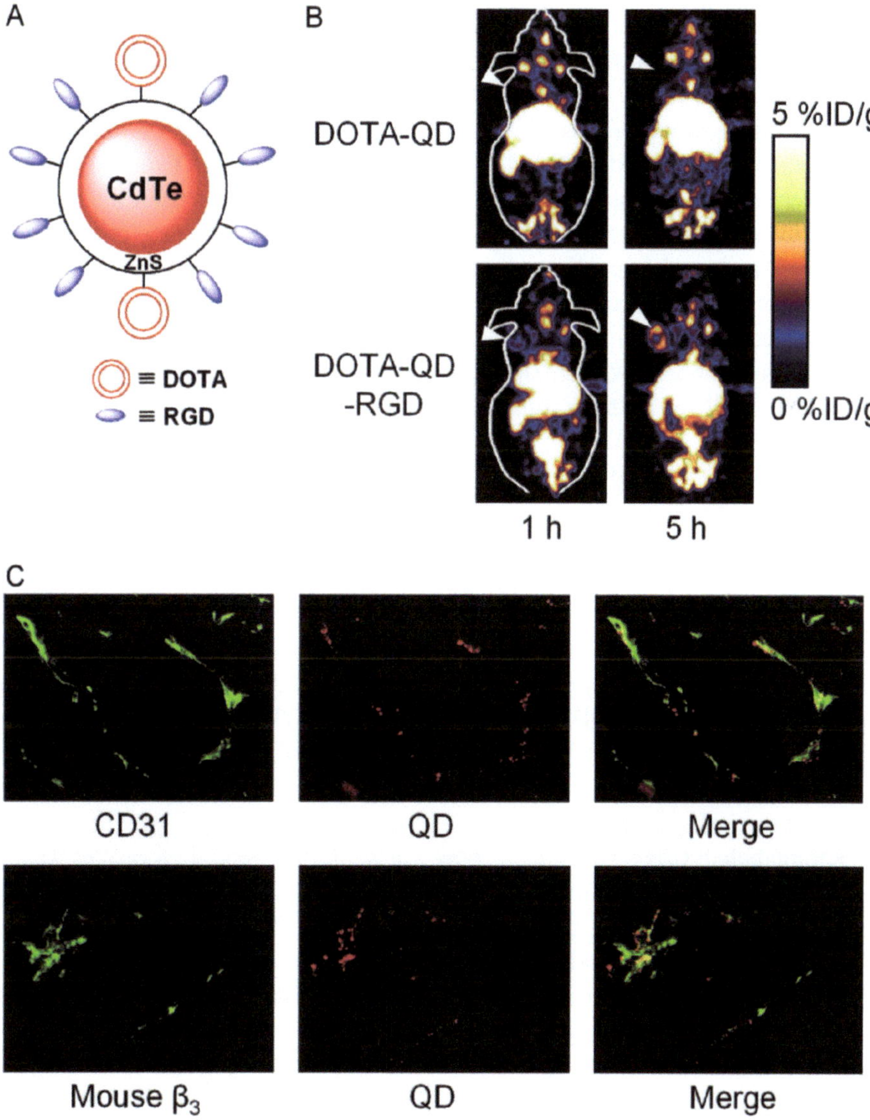

Figure 6.4 Optical/PET imaging of integrin $\alpha_v\beta_3$. (A) Schematic structure of the dual-modality probe. (B) Coronal PET images of tumor-bearing mice at 1 and 5 hours after injection of ^{64}Cu-labeled DOTA-QD or DOTA-QD-RGD. Arrowheads indicate tumors. (C) Excellent overlay between CD31 staining and QD fluorescence (top), as well as between murine β_3 staining and QD fluorescence (bottom), confirmed that DOTA-QD-RGD mainly targeted integrin $\alpha_v\beta_3$ in tumor vasculature. Adapted from Ref. [8].

Combination of optical and PET imaging overcomes the tissue penetration limitation of optical imaging and allows for quantitative *in vivo* targeted imaging in deep tissue, which is crucial for future image-guided surgery through sensitive, specific, and real-time intraoperative visualization of the molecular features of normal and diseased processes. Interestingly, a few other reports have also focused on radiolabeled QDs. In one study, the biodistribution of commercially available CdSe QDs in mice were investigated with PET imaging after ^{64}Cu labeling [106]. Rapid uptake of these radiolabeled QDs by the liver and spleen (organs of the RES) was observed, and it was concluded that the size of QDs had no influence on the biodistribution within the range tested. Subsequently, the influence of particle size, PEGylation, and surface coating on the biodistribution of QDs in mice was studied [107]. It was reported that PEGylation and peptide coating slowed the RES uptake of QDs and small QDs were in part renally excreted. Although these radiolabeled QDs qualify as dual-modality optical/PET imaging agents, the optical properties of these QDs were not utilized in these studies and only PET imaging was carried out. Like the previously mentioned Cd^{125m}Te/ZnS QDs [56, 124], the primary focus of these two studies on ^{64}Cu-labeled QDs was their biodistribution in mice rather than their dual-modality imaging capability.

Fluorine-18 (^{18}F), commonly used for clinical PET imaging, has been used to label phospholipid QD micelles [33]. After injection into mice, whole-body biodistribution and pharmacokinetics of the dual-modality optical/PET agent could be monitored with both PET and *in vivo* fibered confocal fluorescence imaging. This work represents the first report of ^{18}F-labeled QDs and the results differ significantly from previous reports on QD biodistribution in mice. The phospholipid QD micelles were found to be well distributed throughout the body with long circulation half-life and slow RES uptake, which makes them promising agents for future *in vivo* targeted imaging applications.

Radiolabeled QDs represent a new class of dual-modality probes that has enormous potential for clinical applications. Different from other molecular imaging modalities where typically the nanoparticle itself is detected, radionuclide-based imaging detects the radiolabel rather than the QDs. The distribution of QDs is measured indirectly by assessing the localization of the radionuclide, which can provide quantitative measurement of the tumor targeting efficacy and pharmacokinetics only if the radiolabel on the QDs is stable enough under physiological conditions. However, dissociation of the radionuclide (typically metal) from the chelator, and/or the radionuclide-containing polymer coating from the QDs, may occur which can cause significant difference between the QD distribution and the radionuclide distribution. Direct measurement of QDs in various tissues using their intrinsic fluorescence signal, as well as rigorous validation of the stability of the radiolabel on the QDs, should always be carried out to obtain more reliable results. Encapsulating certain PET isotopes inside the QD core (e.g., ^{72}As and ^{74}As in InAs QDs), similar to the Cd^{125m}Te/ZnS QDs for SPECT imaging applications, can solve such *in vivo* stability issue and should be explored in the future.

6.7 CONCLUSION AND FUTURE PERSPECTIVES

The reported multimodality imaging agents to date are mostly optical/MR agents, mainly due to the exploding interest in two types of nanoparticles, IO and QD, for MR and optical imaging, respectively [80, 117]. However, whether optical/MR combination is the most optimal dual-modality approach is debatable since neither modality is quantitative. Other dual-modality agents (optical/SPECT, optical/PET, and PET/MR) have not been widely investigated. Dual-modality agents that combine radionuclide-based imaging (PET or SPECT), which is very sensitive and highly quantitative, and non-radionuclide-based approaches, for example, optical imaging, which can significantly facilitate *ex vivo* validation of the *in vivo* data, and MRI, which can provide high-resolution anatomical information, should be of particular interest for future biomedical applications.

In clinical settings, optical imaging is relevant for tissues close to the surface of the skin, tissues accessible by endoscopy and during intraoperative visualization [16]. One scenario where a dual-modality QD-based optical/PET agent will be particularly useful is that an initial whole-body PET scan can be carried out to identify the location of tumor(s), and optical imaging can be subsequently used to guide tumor resection. The major roadblocks for clinical translation of QD-based agents are inefficient delivery, potential toxicity, and lack of quantification. The use of QD-based multimodality agents can partially overcome these limitations, and various combinations of different imaging modalities can serve different purposes depending on the specific application. We envision that a PET/MR/optical agent may have broad uses in future biomedical/clinical applications since such a combination provides extremely high sensitivity (PET) and quantitation capability (PET), excellent anatomical information and soft tissue contrast (MRI), as well as a means for *ex vivo* validation (optical), which itself can also be useful for highly sensitive imaging in certain sites of the human body.

The use of molecularly targeted nanoparticles (e.g., QDs) affords many advantages over small molecule-based agents. First, hundreds, thousands, or even more imaging labels, or a combination of labels for different imaging modalities, can be attached to a single QD, which can lead to dramatic signal amplification. Second, multiple, potentially different, targeting ligands on the QD can provide enhanced binding affinity/specificity to the target. Third, the ability to integrate various means to bypass the biological barriers may also improve the targeting efficacy. Nanoparticles, in particular QDs, suffer from poorer extravasation compared with small molecules or proteins. Therefore, the most promising applications of QD-based multimodality agents will be in cardiovascular imaging/medicine, where there is much less biological barrier for efficient delivery of QDs, and in oncology, where the leaky tumor vasculature can allow for better tissue penetration than in normal organs/tissues. Mounting evidence has suggested that peptides are more desirable targeting ligands than antibodies for QD conjugation, not only to keep the overall size small but also to fully take advantage of the polyvalency effect [72] since significantly more peptides can be attached to a QD than the macromolecular antibodies.

Several barriers exist for preclinical and clinical *in vivo* applications of QD-based multimodality imaging agents, among which are biocompatibility, *in vivo* kinetics, targeting efficacy, acute and chronic toxicity, the ability to escape the RES, and cost-effectiveness. Recently, nanotoxicology has emerged as a new branch of toxicology for studying the undesirable effects of nanoparticles [54, 92]. Development of QD-based agents for biomedical applications must proceed in tandem with the assessment of any toxicological side effects. With the development of smaller, non-cadmium-based QDs and further improvement on the conjugation strategy [58, 65, 102, 128], it is expected that QD-based agents will achieve optimal tumor targeting efficacy with acceptable toxicity profile for clinical translation in the near future.

References

1. Akerman M. E., Chan W. C. W., Laakkonen P., Bhatia S. N., and Ruoslahti E. (2002) Nanocrystal targeting *in vivo*. *Proc Natl Acad Sci USA* **99**, 12617–12621.
2. Alivisatos A. P., Gu W., and Larabell C. (2005) Quantum dots as cellular probes. *Annu Rev Biomed Eng* **7**, 55–76.
3. Alivisatos P. (2004) The use of nanocrystals in biological detection. *Nat Biotechnol* **22**, 47–52.
4. Bakalova R., Zhelev Z., Aoki I., Masamoto K., Mileva M., Obata T., Higuchi M., Gadjeva V., and Kanno I. (2008) Multimodal silica-shelled quantum dots: direct intracellular delivery, photosensitization, toxic, and microcirculation effects. *Bioconjug Chem* **19**, 1135–1142.
5. Bakalova R., Zhelev Z., Aoki I., Ohba H., Imai Y., and Kanno I. (2006) Silica-shelled single quantum dot micelles as imaging probes with dual or multimodality. *Anal Chem* **78**, 5925–5932.
6. Berman D. S., Kiat H., Van Train K., Friedman J. D., Wang F. P., and Germano G. (1994) Dual-isotope myocardial perfusion SPECT with rest thallium-201 and stress Tc-99m sestamibi. *Cardiol Clin* **12**, 261–270.
7. Bruchez M., Jr., Moronne M., Gin P., Weiss S., and Alivisatos A. P. (1998) Semiconductor nanocrystals as fluorescent biological labels. *Science* **281**, 2013–2016.
8. Cai W., Chen K., Li Z. B., Gambhir S. S., and Chen X. (2007) Dual-function probe for PET and near-infrared fluorescence imaging of tumor vasculature. *J Nucl Med* **48**, 1862–1870.
9. Cai W., and Chen X. (2006) Anti-angiogenic cancer therapy based on integrin $\alpha_v\beta_3$ antagonism. *Anti-Cancer Agents Med Chem* **6**, 407–428.
10. Cai W., and Chen X. (2007) Multimodality imaging of vascular endothelial growth factor and vascular endothelial growth factor receptor expression. *Front Biosci* **12**, 4267–4279.

11. Cai W., and Chen X. (2007) Nanoplatforms for targeted molecular imaging in living subjects. *Small* **3**, 1840–1854.

12. Cai W., and Chen X. (2008) Multimodality molecular imaging of tumor angiogenesis. *J Nucl Med* **49** Suppl 2, 113S-128S.

13. Cai W., and Chen X. (2008) Preparation of peptide conjugated quantum dots for tumour vasculature targeted imaging. *Nat Protoc* **3**, 89–96.

14. Cai W., Gambhir S. S., and Chen X. (2005) Multimodality tumor imaging targeting integrin $\alpha_v\beta_3$. *Biotechniques* **39**, S6-S17.

15. Cai W., Gao T., Hong H., and Sun J. (2008) Applications of gold nanoparticles in cancer nanotechnology. *Nanotechnology, Science and Applications* **1**, 17–32.

16. Cai W., Hsu A. R., Li Z. B., and Chen X. (2007) Are quantum dots ready for *in vivo* imaging in human subjects? *Nanoscale Res Lett* **2**, 265–281.

17. Cai W., Niu G., and Chen X. (2008) Imaging of integrins as biomarkers for tumor angiogenesis. *Curr Pharm Des* **14**, 2943–2973.

18. Cai W., Shin D. W., Chen K., Gheysens O., Cao Q., Wang S. X., Gambhir S. S., and Chen X. (2006) Peptide-labeled near-infrared quantum dots for imaging tumor vasculature in living subjects. *Nano Lett* **6**, 669–676.

19. Chan P., Yuen T., Ruf F., Gonzalez-Maeso J., and Sealfon S. C. (2005) Method for multiplex cellular detection of mRNAs using quantum dot fluorescent *in situ* hybridization. *Nucleic Acids Res* **33**, e161.

20. Chan W. C., Maxwell D. J., Gao X., Bailey R. E., Han M., and Nie S. (2002) Luminescent quantum dots for multiplexed biological detection and imaging. *Curr Opin Biotechnol* **13**, 40–46.

21. Chan W. C., and Nie S. (1998) Quantum dot bioconjugates for ultrasensitive nonisotopic detection. *Science* **281**, 2016–2018.

22. Chen A. A., Derfus A. M., Khetani S. R., and Bhatia S. N. (2005) Quantum dots to monitor RNAi delivery and improve gene silencing. *Nucleic Acids Res* **33**, e190.

23. Chen K., Li Z. B., Wang H., Cai W., and Chen X. (2008) Dual-modality optical and positron emission tomography imaging of vascular endothelial growth factor receptor on tumor vasculature using quantum dots. *Eur J Nucl Med Mol Imaging* **35**, 2235–2244.

24. Clapp A. R., Medintz I. L., Fisher B. R., Anderson G. P., and Mattoussi H. (2005) Can luminescent quantum dots be efficient energy acceptors with organic dye donors? *J Am Chem Soc* **127**, 1242–1250.

25. Cormode D. P., Skajaa T., van Schooneveld M. M., Koole R., Jarzyna P., Lobatto M. E., Calcagno C., Barazza A., Gordon R. E., Zanzonico P., Fisher E. A., Fayad Z. A., and Mulder W. J. (2008) Nanocrystal core high-density lipoproteins: a multimodality contrast agent platform. *Nano Lett* **8**, 3715–3723.

26. Curnis F., Sacchi A., Borgna L., Magni F., Gasparri A., and Corti A. (2000) Enhancement of tumor necrosis factor alpha antitumor immunotherapeutic properties by targeted delivery to aminopeptidase N (CD13). *Nat Biotechnol* **18**, 1185–1190.

27. Dabbousi B. O., RodriguezViejo J., Mikulec F. V., Heine J. R., Mattoussi H., Ober R., Jensen K. F., and Bawendi M. G. (1997) (CdSe)ZnS core-shell quantum dots: Synthesis and characterization of a size series of highly luminescent nanocrystallites. *J Phys Chem B* **101**, 9463–9475.

28. Dahan M., Laurence T., Pinaud F., Chemla D. S., Alivisatos A. P., Sauer M., and Weiss S. (2001) Time-gated biological imaging by use of colloidal quantum dots. *Optics Lett* **26**, 825–827.

29. Dahan M., Levi S., Luccardini C., Rostaing P., Riveau B., and Triller A. (2003) Diffusion dynamics of glycine receptors revealed by single-quantum dot tracking. *Science* **302**, 442–445.

30. de Roos A., Doornbos J., Baleriaux D., Bloem H. L., and Falke T. H. (1988) Clinical applications of gadolinium-DTPA in MRI. *Magn Reson Annu*, 113–145.

31. Derfus A. M., Chan W. C. W., and Bhatia S. N. (2004) Probing the cytotoxicity of semiconductor quantum dots. *Nano Lett* **4**, 11–18.

32. Dubertret B., Skourides P., Norris D. J., Noireaux V., Brivanlou A. H., and Libchaber A. (2002) *In vivo* imaging of quantum dots encapsulated in phospholipid micelles. *Science* **298**, 1759–1762.

33. Duconge F., Pons T., Pestourie C., Herin L., Theze B., Gombert K., Mahler B., Hinnen F., Kuhnast B., Dolle F., Dubertret B., and Tavitian B. (2008) Fluorine-18-labeled phospholipid quantum dot micelles for *in vivo* multimodal imaging from whole body to cellular scales. *Bioconjug Chem* **19**, 1921–1926.

34. Edgar R., McKinstry M., Hwang J., Oppenheim A. B., Fekete R. A., Giulian G., Merril C., Nagashima K., and Adhya S. (2006) High-sensitivity bacterial detection using biotin-tagged phage and quantum-dot nanocomplexes. *Proc Natl Acad Sci USA* **103**, 4841–4845.

35. Efros A. L., and Efros A. L. (1982) Interband absorption of light in a semiconductor sphere. *Sov Phys Semicond* **16**, 772–775.

36. Ekimov A. I., and Onushchenko A. A. (1982) Quantum size effect in the optical-spectra of semiconductor micro-crystals. *Sov Phys Semicond* **16**, 775–778.

37. Felgner P. L., and Ringold G. M. (1989) Cationic liposome-mediated transfection. *Nature* **337**, 387–388.

38. Ferrari M. (2005) Cancer nanotechnology: opportunities and challenges. *Nat Rev Cancer* **5**, 161–171.

39. Gambhir S. S. (2002) Molecular imaging of cancer with positron emission tomography. *Nat Rev Cancer* **2**, 683–693.

40. Gambhir S. S., Czernin J., Schwimmer J., Silverman D. H., Coleman R. E., and Phelps M. E. (2001) A tabulated summary of the FDG PET literature. *J Nucl Med* **42**, 1S-93S.

41. Gao X., Cui Y., Levenson R. M., Chung L. W. K., and Nie S. (2004) *In vivo* cancer targeting and imaging with semiconductor quantum dots. *Nat Biotechnol* **22**, 969–976.

42. Gregoriadis G. (1980) Tailoring liposome structure. *Nature* **283**, 814–815.

43. Gu H., Zheng R., Zhang X., and Xu B. (2004) Facile one-pot synthesis of bifunctional heterodimers of nanoparticles: a conjugate of quantum dot and magnetic nanoparticles. *J Am Chem Soc* **126**, 5664–5665.

44. Han M., Gao X., Su J. Z., and Nie S. (2001) Quantum-dot-tagged microbeads for multiplexed optical coding of biomolecules. *Nat Biotechnol* **19**, 631–635.

45. Hirsch L. R., Gobin A. M., Lowery A. R., Tam F., Drezek R. A., Halas N. J., and West J. L. (2006) Metal nanoshells. *Ann Biomed Eng* **34**, 15–22.

46. Hohng S., and Ha T. (2005) Single-molecule quantum-dot fluorescence resonance energy transfer. *Chemphyschem* **6**, 956–960.

47. Howarth M., Takao K., Hayashi Y., and Ting A. Y. (2005) Targeting quantum dots to surface proteins in living cells with biotin ligase. *Proc Natl Acad Sci USA* **102**, 7583–7588.

48. Ishii D., Kinbara K., Ishida Y., Ishii N., Okochi M., Yohda M., and Aida T. (2003) Chaperonin-mediated stabilization and ATP-triggered release of semiconductor nanoparticles. *Nature* **423**, 628–632.

49. Jackson H., Muhammad O., Daneshvar H., Nelms J., Popescu A., Vogelbaum M. A., Bruchez M., and Toms S. A. (2007) Quantum dots are phagocytized by macrophages and colocalize with experimental gliomas. *Neurosurgery* **60**, 524–529; discussion 529–530.

50. Jaiswal J. K., Mattoussi H., Mauro J. M., and Simon S. M. (2003) Long-term multiple color imaging of live cells using quantum dot bioconjugates. *Nat. Biotechnol.* **21**, 47–51.

51. Jin T., Yoshioka Y., Fujii F., Komai Y., Seki J., and Seiyama A. (2008) Gd^{3+}-functionalized near-infrared quantum dots for *in vivo* dual modal (fluorescence/magnetic resonance) imaging. *Chem Commun*, 5764–5766.

52. Kagan C. R., Murray C. B., and Bawendi M. G. (1996) Long-range resonance transfer of electronic excitations in close-packed CdSe quantum-dot solids. *Physical Review. B. Condensed Matter.* **54**, 8633–8643.

53. Kagan C. R., Murray C. B., Nirmal M., and Bawendi M. G. (1996) Electronic energy transfer in CdSe quantum dot solids. *Phys Rev Lett* **76**, 1517–1520.

54. Kagan V. E., Bayir H., and Shvedova A. A. (2005) Nanomedicine and nanotoxicology: two sides of the same coin. *Nanomedicine* **1**, 313–316.

55. Kawasaki E. S., and Player A. (2005) Nanotechnology, nanomedicine, and the development of new, effective therapies for cancer. *Nanomedicine* **1**, 101–109.

56. Kennel S. J., Woodward J. D., Rondinone A. J., Wall J., Huang Y., and Mirzadeh S. (2008) The fate of MAb-targeted $Cd^{125m}Te/ZnS$ nanoparticles *in vivo*. *Nucl Med Biol* **35**, 501–514.

57. Kim S., Lim Y. T., Soltesz E. G., De Grand A. M., Lee J., Nakayama A., Parker J. A., Mihaljevic T., Laurence R. G., Dor D. M., Cohn L. H., Bawendi M. G., and Frangioni J. V. (2004) Near-infrared fluorescent type II quantum dots for sentinel lymph node mapping. *Nat. Biotechnol.* **22**, 93–97.

58. Kim S. W., Zimmer J. P., Ohnishi S., Tracy J. B., Frangioni J. V., and Bawendi M. G. (2005) Engineering $InAs_xP_{1-x}/InP/ZnSe$ III-V alloyed core/shell quantum dots for the near-infrared. *J Am Chem Soc* **127**, 10526–10532.

59. Kirchner C., Liedl T., Kudera S., Pellegrino T., Javier A. M., Gaub H. E., Stoelzle S., Fertig N., and Parak W. J. (2005) Cytotoxicity of colloidal CdSe and CdSe/ZnS nanoparticles. *Nano Lett* **5**, 331–338.

60. Kobayashi H., Hama Y., Koyama Y., Barrett T., Regino C. A., Urano Y., and Choyke P. L. (2007) Simultaneous multicolor imaging of five different lymphatic basins using quantum dots. *Nano Lett* **7**, 1711–1716.

61. Koole R., van Schooneveld M. M., Hilhorst J., Castermans K., Cormode D. P., Strijkers G. J., de Mello Donega C., Vanmaekelbergh D., Griffioen A. W., Nicolay K., Fayad Z. A., Meijerink A., and Mulder W. J. (2008) Paramagnetic lipid-coated silica nanoparticles with a fluorescent quantum dot core: a new contrast agent platform for multimodality imaging. *Bioconjug Chem* **19**, 2471–2479.

62. Lacerda L., Bianco A., Prato M., and Kostarelos K. (2006) Carbon nanotubes as nanomedicines: from toxicology to pharmacology. *Adv Drug Deliv Rev* **58**, 1460–1470.

63. Lacoste T. D., Michalet X., Pinaud F., Chemla D. S., Alivisatos A. P., and Weiss S. (2000) Ultrahigh-resolution multicolor colocalization of single fluorescent probes. *Proc Natl Acad Sci USA* **97**, 9461–9466.

64. Lakowicz J. R., Gryczynski I., Gryczynski Z., Nowaczyk K., and Murphy C. J. (2000) Time-resolved spectral observations of cadmium-enriched cadmium sulfide nanoparticles and the effects of DNA oligomer binding. *Anal Biochem* **280**, 128–136.

65. Landin L., Miller M. S., Pistol M., Pryor C. E., and Samuelson L. (1998) Optical studies of individual InAs quantum dots in GaAs: few-particle effects. *Science* **280**, 262–264.

66. Larson D. R., Zipfel W. R., Williams R. M., Clark S. W., Bruchez M. P., Wise F. W., and Webb W. W. (2003) Water-soluble quantum dots for multiphoton fluorescence imaging *in vivo*. *Science* **300**, 1434–1436.

67. Li Z. B., Cai W., and Chen X. (2007) Semiconductor quantum dots for *in vivo* imaging. *J Nanosci Nanotechnol* **7**, 2567–2581.

68. Lidke D. S., Nagy P., Heintzmann R., Arndt-Jovin D. J., Post J. N., Grecco H. E., Jares-Erijman E. A., and Jovin T. M. (2004) Quantum dot ligands provide new insights into erbB/HER receptor-mediated signal transduction. *Nat Biotechnol* **22**, 198–203.

69. Lin S., Xie X., Patel M. R., Yang Y. H., Li Z., Cao F., Gheysens O., Zhang Y., Gambhir S. S., Rao J. H., and Wu J. C. (2007) Quantum dot imaging for embryonic stem cells. *BMC Biotechnol* **7**, 67.

70. Liu H. Y., and Vu T. Q. (2007) Identification of quantum dot bioconjugates and cellular protein co-localization by hybrid gel blotting. *Nano Lett* **7**, 1044–1049.

71. Mahtab R., Harden H. H., and Murphy C. J. (2000) Temperature- and salt-dependent binding of long DNA to protein-sized quantum dots: thermodynamics of "inorganic protein"-DNA interactions. *J Am Chem Soc* **122**, 14–17.

72. Mammen M., Chio S., and Whitesides G. M. (1998) Polyvalent interactions in biological systems: implications for design and use of multivalent ligands and inhibitors. *Angew Chem Int Ed Engl* **37**, 2755–2794.

73. Mankoff D. A. (2007) A definition of molecular imaging. *J Nucl Med* **48**, 18N, 21N.

74. Massoud T. F., and Gambhir S. S. (2003) Molecular imaging in living subjects: seeing fundamental biological processes in a new light. *Genes Dev* **17**, 545–580.

75. Medintz I. L., Clapp A. R., Brunel F. M., Tiefenbrunn T., Tetsuo Uyeda H., Chang E. L., Deschamps J. R., Dawson P. E., and Mattoussi H. (2006) Proteolytic activity monitored by fluorescence resonance energy transfer through quantum-dot-peptide conjugates. *Nat Mater* **5**, 581–589.

76. Medintz I. L., Clapp A. R., Mattoussi H., Goldman E. R., Fisher B., and Mauro J. M. (2003) Self-assembled nanoscale biosensors based on quantum dot FRET donors. *Nat. Mater.* **2**, 630–638.

77. Medintz I. L., Konnert J. H., Clapp A. R., Stanish I., Twigg M. E., Mattoussi H., Mauro J. M., and Deschamps J. R. (2004) A fluorescence resonance energy transfer-derived structure of a quantum dot-protein bioconjugate nanoassembly. *Proc. Natl. Acad. Sci. U.S.A.* **101**, 9612–9617.

78. Medintz I. L., Uyeda H. T., Goldman E. R., and Mattoussi H. (2005) Quantum dot bioconjugates for imaging, labelling and sensing. *Nat Mater* **4**, 435–446.

79. Michalet X., Pinaud F., Lacoste T. D., Dahan M., Bruchez M. P., Alivisatos A. P., and Weiss S. (2001) Properties of Fluorescent Semiconductor Nanocrystals and their Application to Biological Labeling. *Single Mol* **2**, 261–276.

80. Michalet X., Pinaud F. F., Bentolila L. A., Tsay J. M., Doose S., Li J. J., Sundaresan G., Wu A. M., Gambhir S. S., and Weiss S. (2005) Quantum dots for live cells, *in vivo* imaging, and diagnostics. *Science* **307**, 538–544.

81. Milanovic V. V., and Ikonic Z. (1989) Intraband absorption of infrared radiation in a semiconductor quantum dot. *Physical Review B. Condensed Matter* **39**, 7982–7984.

82. Mistretta C. A. (2009) Undersampled radial MR acquisition and highly constrained back projection (HYPR) reconstruction: Potential medical imaging applications in the post-Nyquist era. *J Magn Reson Imaging* **29**, 501–516.

83. Mitchell G. P., Mirkin C. A., and Letsinger R. L. (1999) Programmed assembly of DNA functionalized quantum dots. *J Am Chem Soc* **121**, 8122–8123.

84. Miyawaki A., Sawano A., and Kogure T. (2003) Lighting up cells: labelling proteins with fluorophores. *Nat Cell Biol* Suppl, S1–7.

85. Morgan N. Y., English S., Chen W., Chernomordik V., Russo A., Smith P. D., and Gandjbakhche A. (2005) Real time *in vivo* noninvasive optical imaging using near-infrared fluorescent quantum dots. *Acad Radiol* **12**, 313–323.

86. Muhammad O., Popescu A., and Toms S. A. (2007) Macrophage-mediated colocalization of quantum dots in experimental glioma. *Methods Mol Biol* **374**, 161–172.

87. Mulder W. J., Castermans K., van Beijnum J. R., Oude Egbrink M. G., Chin P. T., Fayad Z. A., Lowik C. W., Kaijzel E. L., Que I., Storm G., Strijkers G. J., Griffioen A. W., and Nicolay K. (2009) Molecular imaging of tumor angiogenesis using $\alpha_v\beta_3$-integrin targeted multimodal quantum dots. *Angiogenesis* **12**, 17–24.

88. Mulder W. J., Koole R., Brandwijk R. J., Storm G., Chin P. T., Strijkers G. J., de Mello Donega C., Nicolay K., and Griffioen A. W. (2006) Quantum dots with a paramagnetic coating as a bimodal molecular imaging probe. *Nano Lett* **6**, 1–6.

89. Mulder W. J., Strijkers G. J., Briley-Saboe K. C., Frias J. C., Aguinaldo J. G., Vucic E., Amirbekian V., Tang C., Chin P. T., Nicolay K., and Fayad Z. A. (2007) Molecular imaging of macrophages in atherosclerotic plaques using bimodal PEG-micelles. *Magn Reson Med* **58**, 1164–1170.

90. Murphy C. J. (2002) Optical sensing with quantum dots. *Anal Chem* **74**, 520A-526A.

91. Nath J., and Johnson K. L. (2000) A review of fluorescence *in situ* hybridization (FISH): current status and future prospects. *Biotech Histochem* **75**, 54–78.

92. Oberdorster G., Oberdorster E., and Oberdorster J. (2005) Nanotoxicology: an emerging discipline evolving from studies of ultrafine particles. *Environ Health Perspect* **113**, 823–839.

93. Oostendorp M., Douma K., Hackeng T. M., Dirksen A., Post M. J., van Zandvoort M. A., and Backes W. H. (2008) Quantitative molecular magnetic resonance imaging of tumor angiogenesis using cNGR-labeled paramagnetic quantum dots. *Cancer Res* **68**, 7676–7683.

94. Ornberg R. L., Harper T. F., and Liu H. (2005) Western blot analysis with quantum dot fluorescence technology: a sensitive and quantitative method for multiplexed proteomics. *Nat Methods* **2**, 79–81.

95. Park J. H., von Maltzahn G., Ruoslahti E., Bhatia S. N., and Sailor M. J. (2008) Micellar hybrid nanoparticles for simultaneous magnetofluorescent imaging and drug delivery. *Angew Chem Int Ed Engl* **47**, 7284–7288.

96. Parungo C. P., Colson Y. L., Kim S. W., Kim S., Cohn L. H., Bawendi M. G., and Frangioni J. V. (2005) Sentinel lymph node mapping of the pleural space. *Chest* **127**, 1799–1804.

97. Parungo C. P., Ohnishi S., Kim S. W., Kim S., Laurence R. G., Soltesz E. G., Chen F. Y., Colson Y. L., Cohn L. H., Bawendi M. G., and Frangioni J. V. (2005) Intraoperative identification of esophageal sentinel lymph nodes with near-infrared fluorescence imaging. *J Thorac Cardiovasc Surg* **129**, 844–850.

98. Pathak A. P., Gimi B., Glunde K., Ackerstaff E., Artemov D., and Bhujwalla Z. M. (2004) Molecular and functional imaging of cancer: Advances in MRI and MRS. *Methods Enzymol* **386**, 3–60.

99. Patolsky F., Gill R., Weizmann Y., Mokari T., Banin U., and Willner I. (2003) Lighting-up the dynamics of telomerization and DNA replication by CdSe-ZnS quantum dots. *J Am Chem Soc* **125**, 13918–13919.

100. Patolsky F., Zheng G., and Lieber C. M. (2006) Nanowire sensors for medicine and the life sciences. *Nanomed* **1**, 51–65.

101. Pautler R. G., and Fraser S. E. (2003) The year(s) of the contrast agent - micro-MRI in the new millennium. *Curr Opin Immunol* **15**, 385–392.

102. Pradhan N., Battaglia D. M., Liu Y., and Peng X. (2007) Efficient, stable, small, and water-soluble doped ZnSe nanocrystal emitters as non-cadmium biomedical labels. *Nano Lett* **7**, 312–317.

103. Prinzen L., Miserus R. J., Dirksen A., Hackeng T. M., Deckers N., Bitsch N. J., Megens R. T., Douma K., Heemskerk J. W., Kooi M. E., Frederik P. M., Slaaf D. W., van Zandvoort M. A., and Reutelingsperger C. P. (2007) Optical and magnetic resonance imaging of cell death and platelet activation using annexin a5-functionalized quantum dots. *Nano Lett* **7**, 93–100.

104. Rieger S., Kulkarni R. P., Darcy D., Fraser S. E., and Koster R. W. (2005) Quantum dots are powerful multipurpose vital labeling agents in zebrafish embryos. *Dev. Dyn.* **234**, 670–681.

105. Santra S., Yang H., Holloway P. H., Stanley J. T., and Mericle R. A. (2005) Synthesis of water-dispersible fluorescent, radio-opaque, and paramagnetic CdS:Mn/ZnS quantum dots: a multifunctional probe for bioimaging. *J Am Chem Soc* **127**, 1656–1657.

106. Schipper M. L., Cheng Z., Lee S. W., Bentolila L. A., Iyer G., Rao J., Chen X., Wu A. M., Weiss S., and Gambhir S. S. (2007) microPET-based biodistribution of quantum dots in living mice. *J Nucl Med* **48**, 1511–1518.

107. Schipper M. L., Iyer G., Koh A. L., Cheng Z., Ebenstein Y., Aharoni A., Keren S., Bentolila L. A., Li J., Rao J., Chen X., Banin U., Wu A. M., Sinclair R., Weiss S., and Gambhir S. S. (2009) Particle size, surface coating, and PEGylation influence the biodistribution of quantum dots in living mice. *Small* **5**, 126–134.

108. Selvan S. T., Patra P. K., Ang C. Y., and Ying J. Y. (2007) Synthesis of silica-coated semiconductor and magnetic quantum dots and their use in the imaging of live cells. *Angew Chem Int Ed Engl* **46**, 2448–2452.

109. Smith B. R., Cheng Z., De A., Koh A. L., Sinclair R., and Gambhir S. S. (2008) Real-time intravital imaging of RGD-quantum dot binding to luminal endothelium in mouse tumor neovasculature. *Nano Lett* **8**, 2599–2606.

110. So M. K., Xu C., Loening A. M., Gambhir S. S., and Rao J. (2006) Self-illuminating quantum dot conjugates for *in vivo* imaging. *Nat Biotechnol* **24**, 339–343.

111. Sosnovik D. E., and Weissleder R. (2007) Emerging concepts in molecular MRI. *Curr Opin Biotechnol* **18**, 4–10.

112. Stroh M., Zimmer J. P., Duda D. G., Levchenko T. S., Cohen K. S., Brown E. B., Scadden D. T., Torchilin V. P., Bawendi M. G., Fukumura D., and Jain R. K. (2005) Quantum dots spectrally distinguish multiple species within the tumor milieu *in vivo*. *Nat Med* **11**, 678–682.

113. Stryer L. (1978) Fluorescence energy transfer as a spectroscopic ruler. *Annu Rev Biochem* **47**, 819–846.

114. Tada H., Higuchi H., Wanatabe T. M., and Ohuchi N. (2007) *In vivo* real-time tracking of single quantum dots conjugated with monoclonal anti-HER2 antibody in tumors of mice. *Cancer Res.* **67**, 1138–1144.

115. Tan W. B., and Zhang Y. (2007) Multi-functional chitosan nanoparticles encapsulating quantum dots and Gd-DTPA as imaging probes for bio-applications. *J Nanosci Nanotechnol* **7**, 2389–2393.

116. Thayer A. M. (2007) Building up nanotech research. *Chem Eng News* **85**, 15–21.

117. Thorek D. L., Chen A. K., Czupryna J., and Tsourkas A. (2006) Superparamagnetic iron oxide nanoparticle probes for molecular imaging. *Ann Biomed Eng* **34**, 23–38.

118. Thorne R. G., and Nicholson C. (2006) *In vivo* diffusion analysis with quantum dots and dextrans predicts the width of brain extracellular space. *Proc Natl Acad Sci USA* **103**, 5567–5572.

119. van Tilborg G. A., Mulder W. J., Chin P. T., Storm G., Reutelingsperger C. P., Nicolay K., and Strijkers G. J. (2006) Annexin A5-conjugated quantum dots with a paramagnetic lipidic coating for the multimodal detection of apoptotic cells. *Bioconjug Chem* **17**, 865–868.

120. Voura E. B., Jaiswal J. K., Mattoussi H., and Simon S. M. (2004) Tracking metastatic tumor cell extravasation with quantum dot nanocrystals and fluorescence emission-scanning microscopy. *Nat Med* **10**, 993–998.

121. Wang D., He J., Rosensweig N., and Rosenzweig Z. (2004) Superparamagnetic Fe_2O_3 beads-CdSe/ZnS quantum dots core-shell nanocomposite particles for cell separation. *Nano Lett* **4**, 409–413.

122. Wang S., Jarrett B. R., Kauzlarich S. M., and Louie A. Y. (2007) Core/shell quantum dots with high relaxivity and photoluminescence for multimodality imaging. *J Am Chem Soc* **129**, 3848–3856.

123. Willard D. M., Carillo L. L., Jung J., and van Orden A. (2001) CdSe-ZnS quantum dots as resonance energy transfer donors in a model protein-protein binding assay. *Nano Lett* **1**, 469–474.

124. Woodward J. D., Kennel S. J., Mirzadeh S., Dai S., Wall J. S., Richey T., Avenell J., and Rondinone A. J. (2007) *In vivo* SPECT/CT imaging and biodistribution using radioactive $Cd^{125m}Te$/ZnS nanoparticles. *Nanotechnology* **18**, 175103.

125. Wu S. M., Zhao X., Zhang Z. L., Xie H. Y., Tian Z. Q., Peng J., Lu Z. X., Pang D. W., and Xie Z. X. (2006) Quantum-dot-labeled DNA probes for fluorescence *in situ* hybridization (FISH) in the microorganism Escherichia coli. *Chemphyschem* **7**, 1062–1067.

126. Wu X., Liu H., Liu J., Haley K. N., Treadway J. A., Larson J. P., Ge N., Peale F., and Bruchez M. P. (2003) Immunofluorescent labeling of cancer marker Her2 and other cellular targets with semiconductor quantum dots. *Nat. Biotechnol.* **21**, 41–46.

127. Xia Z., Xing Y., So M. K., Koh A. L., Sinclair R., and Rao J. (2008) Multiplex detection of protease activity with quantum dot nanosensors prepared by intein-mediated specific bioconjugation. *Anal Chem* **80**, 8649–8655.

128. Xie R., Chen K., Chen X., and Peng X. (2008) InAs/InP/ZnSe core /shell /shell quantum dots as near-infrared emitters: bright, narrow-band, non-cadmium containing, and biocompatible. *Nano Res* **1**, 457–464.

129. Xu H., Sha M. Y., Wong E. Y., Uphoff J., Xu Y., Treadway J. A., Truong A., O'Brien E., Asquith S., Stubbins M., Spurr N. K., Lai E. H., and Mahoney W. (2003) Multiplexed SNP

genotyping using the Qbead system: a quantum dot-encoded microsphere-based assay. *Nucleic Acids Res* **31**, e43.

130. 1Yao H., Zhang Y., Xiao F., Xia Z., and Rao J. (2007) Quantum dot/bioluminescence resonance energy transfer based highly sensitive detection of proteases. *Angew Chem Int Ed Engl* **46**, 4346–4349.

131. Yu X., Chen L., Li K., Li Y., Xiao S., Luo X., Liu J., Zhou L., Deng Y., Pang D., and Wang Q. (2007) Immunofluorescence detection with quantum dot bioconjugates for hepatoma *in vivo*. *J Biomed Opt* **12**, 014008.

132. Yutzy S. R., and Duerk J. L. (2008) Pulse sequences and system interfaces for interventional and real-time MRI. *J Magn Reson Imaging* **27**, 267–275.

133. Zhang Z., Nair S. A., and McMurry T. J. (2005) Gadolinium meets medicinal chemistry: MRI contrast agent development. *Curr Med Chem* **12**, 751–778.

134. Zimmer J. P., Kim S. W., Ohnishi S., Tanaka E., Frangioni J. V., and Bawendi M. G. (2006) Size series of small indium arsenide-zinc selenide core-shell nanocrystals and their application to *in vivo* imaging. *J Am Chem Soc* **128**, 2526–2527.

Chapter 7

MULTIMODAL IMAGING OF DENDRIMERS

Michelle R. Longmire, Peter L. Choyke, and Hisataka Kobayashi
Molecular Imaging Program, Center for Cancer Research, National Cancer Institute, National Institutes of Health, 10 Center Dr., Bethesda, MD 20892–1088, USA
kobayashi@mail.nih.gov

7.1 INTRODUCTION

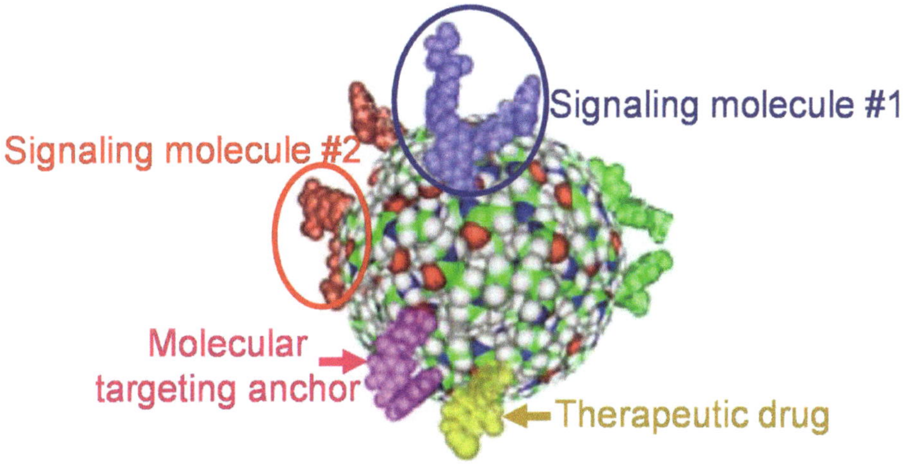

Figure 7.1 Schema for the multiple functionality of the dendrimer, including multimodal imaging, drug delivery and molecular targeting.

Dendrimers represent a subset of nanoparticles with considerable potential in nanomedicine. The precise nanoscale scaffolding, nanocontainer properties, and tunable variation in size, surface chemistry, and interior void space make dendrimers unique nanoparticles with considerable potential as nanoimaging contrast agents as well as targetable nanocontainers for site-specific drug delivery [52]. The word "dendrimer" originates from the Greek word *dendron*, meaning "tree," and refers

Nanoimaging
Edited by Beth Goins and William Phillips

www.panstanford.com

to the basic structure of the particle: a central core that gives rise to numerous branches. The basic dendrimer structure consists of a central core surrounded concentrically by annular interior shells (or branches), which terminate peripherally to provide numerous functional groups on the surface of the dendrimer [52]. These surface groups create an abundance of potential binding sites for targeting ligands, such as monoclonal antibodies or peptides, for site-specific delivery of imaging payloads or therapeutic agents [49]. The branching creates a flexible interior space within the dendrimer that can carry therapeutic payloads (Fig. 7.1). The space within the molecule is an isolated microenvironment, which effectively shields the payload from enzymatic destruction while decreasing its toxicity [49] as well as enabling the controlled release of the payload after target binding. The many highly tunable properties of dendrimers make them attractive nanoparticles for multimodal imaging.

Although the complexities of dendrimer chemistry are beyond the scope of this review, basic aspects of surface and interior chemistry are important. The surface residues vary extensively; however, amine, carboxyl, or alcohol groups, are often utilized due to the ease of conjugation that can be achieved with imaging agents or targeting molecules. The ability to modify and expand dendrimer surface groups permits the rational manipulation of agent biodistribution via changes in agent diameter (through the addition of annular shells or generations) and payload deposition through controlled release by receptor-mediated targeting. Dendrimers are composed of various core types, including ethylene diamine (EDA) and diaminobutyl (DAB), as well as several interior types, such as polyamidoamine (PAMAM) and polypropylimine (PPI). The interiors typically used in imaging agents, such as PAMAM or PPI dendrimers, are highly soluble in aqueous solution and possess a primary amine-rich surface that is totally uncharged at pH greater than 9 or a carboxyl-rich surface that is totally uncharged at pH less than 4. The abundance of primary amine functional groups on the exterior shell serve as optimal attachment sites for chelating agents in the development of macromolecular contrast agents, or for antibody molecules for site-specific targeting. The ability to attach both targeting and chelating moieties to PPI and PAMAM dendrimers has propelled their use as contrast agents for molecular imaging. In comparison to EDA and DAB dendrimer cores, which have diamine core components, newer families of dendrimers possess a cystamine core, which, when subjected to traditional redox reactions, dissociates to produce free sulfhydryl groups and permits the use of maleimide and sulfohydryl coupling reactions, for the conjugation of peptides, monoclonal antibodies, biotin, or other targeting moieties.

The ability to optimize dendrimer biodistribution and pharmacokinetics through modification of particle size offers optimization of biocompatibility and control of agent biodistribution *in vivo*. In general, greater biocompatibility is associated with lower generation numbers and anionic or neutral surface groups compared to similar particles of higher generation or those with cationic surface groups. Increased biocompatibility is also achieved through the attachment of small molecules or

PEGylating the surface, which increases the hydrophilicity and decreases agent immunogenicity. Among the tunable properties, size has been shown to have a tremendous effect on *in vivo* behavior, even when comparing molecules with similar chemical properties [10–12]. Surprisingly small changes (of only several nanometers) in diameter can greatly impact the pharmacokinetics, permeability across the vascular wall, excretion route, and recognition by the reticuloendothelial system [19]. *In vivo* studies have demonstrated that particles with smaller generation numbers (<4 nm diameter; which are found in dendrimers smaller than the generation 3 EDA/PAMAM dendrimer) demonstrate leakage of the agent from the vascular space and rapid renal excretion. Therefore, smaller generation dendrimers that are excreted rapidly through the kidney may be useful in assessing renal function or in rapidly clearing unbound but targeted payload-carrying dendrimers [8, 9, 20, 21, 24]. Intermediate dendrimer generations between 5 and 8 nm (G4 or G7) in diameter leak through hyperpermeable tumor vasculature but are retained within normal vessels [2]. Larger generation dendrimers, greater than 13 nm in diameter (G9 and G10), provide excellent enhancement of the reticuloendothelial system of the liver and spleen. The intermediate generation dendrimers (G6 to G8) demonstrate blood-pool retention of both normal and tumor vessels and are therefore excellent candidates for vascular imaging agents [27, 30, 31].

In addition to size, the chemical characteristics of dendrimers impact *in vivo* behavior. Among the characteristics, particle hydrophilicity greatly impacts agent pharmacokinetics. Increasing the hydrophilicity of G4-PAMAM dendrimers by PEGylation decreases both the renal excretion and the liver accumulation of these particles, increasing particle half-life and improving vascular retention [26]. Conversely, hydrophobic dendrimers, such as those with DAB cores even of small diameter, preferentially accumulate in the liver in contrast to similar-sized PAMAM agents, making them promising hepatic imaging agents [25].

Medical imaging represents a promising application of dendrimers within the broader field of medical nanotechnology. Developments in medical imaging include applications in magnetic resonance (MR) imaging, computer tomography (CT), and radionuclide and optical imaging. Dendrimer-based contrast agents offer considerable advantages compared with conventional agents. Most conventional imaging agents are low in molecular weight and therefore demonstrate rapid clearance from vascular circulation and excretion from the body through renal filtration. The rapid clearance properties of low-molecular-weight (LMW) agents limit their utility in time-dependent imaging studies in which longer particle circulation and retention times are needed. Unlike LMW agents, macromolecular agents, such as dendrimers, display desirable blood-pool properties with longer circulation and retention times. In addition, dendrimer-based agents are able to depict the lymphatic system after interstitial injection due to the preferential retention of such agents within the lymphatic channels [28]. Among the macromolecular agents, dendrimers offer a unique platform for the attachment of imaging and targeting moieties, and as discussed herein, dendrimers represent an optimal particle for multimodal imaging.

7.2 MAGNETIC RESONANCE IMAGING

Two classes of dendrimers with MR capabilities have been developed: (i) dendrimers containing paramagnetic iron oxide particles (magnetodendrimers) for cell labeling and cell tracking, and (ii) dendrimers incorporating high numbers of gadolinium (Gd) chelates [17, 26]. The abundance of binding sites on the exterior shell of the dendrimer particle is ideal for attaching multiple Gd(III)-chelates during the synthesis of dendrimer-based MR contrast agents [52]. In order to achieve sufficient signal for MR imaging, Gd(III) chelates must be able to interact with the water protons. The numerous attachment sites available on the dendrimer surface offer Gd(III) conjugation sites that are exposed to the external environment with a high potential for interaction with water molecules. In order to sequester metal ions at the dendrimer surface, derivatives of the two major classes of MR chelates, diethylene triamine pentaacidic acid (DTPA) and 1,4,7,10-tetraazacyclo dodecane–N,N^{I},N^{II},N^{III}–tetraacidic acid (DOTA), are utilized [12, 20, 40, 42]. Although there are considerable differences in the physicochemical properties of DTPA and DOTA when chelated with Gd(III), the differences in R_1 relaxivity of each type of Gd-chelate are negligible (4.7mM^{-1}s^{-1} at 20 MHz and 25°C in PBS) [20]. Additionally, compared with LMW agents, macromolecular agents, such as dendrimers, demonstrate 5- to 10-fold increased relaxivity at 40–64 MHz and 25°C as a result of slower particle tumbling rates in solution [5, 19]. The relatively higher relaxivity reduces the dose of gadolinium required to achieve sufficient alterations in relaxation rates to be detectable by MRI. This feature of dendrimers for MRI has gained favorable attention due to the growing concern for gadolinium-associated nephrogenic systemic fibrosis, as the lower molar relaxivity of conventional MR agents requires the use of relatively larger doses (0.1mmol/kg) of gadolinium contrast compared with macromolecule-based solutions (0.001–0.03 mmol/kg) [41, 51].

Beginning with the first *in vivo* studies conducted by Wiener *et al.*, dendrimer-based macromolecular contrast agents were found to effectively enhance conventional MR images and 3D time-of-flight MR angiograms, with considerably longer enhancement lives compared with conventional agents [53]. Research in dendrimer-based MR contrast agents has revealed that modification of dendrimer properties, such as size, cores, and interiors, and exterior shell chemistry, permit application-specific customization [19–23]. Much of the data regarding the relationship between dendrimer generation size and *in vivo* particle behavior is derived from dendrimers functionalized for MR imaging. When comparing dendrimer-based contrast agents for MRI that vary only in generation number, size-dependent changes in the pharmacokinetics are observed [31]. The moderately larger conjugates of approximately 10 nm in diameter remain in the blood significantly longer than all of the other preparations, resulting in higher blood-to-organ ratios with increasing molecular size [31]. Smaller generation MR imaging agents (<4 nm in diameter) leak from normal vasculature, similar to conventional Gd(III) agents of less than 1 nm in size, but at a slightly slower rate [31, 56]. These smaller agents are rapidly excreted via the kidneys, primarily by renal filtration, and their utility in renal imaging has been demonstrated [9, 23]. MR imaging of intermediate dendrimers (G5-G8) functionalized with Gd(III) chelates demonstrates retention in normal vessels but leakage from hyperpermeable neoplastic vasculature, while larger generation agents (>G8) mostly remain within the porous tumor vasculature, demonstrating minimal leakage from even diseased vessels [2]. Additionally, as generation number increases, there is a proportional decrease in renal excretion and corresponding increase in liver and spleen uptake [18].

The versatility of dendrimer nanoparticle behavior has led to the development of several interesting dendrimer-based MR contrast agent applications. Macromolecular MR contrast agents are useful for vascular imaging, specifically microvascular imaging, and for determining functional capillary density in tumor tissues [20]. Among the dendrimer-based agents, G7 and G8 agents permit visualization of the blood vessels both within and outside tumors most clearly, and PEG conjugation further improves image quality as well as excretion profiles [20]. Dendrimers with hydrophobic properties have proven useful in hepatic imaging. Utilizing the predisposition of slightly less hydrophilic macromolecules to accumulate in the liver, a G5 DAB/PPI dendrimer with a hydrophobic core was shown to homogenously enhance liver parenchyma [20]. Furthermore, a G5 agent with a DAB/PPI core/interior applied to dynamic micro-MRIs enabled serial visualization of growing metastatic colon cancer foci as small as 0.3 mm in diameter within the livers of living mice [20].

The utility of dendrimer-based MR contrast agents has been demonstrated in the functional assessment of organs as well. Although small dendrimer-based agents are quickly excreted from the kidneys, agents can be designed for retention within blood vessels or the urinary tract, with minimal perfusion into the extravascular tissue. After glomerular filtration, these agents are concentrated at the proximal tubule, forming a high-intensity band [20]. The absence or delayed formation of this high-intensity band was shown to correlate with renal tubular function disturbances associated with acute tubular necrosis (ATN) in a mouse model and the enhancement pattern was predictive of acute and/or chronic renal failure (e.g., heavy metal exposure, ischemia, obstruction, and sepsis) [9, 22]. Functional kidney imaging is readily performed with dendrimer-based MRI agents, and among the numerous agents, the G2 DAB/PPI agent has demonstrated the greatest efficacy in imaging renal function [20].

7.3 COMPUTED TOMOGRAPHY

Conventional iodinated contrast media are routinely used for imaging patients undergoing contrast enhanced computed tomography (CT) scans [12]. However, conventional CT agents suffer from limitations inherent to LMW (<2000 Da) agents. In addition, to obtain sufficient enhancement for detection on CT, as much as 3–4 mmol/kg of organic iodine agent is necessary, which represents a 30–40-fold higher concentration compared with the clinical dose of Gd ions needed for

MRI. Similar to conventional MRI agents, conventional CT contrast media quickly equilibrate between intravascular and extracellular compartments and are rapidly excreted from the body [42]. Additionally, conventional CT contrast media readily leak from hyperpermeable tumor vasculature and have high first-pass extravasation through normal noncerebral vessels [2]. Macromolecular agents, which possess prolonged intravascular half-life and initial volumes of distribution that approximate plasma, may prove superior blood pool–enhancing imaging agents, especially for quantitative characterization of tissues, blood vessels, and pathologic alterations associated with cancer, ischemic injury, and angiogenesis [4, 7, 46, 48].

Although macromolecular CT contrast agents have yet to advance to clinical trials, research efforts to develop large (>30 kDa) macromolecular agents for prolonged blood pool imaging have led to the development of unique agents, including polymeric micelles formed from amphipathic co-polymers and iopromide-containing liposomes [11]. However, these agents demonstrate a high degree of polydispersity, which may compromise imaging accuracy and the ability to interpret results. More effective imaging may be possible using low-polydispersity dendrimers, and therefore further research is needed to explore the potential use of dendrimer-based CT agents as potential blood pool agents.

Similar to Gd(III) dendrimers, prolonged retention of iodinated dendrimer agents may also prove problematic. Due to concern regarding the toxicity of high dose iodinated contrast agent required for CT and concern for repetitive exposure of high dose radiation, there has been relatively little research on the use of dendrimers as CT agents [2]. The main side effects of iodinated agents are anaphlactoid reactions and damage to the kidneys causing acute renal failure [11, 45]. For this reason iodinated agents are contraindicated in patients with existing renal insufficiency or in patients at significant risk for renal disease (e.g., patients with diabetes mellitus, multiple myeloma); patients on dialysis are exempt, as dialysis can be used to remove the agent. However, due to the increased size of macromolecular agents, dialysis may not be as successful as it is for the removal of LMW contrast agents, possibly leading to additional risk for *in vivo* use.

7.4 RADIONUCLIDE IMAGING

Dendrimer-based imaging agents have not been extensively used for radionuclide imaging. The only reported use of dendrimers for radionuclide imaging is to increase the specific activity for labeling monoclonal antibody without conjugating multiple chelates [32, 33, 54]. By employing this method, a single attachment of the dendrimer-radionuclide conjugate could load as much as 49 radionuclides (indium-111 ions) to a single monoclonal antibody molecule without compromising the specific binding to the antigen or *in vivo* targeting ability [32]. However, the pharmacokinetics can be altered by the addition of such a large molecule. Nevertheless, the high sensitivity of radionuclide imaging lends itself to the labeling of a dendrimers. For example, in dual MR/radionuclide dendrimer-based agents, the majority of binding sites are needed for gadolinium for MRI. However, the high sensitivity of radionuclide enables effective dual-imaging with a relatively few number of isotopes required, which are easily added to the remaining few attachment sites.

7.5 OPTICAL IMAGING

Optical imaging offers numerous advantages compared with other modalities, including portability and affordibility, high sensitivity (~pM), and excellent spatial and temporal image quality without exposure to ionizing radiation [13, 14, 39]. Although considerable research has focused on the use of dendrimers for optical imaging, most research has utilized *in vitro* studies. Interestingly, PAMAM dendrimers with carboxyl and amino terminals possess a weak intrinsic fluorescence and oxidation of hydroxyl-terminated PAMAM dendrimers produces intrinsic blue fluorescence with high quantum yield [35, 36]. Utilizing the intrinsic fluorescence of the dendrimer confers the distinct advantage of eliminating the need for additional optical agents such as inorganic dyes or quantum dots with which to label the dendrimer. Moreover, the original size and chemical nature of the dendrimer remains unaltered, unlike other agents in which optically active agents must be attached, thereby changing the inherent properties of the particle. Dendrimer-based optical agents have been used for diverse applications, including assessing metabolic states and cancer and angiogenesis imaging. Ibey *et al.* developed an implantable chemical assay for monitoring glucose levels of diabetic patients using an Alexa Fluor 594-labeled G4-PAMAM dendrimer [16]. Shi *et al.* targeted cancer cells for optical imaging using a folic acid ligand and fluoroscein isothiocyanate (FITC)-labeled G5 dendrimer entrapped in gold nanoparticles [47].The gold nanoparticle potentially allowed for treatment of the cells with light therapy. Hill *et al.* developed a targeted, G5-PAMAM dendrimer conjugated with FITC imaging agent that binds peptide motifs associated with tumor angiogenesis and proliferating endothelial cells [15, 47].

Although limited in number, *in vivo* applications of dendrimer-based optical agents demonstrate the potential of this imaging platform. Scherer *et al.* developed an application for the *in vivo* detection and quantification of matrix metalloproteinase (MMP) activity using a pegylated G4-PAMAM dendrimer core covalently coupled to a Cy5.5-labeled peptide substrate specific for matrix metalloproteinase 7 (MMP-7) activity [44]. MMP-7 is an enzyme involved in the remodeling of neoplastic vessels for various cancers. After binding MMP-7, the peptide component is cleaved yielding a several-fold increase in optical fluorescence [44]. This activatable imaging platform may prove useful in detection of malignancy or monitoring disease recurrence and progression.

7.6 MULTIMODALITY IMAGING

Utilizing the numerous attachment sites available on a single dendrimer molecule, the creation of multifunction imaging agents capable of drug delivery and/or multimodal imaging (such as MR and optical imaging) is possible using the dendrimer platform. The advantages of multimodality agents include (i) the ability to label different components of the nanoparticle to study the degradation pathway of the agent, (ii) the ability to be used with a cocktail of multiple contrast agents for more timely imaging procedures, (iii) the ability to monitor the fate of a therapeutic payload. Clearly, each imaging modality has its distinct limitations, and the simultaneous use of modalities offers the possibility of improving the information obtained during a single imaging procedure. Despite the extensive potential, however, few dual-modality dendrimer-based probes have been developed to date due to their complexity and difficult clinical approval pathway [2].

The first report of a dendrimer-based agent with multiple imaging functions was by Talanov *et al.* [50]. This group created a G6 PAMAM dendrimer-based probe complexed with both Gd(III) and Cy5.5 molecules. Notably, Gd(III) conjugation did not affect the quantum yield of the fluorescence. The dendrimer-based dual-modality probe demonstrated efficient *in vivo* visualization of sentinel lymph nodes (SLNs) in a murine model by both MRI and fluorescence imaging modalities (Fig. 7.2).

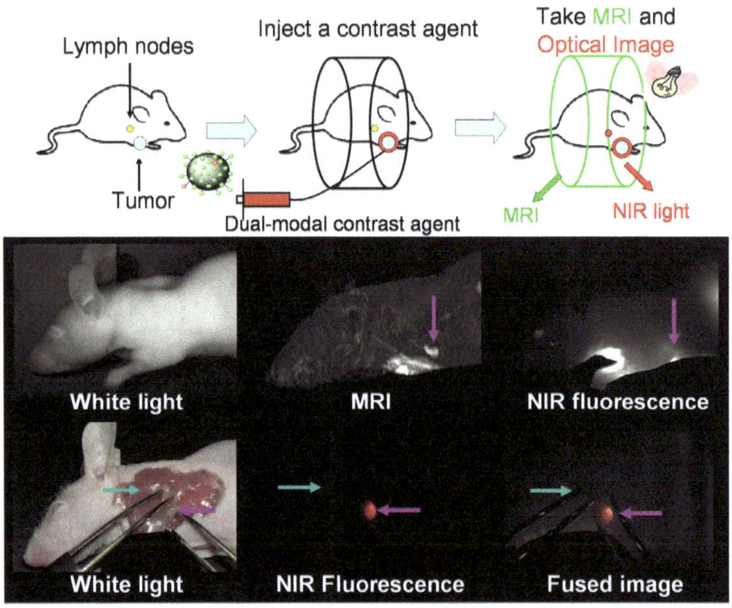

Figure 7.2 Sentinel lymph node detection imaging using G6 PAMAM dendrimer-based MRI/optical multimodal contrast agent. MRI can noninvasively identify the lateral thoracic lymph node in the sentinel lymph node of the breast cancer in this mouse. Near-infrared fluorescence imaging can help with detection and resection of the lateral thoracic lymph node easily by the guide of fluorescence signal.

Koyama *et al.* furthered this work by demonstrating the use of the single hybrid MR/near-infrared (NIR) optical dendrimer-based probe to serve as an intraoperative guide for the identification and resection of SLNs during image-guided surgery [34]. The agent was injected into the mammary glands of a murine model, and MR was used to examine the lymphatic drainage around the breast. NIR optical image-guided surgery was then performed to remove SLNs. All SLNs could be easily identified and resected under fluorescence imaging. Although the high background signal present at the site of injection limited the ability of external NIR optical imaging to identify SLNs in proximity to the site, MR lymphography was able to identify all SLNs regardless of their location. This probe has potential in the detection and removal of SLNs in cases of breast cancer or malignant melanoma.

Another dual-modality application is the dendrimer-based agent combining a radioisotope and multicolor NIR fluorophores also developed for lymphatic imaging [29]. Using a fluorophore-labeled G6-PAMAM dendrimer, an efficient agent requiring a low injected dose (1:100 compared with the previously mentioned MRI-optical imaging agent) was created. However, with this conjugation ratio, the dose of injected reagent was optimal for MRI but still too much for optical imaging. The comparable sensitivity of the radionuclide and fluorescence markers enabled minimization of the dose required (Fig. 7.3). Kobayashi *et al.* have demonstrated that the use of dendrimer-based imaging moieties with radionuclide and fluorescence of similar sensitivities is a useful strategy to reduce required doses of multimodal agents for lymphatic imaging [30].

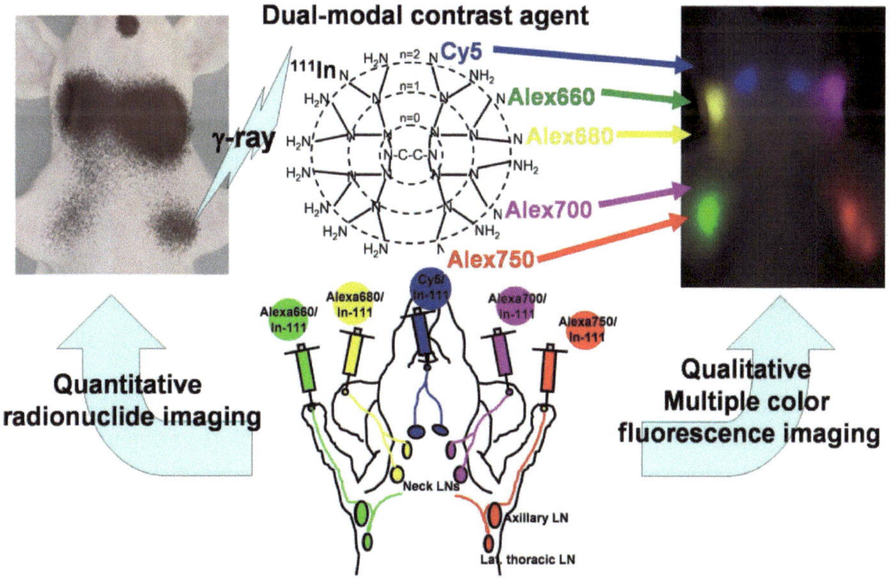

Figure 7.3 Head and neck lymphatic drainage imaging using G6 PAMAM dendrimer-based radionuclide/optical multimodal contrast agent. Radionuclide imaging can provide quantitative measurement of each lymphatic drainage. Near-infrared multiple-color fluorescence spectral imaging can depict five different lymphatic basins in the head and neck region of this mouse.

Multimodal dendrimer-based nanoparticles can be further modified to also include targeting. Boswell *et al.* developed a method of imaging angiogenesis using PAMAM dendrimers covalently targeted with small monovalent peptides specific for $\alpha_v\beta_3$ protein, an integrin expressed on proliferating tumor vasculature [3]. The targeted dendrimers were functionalized with optical agents as well as Gd(III) chelates, to create a multimodal macromolecular agent [3]. Fluorescence microscopy revealed selective binding of the resulting peptide-bearing dendrimer to $\alpha_v\beta_3$-expressing cells, although the selectivity was slightly reduced with Gd(III) conjugation [3]. *In vivo* tissue distribution was evaluated in M21 melanoma tumor-bearing mice and showed primarily renal and reticuloendothelial accumulation, with an impressive tumor-to-blood ratio [3].

In addition to multimodal imaging, dendrimers may be utilized to combine imaging with therapy. The multifunctional property of dendrimers to carry a drug and to monitor its delivery represents a great advantage of these nanoparticles. Criscione *et al.* reported the creation of a multimodal dendrimer-based agent with MR and drug delivery capabilities that spontaneously self-assembles into dense nanoscopic and microscopic particulates that exhibit low pH–induced disassembly over time, enabling controlled release of encapsulated agents [6]. Sarin *et al.* imaged effective transvascular delivery of dendrimer-encased therapeutic agents to malignant gliomas using PAMAM dendrimers functionalized with MR and fluorescence imaging probes [43]. Transvascular transport and accumulation of the nanoparticles in brain tumor tissue was measured *in vivo* with dynamic contrast-enhanced MR imaging [42]. Localization of the nanoparticles within glioma cells was confirmed *ex vivo* with fluorescence imaging [42].

Dendritic nanoparticles have also demonstrated promise as DNA transport vectors for cell-specific delivery in gene therapy [1, 10, 38]. It has been shown that dendrimers, when complexed with DNA, are capable of spontaneously forming a supramolecular assembly that is able to become entrapped in experimental tumors through the enhanced permeability and retention effect [1]. Experiments using small-animal single-photon emission computed tomography/computer tomography (SPECT/CT) and the human Na/I symporter (NIS) as a reporter gene have shown highly specific tumor targeting in a murine model [1]. NIS imaging of transgene expression has been recently validated in humans, highlighting the potential of dendrimers for cancer gene therapy [1].

Another promising direction in dendrimer-based contrast agent research is the development of biodegradable agents [37, 55]. Lu *et al.* introduced disulfide groups close to the gadolinium chelate in the interior of the dendrimer to catabolize and release the chelated gadolinium ions as small molecules that can be rapidly eliminated by renal excretion, which may avoid heavy metal toxicity associated with free gadolinium ions. Almutairi *et al.* have developed a targeted biodegradable positron-emitting dendritic nanoprobe that selectively binds to angiogenic markers for noninvasive imaging of angiogenesis [1]. *In vivo* studies in a murine hindlimb ischemia model revealed highly specific accumulation of the dendritic nanoprobe in angiogenic muscles, suggesting the agent may be an effective method for imaging angiogenesis [1].

7.7 SUMMARY

Dendrimers are a highly versatile subclass of nanoparticles. The numerous functional groups on the outer shell offer the ability to conjugate imaging beacons and targeting moieties, while the highly tunable architecture of the core of these particles enables the tailoring of biodistribution and pharmacokinetics by altering agent size and chemical properties. Research has already demonstrated the efficacy of dendrimers in medical imaging and drug/gene delivery, and the development of multifunctional dendrimers with combined multimodal imaging and drug/gene delivery capabilities represents an exciting and promising area of nanomedicine research. The development of multimodal agents may help overcome limitations of individual imaging techniques and combining imaging with drug delivery may ensure delivery of treatments and permit monitoring of therapeutic response.

References

1. Almutairi A., Rossin R., Shokeen M., Hagooly A., Ananth A., Capoccia B., Guillaudeu S., Abendschein D., Anderson C. J., Welch M. J., and Frechet J. M. (2009) Biodegradable dendritic positron-emitting nanoprobes for the noninvasive imaging of angiogenesis. *Proc Natl Acad Sci U S A* **106**, 685–690.
2. Barrett T., Ravizzini G., Choyke P. L., and Kobayashi H. (2009) Dendrimers in medical nanotechnology. *IEEE Eng Med Biol Mag* **28**, 12–22.
3. Boswell C. A., Eck P. K., Regino C. A., Bernardo M., Wong K. J., Milenic D. E., Choyke P. L., and Brechbiel M. W. (2008) Synthesis, characterization, and biological evaluation of integrin alphavbeta3-targeted PAMAM dendrimers. *Mol Pharm* **5**, 527–539.
4. Brasch R. C. (1991) Rationale and applications for macromolecular Gd-based contrast agents. *Magn Reson Med* **22**, 282–287; discussion 300–283.
5. Bryant L. H., Jr., Brechbiel M. W., Wu C., Bulte J. W., Herynek V., and Frank J. A. (1999) Synthesis and relaxometry of high-generation (G = 5, 7, 9, and 10) PAMAM dendrimer-DOTA-gadolinium chelates. *J Magn Reson Imaging* **9**, 348–352.
6. Criscione J. M., Le B. L., Stern E., Brennan M., Rahner C., Papademetris X., and Fahmy T. M. (2009) Self-assembly of pH-responsive fluorinated dendrimer-based particulates for drug delivery and noninvasive imaging. *Biomaterials* **30**, 3946–3955.
7. Daldrup-Link H. E., and Brasch R. C. (2003) Macromolecular contrast agents for MR mammography: current status. *Eur Radiol* **13**, 354–365.
8. Dear J. W., Kobayashi H., Brechbiel M. W., and Star R. A. (2006) Imaging acute renal failure with polyamine dendrimer-based MRI contrast agents. *Nephron Clin Pract* **103**, c45–49.
9. Dear J. W., Kobayashi H., Jo S. K., Holly M. K., Hu X., Yuen P. S., Brechbiel M. W., and Star R. A. (2005) Dendrimer-enhanced MRI as a diagnostic and prognostic biomarker of

sepsis-induced acute renal failure in aged mice. *Kidney Int* **67**, 2159–2167.

10. Diez S., Navarro G., and de I. C. T. (2009) *In vivo* targeted gene delivery by cationic nanoparticles for treatment of hepatocellular carcinoma. *J Gene Med* **11**, 38–45.

11. Finn W. F. (2006) The clinical and renal consequences of contrast-induced nephropathy. *Nephrol Dial Transplant* **21**, i2–10.

12. Fu Y., Nitecki D. E., Maltby D., Simon G. H., Berejnoi K., Raatschen H. J., Yeh B. M., Shames D. M., and Brasch R. C. (2006) Dendritic iodinated contrast agents with PEG-cores for CT imaging: synthesis and preliminary characterization. *Bioconjug Chem* **17**, 1043–1056.

13. Graves E. E., Ripoll J., Weissleder R., and Ntziachristos V. (2003) A submillimeter resolution fluorescence molecular imaging system for small animal imaging. *Med Phys* **30**, 901–911.

14. Graves E. E., Weissleder R., and Ntziachristos V. (2004) Fluorescence molecular imaging of small animal tumor models. *Curr Mol Med* **4**, 419–430.

15. Hsu A. R., Veeravagu A., Cai W., Hou L. C., Tse V., and Chen X. (2007) Integrin alpha v beta 3 antagonists for anti-angiogenic cancer treatment. *Recent Pat Anticancer Drug Discov* **2**, 143–158.

16. Ibey B. L., Beier H. T., Rounds R. M., Cote G. L., Yadavalli V. K., and Pishko M. V. (2005) Competitive binding assay for glucose based on glycodendrimer-fluorophore conjugates. *Anal Chem* **77**, 7039–7046.

17. Kiessling F., Morgenstern B., and Zhang C. (2007) Contrast agents and applications to assess tumor angiogenesis *in vivo* by magnetic resonance imaging. *Curr Med Chem* **14**, 77–91.

18. Kobayashi H., and Brechbiel M. (2002) Dendrimer-based macromolecular MRI contrast agents: characteristics and application. *Mol Imaging* **2**, 1–10.

19. Kobayashi H., and Brechbiel M. W. (2004) Dendrimer-based nanosized MRI contrast agents. *Curr Pharm Biotechnol* **5**, 539–549.

20. Kobayashi H., and Brechbiel M. W. (2005) Nano-sized MRI contrast agents with dendrimer cores. *Adv Drug Deliv Rev* **57**, 2271–2286.

21. Kobayashi H., Jo S. K., Kawamoto S., Yasuda H., Hu X., Knopp M. V., Brechbiel M. W., Choyke P. L., and Star R. A. (2004) Polyamine dendrimer-based MRI contrast agents for functional kidney imaging to diagnose acute renal failure. *J Magn Reson Imaging* **20**, 512–518.

22. Kobayashi H., Kawamoto S., Brechbiel M. W., Jo S. K., Hu X., Yang T., Diwan B. A., Waldmann T. A., Schnermann J., Choyke P. L., and Star R. A. (2004) Micro-MRI methods to detect renal cysts in mice. *Kidney Int* **65**, 1511–1516.

23. Kobayashi H., Kawamoto S., Jo S. K., Bryant H. L., Jr., Brechbiel M. W., and Star R. A. (2003) Macromolecular MRI contrast agents with small dendrimers: pharmacokinetic differences between sizes and cores. *Bioconjug Chem* **14**, 388–394.

24. Kobayashi H., Kawamoto S., Jo S. K., Sato N., Saga T., Hiraga A., Konishi J., Hu S., Togashi K., Brechbiel M. W., and Star R. A. (2002) Renal tubular damage detected by dynamic micro-MRI with a dendrimer-based magnetic resonance contrast agent. *Kidney Int* **61**,

1980–1985.

25. Kobayashi H., Kawamoto S., Saga T., Sato N., Hiraga A., Ishimori T., Akita Y., Mamede M. H., Konishi J., Togashi K., and Brechbiel M. W. (2001) Novel liver macromolecular MR contrast agent with a polypropylenimine diaminobutyl dendrimer core: comparison to the vascular MR contrast agent with the polyamidoamine dendrimer core. *Magn Reson Med* **46**, 795–802.

26. Kobayashi H., Kawamoto S., Saga T., Sato N., Hiraga A., Ishimori T., Konishi J., Togashi K., and Brechbiel M. W. (2001) Positive effects of polyethylene glycol conjugation to generation-4 polyamidoamine dendrimers as macromolecular MR contrast agents. *Magn Reson Med* **46**, 781–788.

27. Kobayashi H., Kawamoto S., Saga T., Sato N., Hiraga A., Konishi J., Togashi K., and Brechbiel M. W. (2001) Micro-MR angiography of normal and intratumoral vessels in mice using dedicated intravascular MR contrast agents with high generation of polyamidoamine dendrimer core: reference to pharmacokinetic properties of dendrimer-based MR contrast agents. *J Magn Reson Imaging* **14**, 705–713.

28. Kobayashi H., Kawamoto S., Sakai Y., Choyke P. L., Star R. A., Brechbiel M. W., Sato N., Tagaya Y., Morris J. C., and Waldmann T. A. (2004) Lymphatic drainage imaging of breast cancer in mice by micro-magnetic resonance lymphangiography using a nano-size paramagnetic contrast agent. *J Natl Cancer Inst* **96**, 703–708.

29. Kobayashi H., Koyama Y., Barrett T., Hama Y., Regino C. A., Shin I. S., Jang B. S., Le N., Paik C. H., Choyke P. L., and Urano Y. (2007) Multimodal nanoprobes for radionuclide and five-color near-infrared optical lymphatic imaging. *ACS Nano* **1**, 258–264.

30. Kobayashi H., Saga T., Kawamoto S., Sato N., Hiraga A., Ishimori T., Konishi J., Togashi K., and Brechbiel M. W. (2002) Dynamic micro-MRI of liver micrometastasis with a novel liver macromolecular MR contrast agent DAB-Am64-(1B4M-Gd)64. *Acad Radiol* **9** Suppl 2, S452–454.

31. Kobayashi H., Sato N., Hiraga A., Saga T., Nakamoto Y., Ueda H., Konishi J., Togashi K., and Brechbiel M. W. (2001) 3D-micro-MR angiography of mice using macromolecular MR contrast agents with polyamidoamine dendrimer core with reference to their pharmacokinetic properties. *Magn Reson Med* **45**, 454–460.

32. Kobayashi H., Sato N., Saga T., Nakamoto Y., Ishimori T., Toyama S., Togashi K., Konishi J., and Brechbiel M. W. (2000) Monoclonal antibody-dendrimer conjugates enable radiolabeling of antibody with markedly high specific activity with minimal loss of immunoreactivity. *Eur J Nucl Med* **27**, 1334–1339.

33. Kobayashi H., Wu C., Kim M. K., Paik C. H., Carrasquillo J. A., and Brechbiel M. W. (1999) Evaluation of the *in vivo* biodistribution of indium-111 and yttrium-88 labeled dendrimer-1B4M-DTPA and its conjugation with anti-Tac monoclonal antibody. *Bioconjug Chem* **10**, 103–111.

34. Koyama Y., Talanov V. S., Bernardo M., Hama Y., Regino C. A., Brechbiel M. W., Choyke P. L., and Kobayashi H. (2007) A dendrimer-based nanosized contrast agent dual-labeled for magnetic resonance and optical fluorescence imaging to localize the sentinel lymph node in mice. *J Magn Reson Imaging* **25**, 866–871.

35. Larson C. L., and Tcker S. A. (2001) Intrinsic fluorescence of carboxylate-terminated

polyamidoamine dendrimers. *Appl Spectrosc* **55**, 679–683.

36. Lee W. I., Bae Y., and Bard A. J. (2004) Strong blue photoluminescence and ECL from OH-terminated PAMAM dendrimers in the absence of gold nanoparticles. *J Am Chem Soc* **126**, 8358–8359.
37. Lu Z. R., Mohs A. M., Zong Y., and Feng Y. (2006) Polydisulfide Gd(III) chelates as biodegradable macromolecular magnetic resonance imaging contrast agents. *Int J Nanomed* **1**, 31–40.
38. Navarro G., and de Ilarduya C. T. (2009) Activated and non-activated PAMAM dendrimers for gene delivery *in vitro* and *in vivo*. *Nanomedicine* **5**, 287–297.
39. Ntziachristos V. (2006) Fluorescence molecular imaging. *Annu Rev Biomed Eng* **8**, 1–33.
40. Nwe K., Xu H., Regino C. A., Bernardo M., Ileva L., Riffle L., Wong K. J., and Brechbiel M. W. (2009) A new approach in the preparation of dendrimer-based bifunctional diethylenetriaminepentaacetic acid MR contrast agent derivatives. *Bioconjug Chem* **20**, 1412–1418.
41. Prince M. R., Zhang H., Morris M., MacGregor J. L., Grossman M. E., Silberzweig J., DeLapaz R. L., Lee H. J., Magro C. M., and Valeri A. M. (2008) Incidence of nephrogenic systemic fibrosis at two large medical centers. *Radiology* **248**, 807–816.
42. Raymond K. N., and Pierre V. C. (2005) Next generation, high relaxivity gadolinium MRI agents. *Bioconjug Chem* **16**, 3–8.
43. Sarin H., Kanevsky A. S., Wu H., Brimacombe K. R., Fung S. H., Sousa A. A., Auh S., Wilson C. M., Sharma K., Aronova M. A., Leapman R. D., Griffiths G. L., and Hall M. D. (2008) Effective transvascular delivery of nanoparticles across the blood–brain tumor barrier into malignant glioma cells. *J Transl Med* **6**, 80.
44. Scherer R. L., VanSaun M. N., McIntyre J. O., and Matrisian L. M. (2008) Optical imaging of matrix metalloproteinase-7 activity *in vivo* using a proteolytic nanobeacon. *Mol Imaging* **7**, 118–131.
45. Schild H. H., Kuhl C. K., Hubner-Steiner U., Bohm I., and Speck U. (2006) Adverse events after unenhanced and monomeric and dimeric contrast-enhanced CT: a prospective randomized controlled trial. *Radiology* **240**, 56–64.
46. Schwitter J., Saeed M., Wendland M. F., Derugin N., Canet E., Brasch R. C., and Higgins C. B. (1997) Influence of severity of myocardial injury on distribution of macromolecules: extravascular versus intravascular gadolinium-based magnetic resonance contrast agents. *J Am Coll Cardiol* **30**, 1086–1094.
47. Shi X., Wang S., Meshinchi S., Van Antwerp M. E., Bi X., Lee I., and Baker J. R., Jr. (2007) Dendrimer-entrapped gold nanoparticles as a platform for cancer-cell targeting and imaging. *Small* **3**, 1245–1252.
48. Su M. Y., Najafi A. A., and Nalcioglu O. (1995) Regional comparison of tumor vascularity and permeability parameters measured by albumin-Gd-DTPA and Gd-DTPA. *Magn Reson Med* **34**, 402–411.
49. Svenson S., and Tomalia D. A. (2005) Dendrimers in biomedical applications--reflections on the field. *Adv Drug Deliv Rev* **57**, 2106–2129.

50. Talanov V. S., Regino C. A., Kobayashi H., Bernardo M., Choyke P. L., and Brechbiel M. W. (2006) Dendrimer-based nanoprobe for dual modality magnetic resonance and fluorescence imaging. *Nano Lett* **6**, 1459–1463.

51. Thomsen H. S., Marckmann P., and Logager V. B. (2007) Nephrogenic systemic fibrosis (NSF): a late adverse reaction to some of the gadolinium based contrast agents. *Cancer Imaging* **7**, 130–137.

52. Tomalia D. A., Reyna L. A., and Svenson S. (2007) Dendrimers as multi-purpose nanodevices for oncology drug delivery and diagnostic imaging. *Biochem Soc Trans* **35**, 61–67.

53. Wiener E. C., Brechbiel M. W., Brothers H., Magin R. L., Gansow O. A., Tomalia D. A., and Lauterbur P. C. (1994) Dendrimer-based metal chelates: a new class of magnetic resonance imaging contrast agents. *Magn Reson Med* **31**, 1–8.

54. Wu C., Brechbiel M. W., Kozak R. W., and Gansow O. A. (1994) Metal-chelate-dendrimer-antibody constructs for use in radioimmunotherapy and imaging. *Bioorg Med Chem Lett* **4**, 449–454.

55. Xu R., Wang Y., Wang X., Jeong E. K., Parker D. L., and Lu Z. R. (2007) *In Vivo* evaluation of a PAMAM-cystamine-(Gd-DO3A) conjugate as a biodegradable macromolecular MRI contrast agent. *Exp Biol Med (Maywood)* **232**, 1081–1089.

56. Yordanov A., Kobayashi H., English S., Reijnders K., Milenic D., Krishna M., Mitchell J., and Brechbiel M. (2003) Gadolinium-labeled dendrimers as biometric nanoprobes to detect vascular permeability. *J Mater Chem* **13**, 3.

Chapter 8

MULTIFUNCTIONAL, MULTIMODALITY CANCER IMAGING WITH WATER-SOLUBLE SYNTHETIC POLYMER NANOPARTICLES

Marites P. Melancon,[a,b] Xiaoxia Wen,[a] Guodong Zhang,[a] and Chun Li[a]
Departments of [a] Experimental Diagnostic Imaging and [b] Imaging Physics,
The University of Texas M. D. Anderson Cancer Center,
1515 Holcombe Blvd., Houston, TX 77030, USA
cli@mdanderson.org

8.1 INTRODUCTION

Early cancer diagnosis, in combination with efficacious cancer therapies, is critical for achieving the ultimate goal of totally eradicating the disease [43]. Cancer diagnosis is done by taking a biopsy, but if the tumor is inaccessible for biopsy, one has to rely upon the existing imaging techniques, such as magnetic resonance imaging (MRI), computed tomography (CT), positron emission tomography (PET), single-photon emission CT (SPECT), and ultrasonography (US). These imaging modalities have their strengths and limitations, and most of the times, two or more modalities are used to acquire complementary information. Whichever imaging modality is used, sufficient intensity of a corresponding signal from an area of interest must be achieved in order to differentiate the diseased area from the surrounding tissues. Usually, the imaging of different organs and tissues for early detection and localization of numerous pathologies cannot be successfully achieved without appropriate contrast agents. As in the case involving tumor imaging, only selected agents are expected to have the potential of being useful. An ideal cancer imaging agent, therefore, has to meet the prerequisites of (i) selective accumulation in the diseased sites, (ii) low toxicity, (iii) low immunogenicity, (iv) excretability, and (v) high number of reporter groups.

Over the last two decades, synthetic polymers are increasingly studied as a novel

Nanoimaging
Edited by Beth Goins and William Phillips

www.panstanford.com

class of contrast agents. Modern polymer chemistry has led to the generation of a number of biocompatible polymer structures, including branched, graft, multivalent polymers, and dendrimers. Polymeric imaging agents have prolonged plasma half-lives, enhanced stability, reduced toxicity, improved targeting, and a reduced nonspecific binding, thereby allowing for more specific and amplified imaging probes that can significantly differentiate target areas from background. Additionally, multiple reporters and/or homing ligands can be attached to the polymers, allowing multifunctional and multimodality cancer imaging. Common features of this class of contrast agents include (i) prolonged blood circulation time (thus they can serve as blood pool imaging agents), (ii) ability to provide more accurate estimates of the tumor vascular permeability and blood volume using dynamic imaging protocols, and (iii) capability to offer multimodal imaging and combining multiple functions (diagnostic and therapeutic functions) in a single setting.

This chapter examines the application of some key synthetic polymers in cancer imaging. We describe the structure and preparation of these polymers, as well as summarize the *in vitro* and *in vivo* data published in literature. Finally, we discuss the prospects of polymers in the diagnosis and treatment of cancer. The applications of synthetic polymers as drug carriers have been extensively reviewed by many authors [14, 26, 40, 50]. The use of polymeric MR contrast agents for assessing tumor vascular pharmacokinetic parameters (i.e., vascular permeability product, blood volume, transit time) have also been reviewed [8, 19, 38, 39].

8.2 POLY(L-GLUTAMIC ACID) (PG)

8.2.1 PG for MRI

MRI is a noninvasive imaging modality that can provide three-dimensional imaging with high spatial resolution. MRI with blood pool contrast agents can be used to perform noninvasive angiography, assess angiogenesis, quantify and measure the spacing of blood vessels, and measure blood volume and flow [5, 22]. Blood pool contrast agents are generally conjugates of paramagnetic gadolinium chelated to high-molecular-weight polymers that show prolonged blood circulation time and are largely retained within the intravascular space during MR imaging. Polymeric contrast agents can also passively accumulate within solid tumor tissues owing to the enhanced permeability and retention (EPR) effect [31]. However, the clinical applications of many current macromolecular contrast agents are limited by their slow excretion from the body and the potential toxicity of free gadolinium released by the metabolism of the contrast agents [6, 12, 41]. The ideal polymeric contrast agent is degraded and cleared from the body after completion of MR imaging. Poly (L-glutamic acid) (PG) is a polymeric carrier widely used for diagnosis and therapeutics because of its biocompatibility, biodegradability, and water solubility [24].

8.2.1.1 PG-DTPA-GD

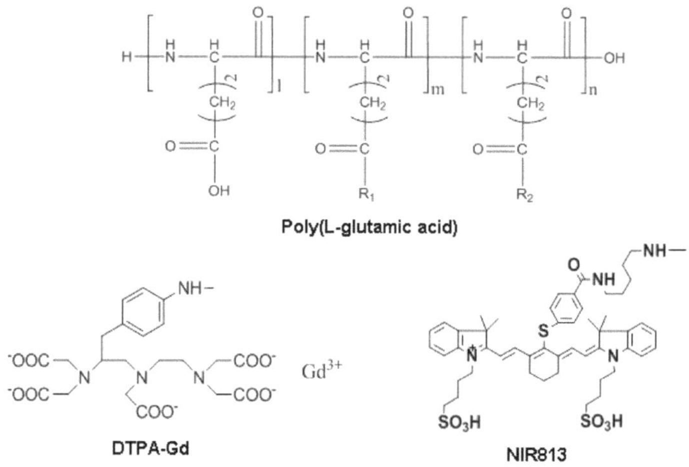

Conjugate	PG-DTPA-Gd	PG-NIR813	PG-DTPA-Gd-NIR813
R_1	DTPA-Gd	NIR813	DTPA-Gd
R_2	OH	OH	NIR813

Figure 8.1 Chemical structures of PG-DTPA-Gd, PG-NIR813, and PG-DTPA-Gd-NIR813.

Poly(L-glutamic acid) conjugated with diethylenetriaminepentaacetic acid gadolinium chelate (PG-DTPA-Gd) (Fig. 8.1) was developed in our laboratory as a biodegradable blood pool MRI contrast agent. It is synthesized through simple two-step reactions. First, *p*-aminobenzyl-DTPA-t-butyl ester is attached to the side chain of PG via an amide bond. Second, the protective t-butyl ester is removed in the presence of trifluoroacetic acid, and gadolinium is chelated in sodium acetate buffer, pH 5.2. PG-DTPA-Gd yields a narrow distribution polymer conjugate with molecular weight of 101 kDa and polydispersity of 1.23.

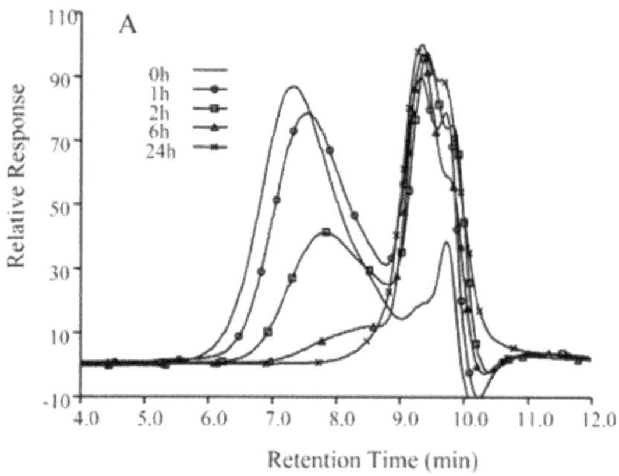

Figure 8.2 Degradation within 24 hours of PG-DTPA-Gd in the presence of cathepsin B. Reproduced with permission from Ref. [53].

PG has been shown to be degraded by cysteine proteases [15]. Conjugates of PG, such as PG-DTPA-Gd, were demonstrated *in vitro* to be rapidly degraded within 24 hours in the presence of cathepsin B (Fig. 8.2).

In vivo, PG-DTPA-Gd, as an MRI blood pool imaging agent, showed enhanced vascular contrast in mice up to 2 hours after contrast injection. By 24 hours postinjection, the blood pool activity had largely returned to the precontrast level (Fig. 8.3A,B). Interestingly, the contrast in certain regions of the tumors was clearly enhanced at 24 hours after contrast injection, reflecting EPR effect [37] of macromolecules in solid tumors. *In vivo* biodistribution in mice confirmed the gradual clearance of the contrast agent from the blood pool and kidney [53]. Intratumoral distribution of PG-DTPA-Gd was further investigated in mice bearing human Colo-205 xenografts or syngeneic murine OCA-1 ovarian tumors.

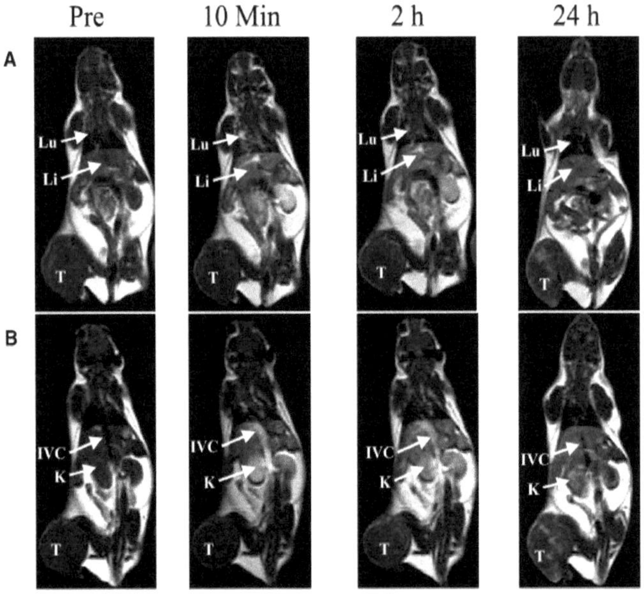

Figure 8.3 (A, B) Whole-body images of mice before and at 10 min, 2 h, and 24 h after intravenous injection of PG-DTPA-Gd at a dose of 0.04 mmol Gd/kg. *Abbreviations*: Lu, lung; Li, liver; IVC, inferior vena cava; K, kidney; T, tumor. Reproduced with permission from Ref. [16].

MR imaging of tumors revealed specific accumulation of the polymeric contrast agent in tumor necrosis tissue (Fig. 8.4A–K). Although the mechanism of the selective localization of PG-DTPA-Gd in areas of necrosis is not entirely clear, the observation that biotin-PG-DTPA-Gd co-localized with macrophages in OCA-1 tumors suggests that the affinity of PG-DTPA-Gd for necrotic tissue may be mediated through macrophages [16]. The contrast enhancement in necrotic tumor tissue was observed with only polymeric contrast agents, not with low-molecular-weight agents. High molecular weight contrast agents have prolonged blood circulation times, which may give them sufficient time to interact with migrating or resident macrophages. The EPR effect of macromolecules also played a role in their transport from the vicinity of tumor blood vessels to the necrotic areas through a diffusion process. These findings suggest that PG is an essential carrier for gadolinium contrast moieties to the tumor necrotic tissue. Consistent with this hypothesis, studies using a monoclonal antibody for paclitaxel (TXL) that specifically recognizes PG-TXL showed that PG-TXL was taken up by macrophages *in vitro* through endocytosis [10].

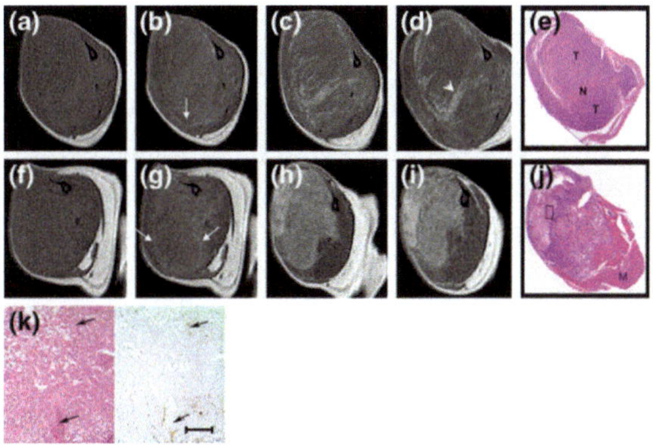

Figure. 8.4 Representative magnetic resonance images of OCA-1 tumors from an untreated mouse (a–e) and a mouse treated with PG-TXL (120 mg equivalent paclitaxel/kg) (f–j). Transverse T_1-weighted spin-echo images before (a, f) and at 10 minutes (b, g), 2 days (c, h), and 4 days (d, i) after intravenous injection of PG-DTPA-Gd (0.04 mmol/kg). (e, j) Macroscopic photographs of the tumors sectioned on the transverse plane showing the necrotic regions in the tumors (N = necrosis; T = tumor; M = muscle). Arrows in b and g: peripheral tumor enhancement owing to blood pool imaging effect of the contrast agents; arrowheads in d: necrotic area. (k) Higher-power photomicrographs (left panel: hematoxylin and eosin stain; right panel: factor VIII immunohistochemical stain) of tumor in (j) from boxed region. Note the scattered factor VIII–positive cells in the necrotic zone (arrows). Scale bar: 100 μm. Reproduced with permission from Ref. [16].

The clear enhancement of vascular contrast for up to 2 hours and the ability to be degraded and cleared from the body after imaging makes PG-DTPA-Gd a favorable blood pool contrast agent for the diagnosis of cardiovascular diseases and cancer detection. PG-DTPA-Gd may also be used to assess chemotherapy-induced necrosis in solid tumors because of its selective localization in necrotic tissue. To further investigate the *in vivo* behavior of PG-DTPA-Gd and identify the optimal dose of PG-DTPA-Gd to be used for clinical trials, rhesus monkeys were intravenously (i.v.) injected with PG-DTPA-Gd at doses ranging from 0.01 to 0.08 mmol Gd/kg, and their pharmacokinetics, biodistribution, and clearance in primates were evaluated. A group of normal Swiss mice also received an i.v. injection of 0.08 mmol Gd/kg of PG-DTPA-Gd in order to compare its *in vivo* behavior in different species. Magnevist, which is a clinically used small-molecular-contrast MR imaging agent, was chosen as a control. Concentrations of PG-DTPA-Gd in blood were determined by quantifying the Gd content using inductively coupled plasma mass spectrometry. T_1-weighted MR images were acquired at 1.5T using a fast spoiled gradient echo imaging protocol. The contrast enhancement of major organs with 0.02 and 0.08 mmol Gd/kg doses was significantly higher than with Magnevist. At 2 days after PG-DTPA-Gd injection, most of the contrast agent was cleared from all major organs, including the kidneys. By 7 days post-injection, no residual contrast could be detected except a little amount in the liver, and the kidneys of the monkey received highest dose of PG-DTPA-Gd (0.08 mmol Gd/kg). This suggests that PG-DTPA-Gd was cleared mainly from the renal routes. The resident time of PG-DTPA-Gd in monkey were 197.8–479.5 hours, compared to 7.69 hours for Magnevist. In comparison, the behavior of PG-DTPA-Gd in mice was different, exhibiting a shorter mean resident time of 50.1 hours [54].

In conclusion, the pharmacokinetics and elimination profile of PG-DTPA-Gd, including vascular retention, renal excretion, and biodegradability of the polymer, make it very promising for use in humans as an intravascular contrast agent for MRI. On the basis of the preclinical studies, 0.02 mmol Gd/kg PG-DTPA-Gd may be used as the initial dose for clinical trials. Biodegradable macromolecular contrast agents such as PG-DTPA-Gd may be a valuable addition to clinically approved low-molecular-weight gadolinium contrast agent.

8.2.1.2 PG-DO3A-Gd

Poly(L-glutamic acid) conjugated with 1,4,7,10-tetraazacyclododecane-1,4,7-trisacetic acid gadolinium chelate (PG-DO3A-Gd) is another PG-based MRI polymeric contrast agent. DO3A is the derivative of DOTA (1,4,7,10-tetraazacyclododecane-1,4,7,10-tetraacetic acid) in which one methylcarboxylic arm is modified by 1,6-hexanediamine, a linker with which Gd-DO3A is attached to PG (Fig. 8.5).

In PG-DO3A-Gd, only three methylcarboxylic arms in DO3A are involved in the chelation with Gd(III), as opposed to four arms in DOTA, which may reduce the stability of the Gd complex. All four of the methylcarboxylic groups of DTPA participated in chelating Gd(III) in PG-DTPA-Gd, which resulted in retention of the stability of the Gd complex. Also, the stability of the DTPA-Gd complex is higher in PG-DTPA-Gd because of the presence of the benzyl group [54]. To investigate the *in vivo* properties of PG-DO3A-Gd, polymer conjugates with molecular weights of 87, 50, and 28 kDa were prepared [56]. MR images of mice bearing MDA-MB-231 human breast cancer xenografts were acquired continuously for the first 20 minutes and then at 30, 60, 120, 180, 240, 360, and 1440 minutes post-injection. The conjugate with the lowest molecular weight (28 kDa) rapidly cleared from the circulation and had lower tumor accumulation, whereas the conjugates with higher molecular weights (50, 87 kDa) exhibited more prolonged blood circulation and higher tumor uptake [56].

PG-DO3A-Gd conjugated with a photosensitizer, mesochlorine$_6$ (Mce$_6$), was reported to be used in monitoring and mediating photodynamic therapy (PDT) [48]. Axial MR images with PG-DO3A-Gd-Mce$_6$ contrast in athymic nude mice bearing MDA-MB-231 xenografts revealed the gradual diffusion of the polymer conjugate from the tumor periphery to the inner tumor tissue owing to the hyperpermeability of the tumor vasculature. More uniform enhancement throughout the tumor was observed at

18 hours postinjection, indicating uniform distribution of the conjugate in the tumor. These findings are consistent with the results obtained with PG-DTPA-Gd [16, 53]. Photodynamic therapy was performed by irradiating the tumor sites with a 650 nm diode laser at 2 and 18 hours post-injection. Tumors in the control group treated with PG-DO3A-Gd grew more rapidly than those treated with Mce_6-conjugated PG-DO3A-Gd. The tumor volume of the mice treated with PG-DO3A-Gd-Mce_6 did not change significantly for at least 30 days. Overall tumor regrowth was significantly delayed in the group treated with the bifunctional conjugate. The results demonstrate that PG-DO3A-Gd-Mce_6, a bifunctional polymer conjugate containing both an MRI contrast agent and a photosensitizer, was effective for cancer imaging (contrast-enhanced MRI) and efficacious for treatment with PDT in an animal model.

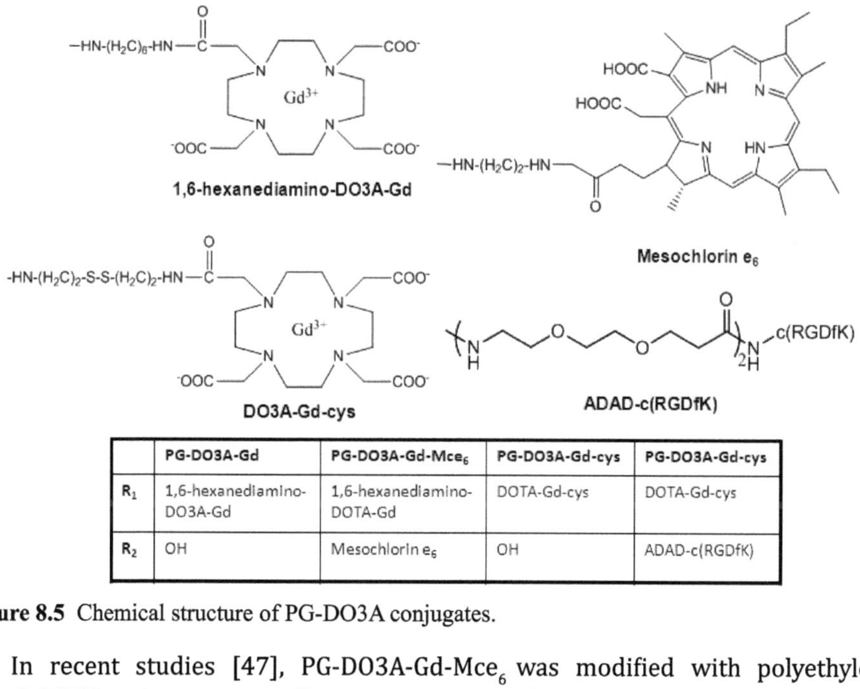

	PG-DO3A-Gd	PG-DO3A-Gd-Mce_6	PG-DO3A-Gd-cys	PG-DO3A-Gd-cys
R_1	1,6-hexanediamino-DO3A-Gd	1,6-hexanediamino-DOTA-Gd	DOTA-Gd-cys	DOTA-Gd-cys
R_2	OH	Mesochlorin e_6	OH	ADAD-c(RGDfK)

Figure 8.5 Chemical structure of PG-DO3A conjugates.

In recent studies [47], PG-DO3A-Gd-Mce_6 was modified with polyethylene glycol (PEG) to improve its efficacy in contrast-enhanced (CE)-MRI-guided PDT by reducing nonspecific liver uptake and increasing tumor accumulation. The pegylated conjugate demonstrated longer blood circulation, less liver uptake, and more tumor accumulation than the nonpegylated conjugate, as shown by MRI. Site-directed laser irradiation of tumor resulted in higher therapeutic efficacy for the pegylated conjugate. Moreover, animals treated with PDT and pegylated PG-DO3A-Gd-Mce_6 showed reduced vascular permeability on dynamic CE-MRI and decreased microvessel density in histological analysis. In conclusion, contrast-enhanced MRI can effectively detect changes in the tissue distribution of polymer conjugates and guide site-directed irradiation of target tissues. Dynamic CE-MRI with the biodegradable macromolecular MRI contrast agent was also effective in noninvasive assessment of tumor response to photodynamic therapy.

In 2003, Lu *et al.* [30] reported their study of a PG-DO3A-Gd conjugate linked with a degradable cystamine spacer. The conjugate was designed to solve the problem of slow excretion of macromolecular contrast agents and potential toxicity due to release of the Gd(III) ion caused by metabolism of the agents. PG-DO3A-Gd-cystamine was prepared by conjugation of DO3A (DOTA) to PG (MW = 50,000) via cystamine, a cleavable disulfide spacer, followed by complexation with $GdCl_3$. The Gd-DO3A chelate was readily released from the polymer conjugate during incubation with cysteine, an endogenous plasma thiol. The conjugate produced significant MRI blood pool contrast enhancement in nude mice bearing OVCAR-3 human ovarian carcinoma xenografts. Less significant contrast enhancement was observed for a small molecular control agent, Gd(DTPA-BMA). The pharmacokinetic MRI study showed that the Gd(III) chelate from the conjugate accumulated in the urinary bladder in a kinetic pattern similar to that of Gd(DTPA-BMA), suggesting that the chelate was released by the endogenous thiols and excreted through renal filtration.

Arginine-glycine-aspartic acid (RGD) peptide was conjugated to PG-DO3A-Gd-cystamine with the aim of detecting $\alpha_v\beta_3$ integrin, an angiogenesis biomarker, in neoplastic tissues. T_1 mapping, a quantitative MRI technique, was used because conventional MRI is not sensitive enough to detect the binding of the targeted contrast agent in the tumor [18]. *In vitro* binding studies revealed that the binding affinity of polymer-bound cyclic-RGD-phenylalanine-lysine, or c(RGDfK), was slightly lower than that of free c(RGDfK) peptide. The RGD-bound conjugate had higher binding affinity to DU145 cells, which highly express $\alpha_v\beta_3$ integrin, than to SLK cells, which was consistent with free c(RGDfK). *In vivo* MR imaging showed that RGD-bound PG-DO3A-Gd-cystamine had higher binding affinity to DU145 xenografts than to SLK xenografts, resulting in a significant decrease in the T_1 values of water protons in the periphery of DU145 tumors, as shown on the MR T_1 maps. No significant decrease in T_1 values was observed in SLK tumors with the targeted conjugate and in both tumors with the nontargeted conjugate. The conjugation of RGD peptide to PG-DO3A-Gd-cystamine resulted in a specific affinity for the human prostate carcinoma DU145 cell line and was effective for the detection of the angiogenesis biomarker $\alpha_v\beta_3$ integrin in the DU145 tumor with quantitative T_1 mapping. The targeted polymeric Gd(III) chelate conjugate with a degradable spacer has the potential to be a new paradigm for safe and effective probes in molecular imaging with quantitative MR T_1 mapping.

8.2.2 PG for Optical Imaging

Optical imaging is an emerging method for the detection and monitoring of cancer because of its high sensitivity, rapid data acquisition, and relatively inexpensive instrumentation. In particular, the near-infrared (NIR) region of the electromagnetic spectrum is highly desirable because most biological substances and impurities in bioprocesses absorb and fluoresce between 190 and 650 nm. The relative sensitivity to detect biomolecules at these wavelengths is greatly reduced owing to the high background noise caused by the molecules themselves (autofluorescence) and the impurities within the ultraviolet-visible (UV-vis) region [23]. Since the NIR region lies in a spectrum beyond these effects, it provides high sensitivity for the detection of biomolecules and an improved signal-to-noise ratio.

PG is known to be a good macromolecular carrier for anticancer drugs, including paclitaxel [26]. PG conjugated with paclitaxel (PG-TXL) has been shown, in a number of preclinical animal tumor models, to have significantly higher uptake in the tumors and greater antitumor activity than the parent drug paclitaxel [25, 28]. PG-TXL is currently in phase III clinical trials. One of the proposed mechanisms for the increased site-specific delivery and enhanced antitumor activity of PG-TXL is the degradation of the PG backbone, through the action of cellular proteases in tumors, which releases the paclitaxel into the tumors. To understand PG's *in vivo* degradation, we conjugated an inexpensive NIR fluorescent dye containing a primary amine to the pendant –COOH groups of the PG (Fig. 8.1) [33]. Interestingly, the fluorescence intensity decreased with increasing amounts of NIR dye (Fig. 8.6).

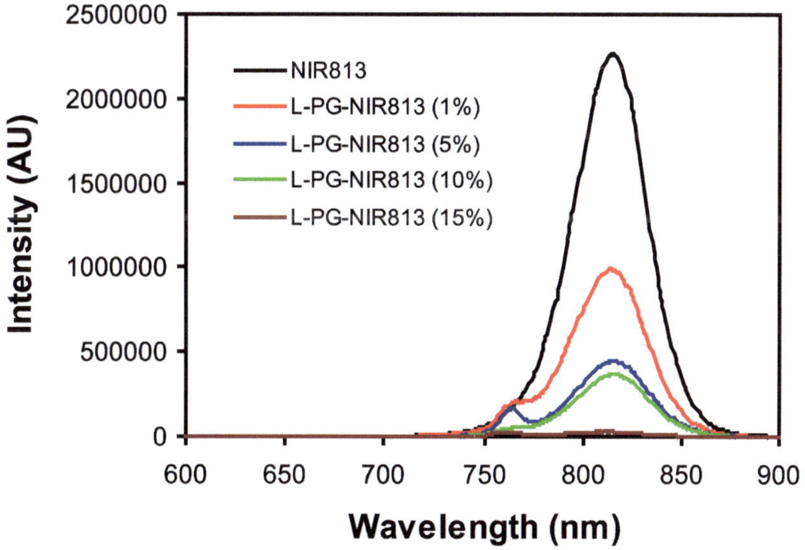

Figure 8.6 Effect of NIR813 loading on the quenching efficiency of L-PG-NIR813. The number-average molecular weight of L-PG was 17,500. All conjugates had the same equivalent concentration of NIR813 (1 mM). Fluorescence intensity decreased with increasing payload of NIR813. Reproduced with permission from Ref. [33].

NIR loading of 8% was optimal to start with an optically silent PG-NIR, but upon exposure to cathepsin B (an enzyme that is overexpressed in most tumors), the fluorescence intensity increased 10-fold (Fig. 8.7).

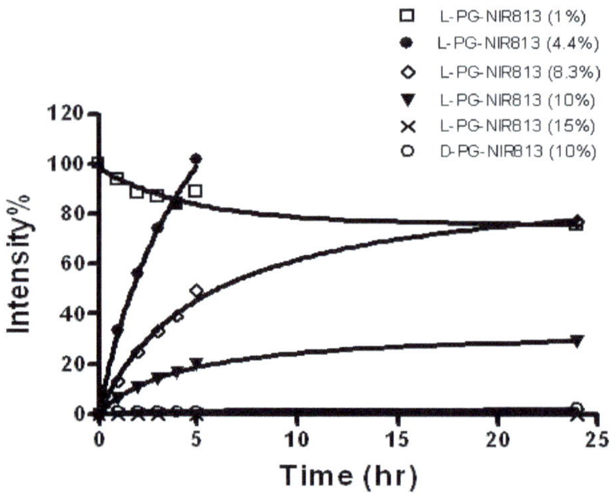

Figure 8.7 Effect of NIR813 loading on degradation of L-PG-NIR813 in the presence of CB. Aliquots (100 mL) of each L-PG-NIR813 solution (10 mM equivalent NIR813) in 20 mM sodium acetate buffer (pH 5.2) were incubated with CB (0.4 units/ml) at 37 °C for 24 hours in triplicate. Reproduced with permission from Ref. [33].

This phenomenon was tested in mice bearing U87 human glioma tumors. (Fig. 8.8) shows that L-PG-NIR, but not D-PG-NIR, is degraded into the tumor that has overexpression of cathepsin B (CB). These results indicate that PG-NIR may be used to monitor the *in vivo* degradation of L-PG-based polymeric drugs and that this agent may be useful in noninvasive imaging of protease activity, particularly that of cysteine proteases [33].

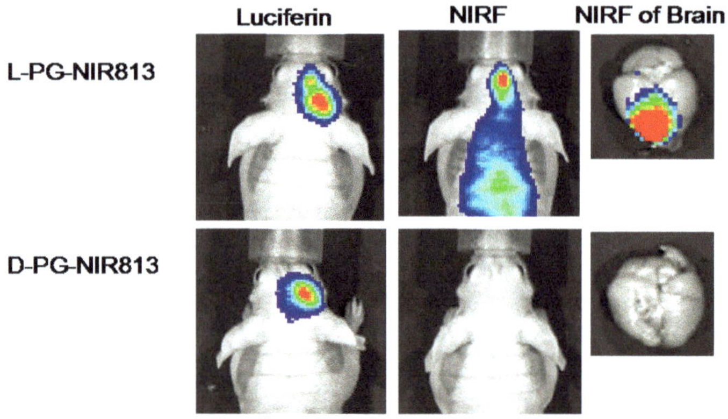

Figure 8.8 Representative *in vivo* NIRF images of CB activity in intracranially inoculated U87/TGL tumors. The mice used in the study had the same tumor burden as indicated by bioluminescence signal generated after intravenous injection of luciferin. NIRF Images were acquired 24 hours after intravenous injection of L-PG-NIR813 (50 nmol/mouse) or D-PG-NIR813 (50 nmol/mouse) with the same NIR dye loading (10%) and the same molecular weight (17,500). NIRF signal was visualized in the mouse injected with L-PG-NIR813 but not the mouse injected with D-PG-NIR813. Reproduced with permission from Ref. [33]

8.2.3 PG for Multimodality Imaging

MR imaging offers excellent spatial resolution and intrinsic soft-tissue contrast, while optical imaging offers high sensitivity. These two modalities when combined together can have significant impact in sentinel lymph node (SLN) mapping and biopsy. Currently, SLN mapping is routinely done to determine whether the cancer has spread beyond the primary tumor. However, the current protocol for SLN mapping exposes the patients and healthcare professionals to ionizing radiation. In addition, patients with head and neck cancer have very complex structures that necessitates the use of a high-resolution, three-dimensional imaging of the anatomical structure.

We have developed a dual-modality MR/optical imaging agent based on PG [34]. We conjugated DTPA-Gd and NIR813 dye on the pendant carboxylic acid groups of PG to obtain PG-Gd-NIR. NIR813 is a dye that fluoresces and absorbs at 813 nm. The conjugate, PG-Gd-NIR813, co-localized with isosulfan blue dye (the gold standard for SLN mapping), and its drainage into the SLN was confirmed by using optical imaging. When injected intralingually in mice bearing DM 14 oral squamous cell carcinoma, MRI could detect not only the presence of PG-Gd-NIR813 but also the pattern of contrast agent distribution, which was different in the metastatic and normal lymph nodes (Fig. 8.9A–K) [34].

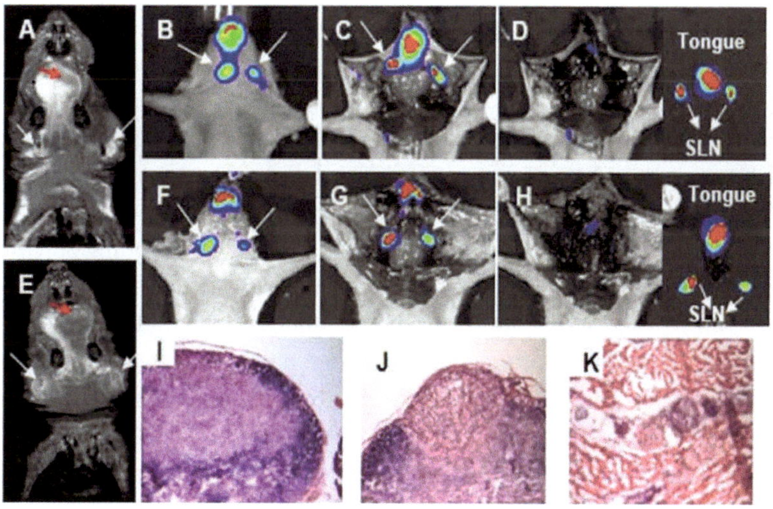

Figure 8.9 Visualization of cervical lymph nodes after interstitial injection of PG-DTPA-Gd-NIR813 (0.02 mmol Gd/kg) into the tongue of a normal mouse (A–D) and a mouse with a human DM14 squamous cell carcinoma xenograft in the tongue (E–H). (A and E) T_1-weighted coronal image acquired 2 hours after contrast injection. (B and F) Overlay of white light and NIRF images 24 hours after contrast injection. (C and G) NIRF images of mice without skin. (D and H) NIRF images of resected lymph nodes. The white arrows indicate SLNs, and the red arrow indicates the primary tumor. (I–K) Microphotographs of H&E-stained sections of (I) normal lymph node, (J) metastatic lymph node, and (K) tumor-bearing tongue indicating the presence of micrometastases, presumably in-transit metastases in a lymphatic duct. Note the difference in the pattern of contrast distribution in the lymph nodes between normal and tumor-bearing mice in the MR images. Reproduced with permission from Ref. [34].

These results suggest that MRI could be used clinically to assess the presence of metastasis. If MRI detects the presence of lymph node metastasis, surgery would have to be performed to remove the entire nodal basin, which would eliminate the need for SLN biopsy and the associated waiting time. If MRI does not detect lymph node metastasis, SLN mapping would have to be performed using pre- and intra-operative near-infrared fluorescent (NIRF) imaging, without the use of an ionizing radiotracer or cumbersome interventional MRI. Because of its high detection sensitivity, NIRF imaging could also be used to monitor the success of complete resection of SLNs during surgery.

8.3 POLY(L-LYSINE)

8.3.1 PLL for MRI

Poly(L-lysine) (PLL) is the first synthetic polymer to be used as an MRI contrast agent [32, 42]. Toward the end of 1980s, polymeric MRI contrast agents such as albumin-DTPA-Gd and dextran-DTPA-Gd demonstrated promising potential for blood pool MR imaging because of their apparently improved r_1 relaxivity and prolonged circulating time *in vivo*. For instance, the r_1 of albumin-DTPA-Gd was reported to be 14.9 $mM^{-1}s^{-1}$ per gadolinium at 0.23 T [37], which is a threefold increase in relaxation enhancement compared with DTPA-Gd (4.9 $mM^{-1}s^{-1}$). However, some issues precluded the clinical development of albumin-DTPA-Gd, including the immunogenic potential of albumin and the substantial retention of gadolinium metabolites in the liver and bone. To circumvent these problems, the search for a polymeric carrier became focused on the synthetic PLL, which can be prepared by ring-opening polymerization of *N*ε-carbobenzyloxy-L-*lysine*-*N*-carboxyanhydride (Z-Lys-*NCA*), and the polymer structure can be tailored by adjusting the polymerization conditions. Furthermore, PLLs with different molecular weights are commercially available. DTPA, the chelator for gadolinium that is currently used in the clinic, is usually conjugated to the primary amine of PLL through a reaction with a commercial reagent such as DTPA-anhydride or DTPA-isothiocyanate [60]. Typically, PLL-DTPA-Gd, which has an average molecular weight of 48,700, would contain ~380 repeating units of lysine in each polymer chain, with 60 to 70 of them covalently linked to the chelator DTPA, and a gadolinium content of 21.8% (w/w) after labeling. *In vitro*, the relaxivity, r_1, of PLL-DTPA-Gd is 13.1 $mM^{-1}s^{-1}$ per gadolinium performed at 20 MHz. This r_1 is more than three times higher than that of DTPA-Gd under similar conditions. No biodegradation or metabolization was observed, and 7 days after i.v. injection about 6% of the Gd ion dose administered was retained in the body. Though PLL-DTPA-Gd is still on the list of developing MRI agents, its potential long-term toxicity undoubtedly is the main concern and may hamper its use in the clinic.

To achieve a longer circulating time and lower toxicity and immunogenicity, PLL carrier was replaced by poly(ethylene glycol)-grafted PLL (PLL-PEG) [1] owing to the outstanding behavior of PEG in PEG-modified protein *in vivo*. Methoxy-PEG with an active ester OSu terminal was first conjugated to the primary amines of PLL, and then DTPA anhydride was reacted with the residual primary amines of the resulting PEG-graft-PLL (PLL-PEG). The conjugation of PEG made a significant difference in the *in vivo* characteristics compared with PLL-DTPA-Gd alone. First, the grafting of PEG to PLL significantly extended the blood half-life of the modified polymeric DTPA-Gd contrast agent. It was indicated that the prolonged circulation time correlated with the chain length of PEG: the greater the molecular weight of PEG, the longer the circulation time *in vivo*; for example, the half-lives of PLL-DTPA-Gd grafted with 0.75 k, 2 k, and 5 k MPEG were 0.36, 1.1, and 14.1 hours, respectively. Second, MPEG grafted to PLL had an apparent effect on the biodistribution of the polymeric DTPA-Gd system *in vivo*, especially uptake in the kidney. When only 10% of amino groups of PLL were modified by 5 k MPEG, their uptake in the kidney decreased by at least fourfold. Third, contrary to the very short timing-frame of MR imaging, PLL-PEG-DTPA-Gd resulted in a maximum signal intensity pixel projection image of a rat head at 40 minutes post-i.v. injection, and small arteries and veins (0.25–1 mm in diameter) became clearly discerned with such a long-circulating contrast agent. On the basis of PLL-PEG-DTPA-Gd–enhanced MR angiography, Bogdanov *et al.* have studied the formation of neovasculature in a rat mammary adenocarcinoma model, which cannot be visualized using small molecular agent-based MR angiography techniques because of the slow and nonlaminar blood flow in the tumors. For example, at 6–8 days after tumor implantation vessels at the tumor/host interface became detectable; after 15 days, the formation of numerous loops and self loops of tumor vessels located inside the tumor was clearly visible. Thus, it was concluded that the combination of MR angiography and MPEG-PLL-DTPA-Gd is an effective technique for studying the development of vascular tumor architecture *in vivo* [6, 7].

8.3.2 PLL for Nuclear Imaging

PLL-PEG-DTPA labeled with ^{99m}Tc has been proven to be a successful blood imaging agent. According to theoretical calculation and experimental data, the molecular structure of the polymeric chelator is PLL_{279}–89PEG_{114}–190DTPA, which means that each PLL with about 279 repeating units was modified by 89 PEG and 190 or so DTPA units. Biodistribution experiments in rats (n = 6) exhibited excellent retention of PLL-PEG-DTPA in the blood pool, and 50% of the dose was detected in the circulation at 6 hours post-injection. The accumulation of the agent in the liver and kidney amounted to less than 10% of the total dose in all the time points measured. Approximately

25% of the dose was excreted in the urine and feces within 24 hours postinjection. Importantly, the synthetic polymer blood pool agent produced satisfactory blood pool imaging, as evidenced by its ability to label red blood cells (RBCs) in a rhesus monkey.

PLL-PEG-DTPA's attractive features, including excellent blood-pool retention, simple labeling procedure, and acceptable stability, led to preclinical evaluation and a phase I clinical trial in 1998. In animal studies using rats and rabbits, there were no signs of acute or sub-acute toxicity at doses 280 times the proposed dose for humans. In clinical trials in healthy humans (n = 6), no significant abnormalities of laboratory values, ECG findings, or hemodynamic parameters were observed. Four of the volunteers studied exhibited typical blood pool biodistribution, with a blood half-life of 20.6±2.3 hours. Within 24 hours post-injection, 22.1%±2.5% of the injected dose had been excreted through the kidneys. Both anterior and posterior whole-body images demonstrated a typical blood pool distribution, and the heart and major blood vessels were clearly visible after administration of PLL-PEG-DTPA-^{99m}Tc with radiochemical purity of greater than 95%. It was concluded that the synthetic PLL-PEG-DTPA-^{99m}Tc possesses the ability to produce high-quality blood pool images in humans [4, 9].

8.3.3 PLL for Optical Imaging

Weissleder *et al.* [44, 52] have developed a method for *in vivo* imaging of tumors with protease-luminated near-infrared fluorescent (NIRF) probes. This method is based on the following facts: (i) PLL-PEG exhibits very high accumulation in tumor, reaching up to 6% injected dose/g tumor tissue; (ii) the residual primary amines can provide sufficient chances to conjugate NIRF dyes, directly or through a protease-cut peptide linker; and (iii) the NIRF dyes can be turned "off or on," depending on whether they are quenched within the polymer chain or released from the polymer chain. In the case of direct conjugation, 11 molecules of Cy5.5 were directly conjugated to PLL-PEG, and each PLL contained about 92 PEG, with molecular weights of PEG and PLL of 5 and 35.5 k, respectively. *In vitro*, after the polymer probe was activated by an enzyme, a 12-fold increase in NIRF signals was observed. Inhibitor experiments showed that the release of dyes from the polymer backbone resulted from the activity of several proteases, including cysteine protease, leupeptin, and trypsin, but not pepstatin, suggesting that NIRF signals are related to lysosomal cysteine/serine protease activity. Using this polymer fluorescence probe, all implanted breast tumors were clearly detected in tumor-bearing mice 24 hours after i.v injection, even microscopic tumors with diameters less than 300 μm. Thus, the proposed imaging strategy has the potential to detect early-stage tumors with submillimeter diameters.

The same research group studied the PLL-PEG-peptide-NIR system, in which the peptide can be specifically cleaved by cathepsin D to evaluate the detection of cathepsin D alone as a model target protease. This cathepsin D-substrate peptide consists of a Gly-Pro-Ile-Cys (Et)-Phe-Arg-Leu-Gly(FITC)-Cys-NH_2 sequence, in which the thiol terminal was used to conjugate with iodol groups of the graft copolymer, prepared by the reaction of the primary amines of PLL-PEG with iodoacetyl anhydride. Before linker peptide cleavage by cathepsin D, NIR dyes were locked along the polymer chains and autoquenched to be silent in that state; thus, the amount of dyes is crucial to the quenching efficiency and the subsequently released luminescence intensity. The current PLL-PEG probe was optimized such that each polymer contains about 22 NIRF reporter groups. *In vitro*, the prepared polymer probe activated by cathepsin D generated fluorescence signals more than 60-fold higher than those generated by the probe in the silent state before activation. The *in vivo* results demonstrated that there is a significant difference in fluorescence intensity between cathepsin D(+) tumor-bearing mice and cathepsin D(–) mice post-i.v injection, and the heterogeneous distribution of NIRF in tumor tissue was also clearly observed.

8.3.4 PLL for Multimodality Imaging

PLL-based fluorescence and MRI multimodal imaging probes have been developed for novel enzyme/pro-drug cancer therapy [24, 27] and for tumor localization [46]. Enzyme/pro-drug therapy is a promising strategy to eliminate the side effects of chemotherapy for cancer [3]. One of the key elements in this type of cancer therapy is to determine the optimal time window for pro-drug injection [36]. PLL-based fluorescence and MRI multi-modal imaging probes provide a chance to image the delivery of a drug-activating enzyme. In a promising example, bacterial cytosine deaminase (bCD), which can convert the nontoxic prodrug 5-fluorocytosine (5FC) to the anticancer drug 5-fluorouracil (5FU), was conjugated to PLL (Mr 5.6 kDa, repeating unit about 44) with DOTA-Gd as a reporter for MRI and rhodamine as the fluorescing agent. Due to the EPR effect, the resulting conjugate, with a molecular weight of about 300 kDa, passively targets the tumor. Dynamic light scattering revealed that the conjugate has an average hydrodynamic radius of 9.4 nm. Both *in vivo* MRI studies and optical imaging studies exhibited that uptake of bCD-PLL in the tumor was evident at 2 hours after injection and still present 24 hours after i.v. injection, whereas most of the conjugate had cleared from the blood and other tissues; thus, 24 hours post-i.v. injection is the desired time for 5-FC delivery. The following strategy not only monitors the production of 5-FU in the tumor by ^{19}F MR spectroscopy in real time but also monitors the treatment efficiency, which shows its great potential in cancer diagnosis and therapy. As the authors pointed out, replacing rhodamine with NIR dyes will enforce its application.

Another PLL-based dual-modality imaging probe is PLL-DTPA-Gd-conjugated with an NIR dye, which has been designed and prepared for preoperative tumor localization and intraoperative tumor margin delineation. Except for the expected MRI imaging, NIR fluorescence from preclinical subcutaneous and orthotopic mammary gland tumors is 4.5 times higher than background signals at 12 hours post-i.v. injection of this dual agent. The author also studied the effect of the conformation of PLL-DTPA-Gd on its uptake in tumor tissue and found that the extended chain, resulting from more substituted DTPA, more easily reached the tumor [45].

8.4 *N*-(2-HYDROXYPROPYL)METHACRYLAMIDE (HPMA)

N-(2-hydroxypropyl)methacrylamide (HPMA) copolymer has been extensively studied as a polymeric carrier for chemotherapy drugs and imaging agents, and some of the copolymers are being studied in clinical trials for the treatment of several different malignancies [21]. The biocompatibility of HPMA copolymers has been validated in clinical trials [14]. Importantly, the chemical versatility of HPMA copolymers can be achieved by copolymerization of designed monomers and post-polymerization modification. Unlike PLL-DTPA-Gd, most of the HPMA copolymer-based MRI systems prefer DOTA as the chelator owing to the kinetic inertness of DOTA-Gd in the blood pool, which guarantees the least toxicity from free gadolinium *in vivo* [35].

To study the features of HPMA copolymers *in vivo* by using contrast-enhanced magnetic resonance imaging, DOTA-Gd-conjugated HPMA copolymers were prepared and evaluated. Lu *et al.* [51] first investigated the effect of molecular weight on the biodistribution, tumor accumulation, and MR imaging in tumor-bearing mice. HPMA copolymers with different molecular weights, including 28, 60, and 121 kDa, were prepared by using the gel permeation chromatography fraction method, and the DOTA-Gd content was basically similar. Taking into account the threshold of renal glomerular filtration (ca. 45 kDa), among the three copolymer conjugates, the smallest one should be easily filtered in the kidneys. In fact, the MRI signal intensity of 28 kDa conjugates showed a continued increase in the urinary bladder for 4 hours post-i.v injections, indicating its rapid clearance from the blood circulation. Clearly, the circulation life of the three conjugates was related to their polymer size, and the one with the highest molecular weight (121 kDa) exhibited the longest circulation life in blood. The dynamic MR images (axial 2D T_1-weighted spin-echo images) also demonstrated a gradual, size-dependent accumulation of conjugates in tumor, that is, the higher the molecular weight, the higher the passive tumor targeting efficiency. For example, the conjugates with 121 kDa exhibited more significant and prolonged tumor enhancement, even in the tumor interstitium at 24 hours post-injection. As the author mentioned, dual-imaging modalities of polymeric conjugate-enhanced MRI and SPECT or PET would provide a noninvasive imaging evaluation with high spatial resolution and high sensitivity (accuracy). In addition, Ghandehari *et al.* [59] prepared HPMA copolymer-doxorubicin-DOTA-gadolinium conjugates to monitor the *in vivo* fate of HPMA-based drug delivery systems by using the MRI method and found that the relaxivity, r_1, of the polymeric complex was 1.6 times higher with than without doxorubicin in *in vitro* studies. This improvement in r_1 may have resulted from the aggregation of polymer conjugates after the introduction of doxorubicin.

Targeted MRI *in vivo* has been limited by the lower sensitivity of conventional gadolinium-enhanced MRI, [2] and polymeric contrast agents seem to be a solution to address this issue, based on a high local gadolinium concentration, improved relaxivity, and easily adjusted targeting moiety amounts [57–58]. One such agent, HPMA copolymer-(DOTA-Gd)-RGDfK conjugate, contains about 17 RGDfK per polymer. In endothelial cell binding studies of $\alpha_v\beta_3$ integrin-positive HUVEC and MDA-MB-231 cells, it was demonstrated that the copolymer with RGDfK does not possess any improved binding ability compared with free peptides, suggesting that steric hindrance clearly prevents multivalent interaction in the current system. The *in vivo* MR imaging was performed on MDA-MB-231 tumor-bearing SCID mice by using T_1 mapping, a quantitative MRI technique allowing assessment of changes in the relaxation time after administration of the contrast agent. Among the different time points measured (10 minutes, 30 minutes, and 1, 2, 6, and 24 hours post-injection), only at the 2 hours time point did the targeted conjugate exhibit a significant decrease in T_1 values in the tumor, compared with conjugate without targeting moieties. This enhancement can be successfully blocked by using targetable conjugate without Gd. Therefore, the current conjugate system has the potential to produce effective targetable MR agents for tumor imaging.

Another example of a targeted HPMA-based MRI contrast agent is HPMA-DOTA-Gd copolymer containing mannosamine and FITC on the side chain (HPMA-DOTA-Gd-ManN-FITC), which is prepared by copolymerization of HPMA, (5-[3-(methacryloylaminopropyl) thioureidyl] fluorescein (MA-AP-FITC), methacrryloylglycylglycyl mannosamine (MA-GG-ManN), and aminopropylmethacrylamide-benzyl-1,4,7,10-tetraazacyclododecane-1,4,7,10-tetraacetic acid (APMA-benzyl-DOTA) and subsequent chelation with gadolinium, which is designed for macrophage targeting [58]. Here, the targeting ligand for activated macrophages is mannosamine, since there is enhanced expression of mannose receptors in activated macrophages. FITC conjugated in copolymer acts as a fluorescent probe to evaluate the uptake of the various mannose-containing polymeric conjugates. In a model of THP-1 human monocytes, it was found that polymeric conjugates with a higher percentage of ligands, such as 4% and 8% (mole ratio), exhibited significantly higher uptake compared to conjugates with 2% targeting ligands or without ligands, which is consistent with the Gd uptake results.

8.5 OTHER POLYMERS

Several other synthetic polymers have generated much interest in the field of cancer imaging due to their favorable properties. These polymers are dendrimers, poly(ethyleneimine) [20], diblock copolymers of poly(ethylene oxide) and polybutadiene, bisamide copolymers [13], and polylactide, just to name a few.

Poly(ethyleneimine) modified with deoxycholic acid hydrosuccinimide ester and then conjugated with Cy5.5 and DEVD has been used in optical imaging to target apoptosis in live cells [20]. Polylactic-*co*-glycolic acid (PLGA) is the most commonly used and investigated material for ultrasound imaging [17, 55].

Dendrimers are another very interesting type of polymer because of their highly branched spherical structure made up of different types of chemical structures and functional groups. The application of dendrimers in cancer diagnosis and therapy has been extensively reviewed [11, 29, 49]. These dendrimers are a prime example of a synthetic polymer that has been used for a multimodal, multifunctional approach [49]. However, these dendrimers are non-biodegradable, which poses concerns that incomplete metal clearance because of the polymer conjugate's slow excretion may result in the presence of toxic metals from the chelates through metabolism and accumulation of the ions in the bone. To resolve these problems, biodegradable polymers, such as PG, are more favorable.

8.6 SUMMARY AND OUTLOOK

The role of any multimodal imaging approach in the detection, staging, and treatment planning of cancer is rapidly growing. Multimodality imaging involves the incorporation of two or more imaging modalities, using, for example, dual- or triple-labeled or "reporter" agents and/or homing ligands. This approach has been used to provide the exact localization, extent, and metabolic activity of the target tissue and to show functional changes within the surrounding tissues for staging and for cancer screening and detection. The complementary information gathered from the multimodal approach raises the exciting possibility of theranostic and personalized oncology.

Tables 8.1–8.4 summarize the major polymers used as carriers for cancer imaging. These polymers have been investigated both *in vitro* and *in vivo* and have shown promise because of their enhanced biodistribution and pharmacokinetics, ability to be functionalized with a wide variety of reporter groups and homing ligands, biocompatibility, and degradability. However, despite these advantages, there are still some challenges in translating these polymers into clinical use. To be clinically relevant, the high cost of preparation has to be reduced, the polydispersity of these polymers has to be carefully controlled, and quality control measures for the synthesized materials have to be devised. Nonetheless, the many beneficial attributes of these synthetic polymers described here make them very promising for use in clinical cancer imaging.

Table 8.1 Summary of polymer PG carriers for cancer imaging

Reporter	*Modality*	*Ligand*	*Physicochemical Properties*	*In vitro*	*In vivo*
DTPA-Gd	MRI	None	Mn = 101 kDa; %Gd = 12.3% (wt/wt); r_1 = 21 $mM^{-1}s^{-1}$, r_2 = 27 $mM^{-1}s^{-1}$ at 1.5T	Degradation by cathepsin B	Biodistribution, pharmacokinetics, and MR imaging in mice and monkeys. Enhanced vascular contrast for up to 2 hours was observed. PG-DTPA-Gd cleared faster in mice than in monkeys. $T_{1/2}$ (h): 50.1 in mice, 197.8–479.5 in monkeys; MRI showed accumulation of PG-DTPA-Gd in necrotic area in tumors [19, 20, 22].
DOTA-Gd	MRI	None	Mn = 28~87 kDa; %Gd = 30–55; r_1 = 9.2~10.6 $mM^{-1}s^{-1}$ at 3T	None	MRI revealed the size effect of the contrast agent on blood pharmacokinetics and distribution in mice. The contrast agents with higher molecular weights showed longer *in vivo* retention and more accumulation across tumor tissue [23].
		Mce_6	Mn = 94 kDa; %Gd = 0.62 mmol/g; Mce_6 = 0.078 mmol /g; r_1 = 8.46 $mM^{-1}s^{-1}$ at 3T	None	Tumors were clearly visualized by MRI. Photodynamic therapy showed significant therapeutic response to cancer treatment [24, 25]
		Cys	Mn = 50 kDa; %Gd = 0.5 mmol/g; r_1 = 2.5 $mM^{-1}s^{-1}$ at 1.5T	None	MRI showed contrast enhancement in heart and blood vessels for up to 1 hours. Gd chelate was released from the conjugate by endogenous thiols, excreted through renal filtration [26].
		RGD	Mn = 49 kDa; %Gd = 9.27% (w/w); RGD= 13% (mol/mol); r_1= 9.7 $mM^{-1}s^{-1}$ at 3T.	Specific binding was observed in DU145 (prostate) and SLK (Kaposi's sarcoma) cell lines	T_1 mapping showed a significant decrease of T_1 values at tumor periphery of DU145 xenografts, which had a high binding affinity to the RGD-bound contrast [27].
NIR813	Optical	None	Mn = 60 kDa; λ_{em}= 813 nm; fluorescence is quenched at %NIR loading = 8	Degradable by cathepsin B and D, and U87/TGL cells (high expression of cathepsin B and D)	Increased fluorescence in the brain of mice bearing U87/TGL human glioma when PG-DTPA-Gd was i.v. injected [31].
DTPA-Gd -NIR813	MRI and Optical	None	Mn = 101 kDa; %Gd = 12.3 (wt/wt); r_1=21 $mM^{-1}s^{-1}$ at 4.7T ; λ_{em} = 813 nm	None	Optical imaging showed drainage into the lymph nodes in mice injected subcutaneously. Metastatic and nonmetastatic lymph nodes can be differentiated in MRI when s.c. injected in mice bearing OSC-19 oral carcinoma [32].

Table 8.2 Summary of polymer PLL carriers for cancer imaging

Reporter	*Modality*	*Ligand*	*Physiochemical Properties*	*In vitro*	*In vivo*
DTPA-Gd	MRI	None	Mn = 48.7 kDa; Gd=21.8; T_1 = 13.1±0.36 /mmol^{-1}s^{-1}, at 20 MHz MRI	None	LD_{50}: 17 mmol Gd/KG in mice. $t_{1/2}\alpha$ (h): 0.134±0.002, $t_{1/2}$ β (h):1.90 ±0.102 [32].
DTPA-Gd-Cy5	MRI and Optical	None	Mw = 270 kDa; ~95% of amino groups attached with DTPA-Gd; λ_{em} = 665 nm; 1 dye per polymer	None	NIR fluorescence could be detected with a signal-to-background ratio of as high as 4.5 at 12 hours post-agent injection at a dye dose of 125 nmol/kg [46].
DOTA-Gd-Rhodamine	MRI and Optical	None	Mn = 5.6 kDa(PLL), Mw of bCD-PLL conjugates >300 kDa; r_1=8.6 mM^{-1}s^{-1} at 4.7T; 14 DOTA-Gd/ conjugate and 8 rhodamine /10 conjugates; hydrodynamic radius of 9.4 nm; λ_{em} = 590 nm	None	Both *in vivo* intratumoral T_1 values and *in vivo* optical imaging showed significant uptake of bCD-PLL conjugates in tumor was evident at 24 hours post-injection [42].

Table 8.3 Summary of polymer PLL-PEG carriers for cancer imaging

Reporter	*Modality*	*Ligand*	*Physiochemical Properties*	*In vitro*	*In vivo*
DTPA-Gd	MRI	None	Mn = 25kD; %Gd=5.5 (wt/wt); r_1 = 18 mM^{-1}s^{-1} at 1.5T	None	5k PEG-grafted-PL-DTPA, $t_{1/2}$ β (h):14.1, and MPEG apparently decrease uptake in kidney [37].
Cy5.5	Optical	None	92MPEG-PLL$_{147-}$11Cy5.5; λ_{ex}= 673 nm; λ_{em}= 689 nm	NIR signals resulted from lysosomal cysteine/serine protease.	*In vivo* imaging exhibited a 12-fold increase in NIRF signals, allowing for the detection of tumors with submillimeter-sized diameters [40].
DTPA-^{99}Tc	Nuclear	None	89MPEG$_{114}$-PLL$_{279-}$190DTPA,	None	Approximately 25% was excreted during 24 hours post-injection. In rat and rabbit studies, no signs of acute or sub-acute toxicity at doses 280 times the proposed dose for humans. Four volunteers (n = 6) exhibited typical blood pool biodistribution [38].

Table 8.4 Summary of polymer HPMA carriers for cancer imaging

Reporter	*Modality*	*Ligand*	*Physiochemical Properties*	*In vitro*	*In vivo*
DOTA-Gd	MRI	1. None	1. Mw of PLL conjugates: 28, 60, 121 kDa; %Gd = 7 (wt/wt); r_1 =10.7–11.7 mM^{-1}s^{-1} at 3.0T.	None	1. The one with Mw of 121 kDa exhibited the longest circulation life in blood and more significant and prolonged MRI signal in tumor, even at 24 hours post-injection [49].
		2. RGDfK	2. Mw 43kDa; 17 RGDfK per polymer, r_1= 20.6 mM^{-1}s^{-1} at 1.5 T.		2. Only at 2 hours time point did the targeted conjugate exhibit significant decrease in T_1 value in tumor [52].

Acknowledgments

The authors thank Dawn Chalaire for editing the manuscript. This work is supported in part by the National Institutes of Health (grants R01 CA119387), a Seed Grant through the Alliance for NanoHealth by the Department of Army Telemedicine and Advanced Technology Research Center (W81XWH-07-2-0101), the John S. Dunn Foundation, and Odyssey Fellowship (to M.P.M.). Funding as an Odyssey Fellow is supported by the Odyssey Program and The Cockrell Foundation Award for Scientific Achievement at the University of Texas M. D. Anderson Cancer Center.

References

1. Abuchowski A., Mccoy J. R., Palczuk N. C., Vanes T., and Davis F. F. (1977) Effect of covalent attachment of polyethylene-glycol on immunogenicity and circulating life of bovine liver catalase. *J Biol Chem* **252**, 3582–3586.
2. Andersonberg W. T., Strand M., Lempert T. E., Rosenbaum A. E., and Joseph P. M. (1986) Nuclear-magnetic-resonance and gamma-camera tumor imaging using gadolinium-labeled monoclonal-antibodies. *J Nucl Med* **27**, 829–833.
3. Bagshawe K. D., Sharma S. K., Burke P. J., Melton R. G., and Knox R. J. (1999) Developments with targeted enzymes in cancer therapy. *Curr Opin Immunol* **11**, 579–583.
4. Bogdanov A. A., Callahan R. J., Wilkinson R. A., Martin C., Cameron J. A., Fischman A. J., Brady T. J., and Weissleder R. (1994) Synthetic copolymer kit for radionuclide blood-pool imaging. *J Nucl Med* **35**, 1880–1886.
5. Bogdanov A. A., Lewin M., and Weissleder R. (1999) Approaches and agents for imaging the vascular system. *Adv Drug Delivery Rev* **37**, 279–293.
6. Bogdanov A. A., Weissleder R., and Brady T. J. (1995) Long-circulating blood pool imaging agents. *Adv Drug Delivery Rev* **16**, 335–348.
7. Bogdanov A. A., Weissleder R., Frank H. W., Bogdanova A. V., Nossif N., Schaffer B. K., Tsai E., Papisov M. I., and Brady T. J. (1993) A new macromolecule as a contrast agent for MR-angiography: preparation, properties, and animal studies. *Radiology* **187**, 701–706.
8. Brasch R., Pham C., Shames D., Roberts T., Van Dijke K., Van Bruggen N., Mann J., Ostrowitzki S., and Melnyk O. (1997) Assessing tumor angiogenesis using macromolecular MR imaging contrast media. *J Magn Reson Imaging* **7**, 68–74.
9. Callahan R. J., Bogdanov A., Fischman A. J., Brady T. J., and Weissleder R. (1998) Preclinical evaluation and phase I clinical trial of a Tc-99m-labeled synthetic polymer used in blood pool imaging. *Am J Roentgenol* **171**, 137–143.
10. Chipman S., Rosler R., and Bonham L. (2006) Energy dependent uptake of paclitaxel poliglumex by human NSCLC tumor and murine macrophage-like cell lines. 18th EORTC-NIC-AACR Symposium on Molecular Targets and Cancer Therapeutics Prague, Czech Republic.
11. Cloninger M. J. (2002) Biological applications of dendrimers. *Curr Opin Chem Biol* **6**, 742–748.

12. Franano F. N., Edwards W. B., Welch M. J., Brechbiel M. W., Gansow O. A., and Duncan J. R. (1995) Biodistribution and metabolism of targeted and nontargeted protein-chelate-gadolinium complexes: evidence for gadolinium dissociation *in vitro* and *in vivo*. *Magn Reson Imaging* **13**, 201–214.

13. Ghoroghchian P. P., Frail P. R., Susumu K., Blessington D., Brannan A. K., Bates F. S., Chance B., Hammer D. A., and Therien M. J. (2005) Near-infrared-emissive polymersomes: self-assembled soft matter for *in vivo* optical imaging. *Proc Natl Acad Sci U S A* **102**, 2922–2927.

14. Haag R., and Kratz F. (2006) Polymer therapeutics: concepts and applications. *Angew Chem, Int Ed* **45**, 1198–1215.

15. Chiu H.-C., Kopečková P., Deshmane S. S., Kopeček J. (1997) Lysosomal degradability of poly(α-amino acids). *J Biomed Mater Res* **34**, 381–392.

16. Jackson E. F., Esparza-Coss E., Wen X., Ng C. S., Daniel S. L., Price R. E., Rivera B., Charnsangavej C., Gelovani J. G., and Li C. (2007) Magnetic resonance imaging of therapy-induced necrosis using gadolinium-chelated polyglutamic acids. *Int J Radiat Oncol Biol Phys* **68**, 830–838.

17. Jain R. A. (2000) The manufacturing techniques of various drug loaded biodegradable poly(lactide-*co*-glycolide) (PLGA) devices. *Biomaterials* **21**, 2475–2490.

18. Ke T., Jeong E. K., Wang X., Feng Y., Parker D. L., and Lu Z. R. (2007) RGD targeted poly(L-glutamic acid)-cystamine-(Gd-DO3A) conjugate for detecting angiogenesis biomarker alpha(v) beta3 integrin with MRT, mapping. *Int J Nanomed* **2**, 191–199.

19. Kiessling F., Morgenstern B., and Zhang C. (2007) Contrast agents and applications to assess tumor angiogenesis *in vivo* by magnetic resonance imaging. *Curr Med Chem* **14**, 77–91.

20. Kim K., Lee M., Park H., Kim J.-H., Kim S., Chung H., Choi K., Kim I.-S., Seong B. L., and Kwon I. C. (2006) Cell-permeable and biocompatible polymeric nanoparticles for apoptosis imaging. *J A Chem Soc* **128**, 3490–3491.

21. Kopecek J., Kopeckova P., Minko T., and Lu Z. R. (2000) HPMA copolymer-anticancer drug conjugates: design, activity, and mechanism of action. *Eur J Pharm Biopharm* **50**, 61–81.

22. Kroft L. J., and de Roos A. (1999) Blood pool contrast agents for cardiovascular MR imaging. *J Magn Reson Imaging* **10**, 395–403.

23. Lacowicz J. (1991) *Topics in Fluorescence Spectroscopy: Priciples*. Plenum Press, New York.

24. Li C., Penet M. F., Winnard P., Artemov D., and Bhujwalla Z. M. (2008) Image-guided enzyme/prodrug cancer therapy. *Clin Cancer Res* **14**, 515–522.

25. Li C., Price J. E., Milas L., Hunter N. R., Ke S., Yu D. F., Charnsangavej C., and Wallace S. (1999) Antitumor activity of poly(L-glutamic acid)-paclitaxel on syngeneic and xenografted tumors. *Clin Cancer Res* **5**, 891–897.

26. Li C., and Wallace S. (2008) Polymer-drug conjugates: recent development in clinical oncology. *Adv Drug Delivery Rev* **60**, 886–898.

27. Li C., Winnard P. T., Takagi T., Artemov D., and Bhujwalla Z. M. (2006) Multimodal image-guided enzyme/prodrug cancer therapy. *J Am Chem Soc* **128**, 15072–15073.

28. Li C., Yu D.-F., Newman R. A., Cabral F., Stephens L. C., Hunter N., Milas L., and Wallace S. (1998) Complete regression of well-established tumors using a novel water-soluble poly(L-glutamic acid)-paclitaxel conjugate. *Cancer Res* **58**, 2404–2409.

29. Longmire M., Choyke P. L., and Kobayashi H. (2008) Dendrimer-based contrast agents for molecular imaging. *Curr Top Med Chem* **8**, 1180–1186.

30. Lu Z. R., Wang X., Parker D. L., Goodrich K. C., and Buswell H. R. (2003) Poly(L-glutamic acid) Gd(III)-DOTA conjugate with a degradable spacer for magnetic resonance imaging. *Bioconjug Chem* **14**, 715–719.

31. Maeda H., Seymour L. W., and Miyamoto Y. (1992) Conjugates of anticancer agents and polymers: advantages of macromolecular therapeutics *in vivo*. *Bioconjug Chem* **3**, 351–362.

32. Marchal G., Bosmans H., Vanhecke P., Speck U., Aerts P., Vanhoenacker P., and Baert A. L. (1990) MR angiography with gadopentetate dimeglumine polylysine: evaluation in rabbits. *Am J Roentgenol* **155**, 407–411.

33. Melancon M. P., Wang W., Wang Y., Shao R., Ji X., Gelovani J. G., and Li C. (2007) A novel method for imaging *in vivo* degradation of poly(L-glutamic acid), a biodegradable drug carrier. *Pharm Res* **24**, 1217–1224.

34. Melancon M. P., Wang Y., Wen X., Bankson J. A., Stephens L. C., Jasser S., Gelovani J. G., Myers J. N., and Li C. (2007) Development of a macromolecular dual-modality MR-optical imaging for sentinel lymph node mapping. *Invest Radiol* **42**, 569–578.

35. Mohs A. M., and Lu Z. R. (2007) Gadolinium(III)-based blood-pool contrast agents for magnetic resonance imaging: status and clinical potential. *Expert Opin Drug Delivery* **4**, 149–164.

36. NiculescuDuvaz I., and Springer C. J. (1997) Antibody-directed enzyme prodrug therapy (ADEPT): a review. *Adv Drug Delivery Rev* **26**, 151–172.

37. Ogan M. D., Schmiedl U., Moseley M. E., Grodd W., Paajanen H., and Brasch R. C. (1987) Albumin labeled with Gd-DTPA—an intravascular contrast-enhancing agent for magnetic-resonance blood pool imaging: preparation and characterization. *Invest Radiol* **22**, 665–671.

38. Padhani A. R. (2002) Dynamic contrast-enhanced MRI in clinical oncology: current status and future directions. *J Magn Reson Imaging* **16**, 407–422.

39. Pham C. D., Roberts T. P. L., van Bruggen N., Melnyk O., Mann J., Ferrara N., Cohen R. L., and Brasch R. C. (1998) Magnetic resonance imaging detects suppression of tumor vascular permeability after administration of antibody to vascular endothelial growth factor. *Cancer Invest* **16**, 225–230.

40. Putnam D., and Kopecek J. (1995) polymer conjugates with anticancer activity. *Biopolymers II*, vol. 122. Springer-Verlag, Berlin, pp. 55–123.

41. Rebizak R., Schaefer M., and Dellacherie E. (1998) Polymeric conjugates of Gd(3+)-diethylenetriaminepentaacetic acid and dextran. 2. Influence of spacer arm length

and conjugate molecular mass on the paramagnetic properties and some biological parameters. *Bioconjug Chem* **9**, 94–99.

42. Schuhmann-Giampieri G., Schmitt-Willich H., Frenzel T., Press W., and Weinmann H. (1991) *In vivo* and *in vitro* evaluation of gadolinium-DTPA-polylysine as a macromolecular contrast agent for magnetic resonance imaging. *Invest Radiol* **26**, 969–974.

43. Sullivan D. C., and Ferrari M. (2004) Nanotechnology and tumor imaging: seizing an opportunity. *Mol Imaging* **3**, 364–369.

44. Tung C. H., Mahmood U., Bredow S., and Weissleder R. (2000) *In vivo* imaging of proteolytic enzyme activity using a novel molecular reporter. *Cancer Res* **60**, 4953–4958.

45. Uzgiris E. (2004) The role of molecular conformation on tumor uptake of polymeric contrast agents. *Invest Radiol* **39**, 131–137.

46. Uzgiris E. E., Sood A, Bove K., Grimmond B, Lee, D, and Lomnes, S. (2006) A multimodal contrast agent for preoperative MR iamging and intraoperative tumor margin delineation. *Technol Cancer Res Treat* **5**, 301–309.

47. Vaidya A., Sun Y., Feng Y., Emerson L., Jeong E. K., and Lu Z. R. (2008) Contrast-enhanced MRI-guided photodynamic cancer therapy with a pegylated bifunctional polymer conjugate. *Pharm Res* **25**, 2002–2011.

48. Vaidya A., Sun Y., Ke T., Jeong E. K., and Lu Z. R. (2006) Contrast enhanced MRI-guided photodynamic therapy for site-specific cancer treatment. *Magn Reson Med* **56**, 761–767.

49. Venditto V. J., Regino C. A. S., and Brechbiel M. W. (2005) PAMAM Dendrimer based macromolecules as improved contrast agents. *Mol Pharm* **2**, 302–311.

50. Vicent M. J., Duncan R. (2006) Polymer conjugates: nanosized medicines for treating cancer. *Trends Biotechnol* **24**, 39–47.

51. Wang Y. L., Ye F. R., Jeong E. K., Sun Y., Parker D. L., and Lu Z. R. (2007) Noninvasive visualization of pharmacokinetics, biodistribution and tumor targeting of poly[*N*-(2-hydroxypropyl)methacrylamide] in mice using contrast enhanced MRI. *Pharm Res* **24**, 1208–1216.

52. Weissleder R., Tung C. H., Mahmood U., and Bogdanov A. (1999) *In vivo* imaging of tumors with protease-activated near-infrared fluorescent probes. *Nat Biotechnol* **17**, 375–378.

53. Wen X., Jackson E. F., Price R. E., Kim E. E., Wu Q., Wallace S., Charnsangavej C., Gelovani J. G., and Li C. (2004) Synthesis and characterization of poly(L-glutamic acid) gadolinium chelate: a new biodegradable MRI contrast agent. *Bioconjug Chem* **15**, 1408–1415.

54. Wen X., Jackson E. F., Uthamanthil R., Ng C. S., Gelovani J. G., Tian M., and Li C. (2009) Pharmacokinetics and MR imaging of rhesus monkeys with PG-Gd, a blood pool imaging agent. World Molecular Imaging Conference Montreal, Canada.

55. Wheatley M. A., Forsberg F., Oum K., Ro R., and El-Sherif D. (2006) Comparison of *in vitro* and *in vivo* acoustic response of a novel 50:50 PLGA contrast agent. *Ultrasonics* **44**, 360–367.

56. Ye F., Ke T., Jeong E. K., Wang X., Sun Y., Johnson M., and Lu Z. R. (2006) Noninvasive visualization of *in vivo* drug delivery of poly(L-glutamic acid) using contrast-enhanced MRI. *Mol Pharm* **3**, 507–515.

57. Zarabi B., Borgman M. P., Zhuo J. C., Gullapalli R., and Ghandehari H. (2009) Noninvasive Monitoring of HPMA Copolymer-RGDfK Conjugates by Magnetic Resonance Imaging. *Pharm Res* **26**, 1121–1129.

58. Zarabi B., Nan A., Zhuo J., Gullapalli R., and Ghandehari H. (2006) Macrophage targeted *N*-(2-hydroxypropyl)methacrylamide conjugates for magnetic resonance imaging. *Mol Pharm* **3**, 550–557.

59. Zarabi B., Nan A. J., Zhuo J. C., Gullapalli R., and Ghandehari H. (2008) HPMA copolymer-doxorubicin-gadolinium conjugates: Synthesis, characterization, and *in vitro* evaluation. *Macromol Biosci* **8**, 741–748.

60. Zhang Z. D., Nair S. A., and McMurry T. J. (2005) Gadolinium meets medicinal chemistry: MRI contrast agent development. *Curr Med Chem* **12**, 751–778.

Chapter 9

PLGA-BASED OPTICAL IMAGING IN BREAST CANCER

Hareesh B. Nair[a] and Rajeshwar R. Tekmal[b]
[a] *Southwest National Primate Research Center, Texas Biomedical Research Institute, San Antonio TX 78227, USA*
[b] *Department of Obstetrics and Gynecology, University of Texas Health Science Center at San Antonio, San Antonio TX 78229, USA*
hnair@TxBiomedGenetics.org

9.1 INTRODUCTION

Cancer is multifactorial in origin with diverse pathologies, and its treatment is highly challenging. Since cancer cells have tremendously greater replication potential than nontumoral cells, the majority of anticancer drugs have been developed to reduce cell proliferation. In recent years, more selective targeted approaches have been developed by using cancer cell–specific ligands and receptors. Early-generation nanoparticles utilized the smaller size (20–400 nm) of particles to take advantage of increased vascular permeability of tumor blood vessels and impaired lymphatic drainage (passive targeting). More recently, researchers have investigated targeted approaches with the goal of increasing treatment efficacy and decreasing side effects in patients.

Polymer nanoparticles have received enormous attention in modern medicine for their ability to deliver an optimum dosage of drugs, minimize side effects of systemic delivery, and enhance therapeutic efficacy [11, 16, 31]. The efficacy of current cancer therapeutics is limited by the high incidence of toxicity, reduced half-life of the active compound, lack of specificity, and acquired drug resistance [2, 17, 21]. It has been well established that the clinical success of a drug correlates with pharmacological and toxicological parameters, selectivity of a targeted population of cells, circulation time in the blood stream, and maximum tolerated dose (MTD). These factors together

Nanoimaging
Edited by Beth Goins and William Phillips

www.panstanford.com

determine the therapeutic index of the drug of formulation. Many polymers have been investigated for research and clinical use including polylactic-co-glycolic acid (PLGA), dextran, *N*-(2-hydroxypropyl)methacrylamide (HPMA), polyglutamate, and polyethylene glycol (PEG).

Several characteristics of polymeric nanoparticles are needed for effective cancer therapy. One of the key features of anticancer drug-loaded nanoparticles is their smaller size. The optimum size for therapeutic nanoparticles is considered to be 10– 100 nm [12]. Since scientific evidence suggests that the enhanced permeability and retention effect (EPR) is present in humans, these smaller nanoparticles can leak out of the blood stream and accumulate in tumors. However, evidence also suggests that large macroparticles (200–700 nm) could have limited diffusion in the extracellular space [1].

9.2 TARGETING: PASSIVE AND ACTIVE

Nanoparticles with longer circulation times are suitable for passive targeting. Passive targeting takes advantages of the inherent size of nanoparticles and the unique properties of tumor vasculature to enhance the efficacy of drugs. The leaky fenestrations in inflamed tumor blood vessels allow this method of targeting. Pegylation (surface coating of nanoparticles by PEG) has been shown to increase the circulation time and reduce opsonization issues. Opsonization by complement proteins, vitronectin, fibronectin, immunoglobulins, and lipoproteins with nanoparticles as foreign objects remains the main challenge for intravenously injected nanoparticles. The opsonized nanoparticles are rapidly cleared by the reticuloendothelial system (RES) of liver and spleen or excretion by kidneys [3]. The bigger the size of the particles, the higher the risk of opsonization. In addition, hydrophobic nanoparticles have a greater chance to be removed from the circulation than nanoparticles with hydrophilic surface. Surface charge, charge density, lipophilic/hydrophilic ratio, and presence of functional groups are key physicochemical properties that determine the successful drug delivery of polymeric nanoparticles. In a recent study using doxorubicin, researchers designed biodegradable polymeric (PLGA) nanoparticles for effective oral delivery [18]. Even though these nanoparticles are passively targeted, oral chemotherapy is a great challenge for doxorubicin, and the availability of an oral formulation of doxorubicin could revolutionize cancer chemotherapy.

Although doxorubicin is a widely accepted anthracyclin antibiotic for cancer chemotherapy, its effects are limited because of dose-dependent cardiotoxicity and myelosuppression [18]. These doxorubicin nanoparticles were prepared by the double emulsion method using PLGA in ethyl acetate and 3% polyvinyl alcohol (PVA) as surfactant. In animal studies these nanoparticles showed promising oral bioavailability of doxorubicin and reduced cardiotoxicity. There are several other types of polymeric passively targeted nanoparticles that have been tested in animals models, many of which are listed in Tables 9.1 and 9.2 and are under clinical development [32].

Table 9.1 Common biocompatible polymers

Agents	*Action*
PEG (poly ethylene glycol)	Increases circulation and time, decreases uptake by macrophages
Chitosan	Increases hydrophilicity, immunogenicity
PLGA (poly lactic-co-glycolide)	Stability, biocompatibility, decreases uptake by RES
PVA (poly vinyl alcohol)	Decreases aggregation
PLL (poly-L-lactide)	Stability, decreases uptake by RES
Dextran	Increases circulation time
PEO-PCL (Poly(ethylene oxide)-modified poly(ε-caprolactone))	Stealth nanoparticles
P(NIPAAm-co-AA-co-CHA)	pH-responsive nanoparticles

Table 9.2 Polymeric nanoparticles in clinical trials [32]

Composition/trade name	*Trial status*	*Treatment*
Dextran-Camptothecin	Phase I/II	Cancer, head and neck tumor
PEG-aspartic acid doxorubicin (NK911)	Phase I	Cancer
Polyglutamate-paclitaxel (CT-2103, XYOTAX™)	Phase II/III	Breast, ovarian
HPMA copolymer-doxorubicin (PK1; FCE28068)	Phase II	Lung and breast cancer
HPMA copolymer palatinate (AP5280)	Phase II	Cancer
HPMA copolymer-doxorubicin-galactosamine (PK2; FCE28069)	Phase I/II	Hepatocellular carcinoma
Polyglutamate-camptothecin (CT-2106)	Phase I	Cancer
PEG-camptothecin (PROTHECAN™)	Phase II	Cancer
Pluronic doxorubicin (SP1049C)	Phase II	Cancer
HPMA-DACH-platinate (ProLindac)	Phase II	Cancer

A wide array of nanoparticle therapeutics that have been approved for clinical use are PEGylated proteins. PEGylation has been applied to various proteins, including enzymes, antibodies, affibodies, and cytokines. PEGylation can offer increased protein stability, prevent rapid kidney clearance, and provide enhanced plasma life [20]. Oncospar by Enzon Pharma, a PEG-L-asparaginase, was approved by FDA in 1994 to treat acute lymphoblastic leukemia. Oncospar showed an increased plasma half-life and reduced hypersensitivity reaction induced by the naked enzyme in phase I studies [8]. PEGylated cytokines have shown superior activity in clinical trials of melanoma and renal cell carcinoma. For example, Schering's "Pegintron" for interferon alpha-2b and Roche's "Pegasys" have showed superior antiviral activity and have been approved for treating hepatitis C, whereas various PEGylated chemotherapeutics, including PEG-irinotecan, PEG-SN-38, and PEG-paclitaxel, are currently in clinical trials [30]. Nanoparticles with targeting ligands have become more attractive because they are envisioned to provide more efficacious therapy. In this chapter we will focus on the specific ligand poly(lactic-co-glycolic acid) (PLGA) nanoparticles we have developed in our laboratory for targeting breast cancer using *in vitro* and *in vivo* models. PLGA has been used for a wide variety of medical applications, from resorbable sutures to bone screws and microspheres for drug delivery. This biodegradable and biocompatible polymer is safe to use in the body and is hydrolyzed into metabolic by-products. Recently, these polyester materials have been widely used to deliver peptide or protein drugs, which have unusual physicochemical properties compared with low-molecular-weight drugs. Degradation of PLGA *in vitro* as well as *in vivo* mainly takes place through hydrolysis of the ester linkages and/or enzymatic degradation. Degradation behavior of polymeric microspheres is crucial for the successful application of such devices in controlled drug delivery.

Attractive breast cancer therapies for the targeting of PLGA nanoparticles can be based on estrogen/estrogen receptor–mediated actions. Estrogen/estrogen receptor–mediated actions play an important role in the normal development and pathophysiology of the breast. Approximately 75% of breast cancer patients respond well to endocrine therapies at the time of diagnosis. Even though systemic hormone therapies that either block local estrogen production by aromatase inhibitors (AIs) or block actions of estrogen/estrogen receptor (ER) by antiestrogens (AEs) is well tolerated, the development of resistance to these agents is still a major concern. The transition of an ER-positive hormone responsive state to an unresponsive state following AE or AI therapy is associated with molecular adaptation that involves loss of ER (ERα), activation of HER2/Raf/MAPK signaling pathways, epigenetic modifications, and alterations in the levels and functions of ER coregulators. Since multiple signaling pathways are involved in ER activation, combination of endocrine and non-endocrine agents that block these signaling pathways may delay the onset of resistance to hormonal therapy.

The current hormone therapies involve blocking of estrogen from binding to the ER using AEs or selective estrogen receptor modulators (SERMS) such as tamoxifen or by inhibiting estrogen synthesis using AIs. Tamoxifen was a widely accepted therapy, until its estrogen-like activity was revealed to significantly increase the risk of endometrial cancer and stroke [14]. To some extent it is still used despite these actions, and the superiority of AIs over tamoxifen has reduced its use. Letrozole is a third-generation AI found to be more active than tamoxifen in neoadjuvant, early- and extended-adjuvant, as well as metastatic disease settings of postmenopausal breast cancer patients [23]. Letrozole is a superior therapy for treating breast cancer; however, some patients may develop acquired resistance during the treatment period. To increase the therapeutic efficiency and reduce side effects, much research has been devoted to the development of a biodegradable delivery system capable of increasing the localized dose of drug to the tumor site. Nanoparticles composed of biodegradable polymers have attracted tremendous interest in recent years for clinical administration of anticancer drugs.

A major limitation in cancer chemotherapy is the development of acquired drug resistance. Chemoresistance is mainly generated in two ways: (i) physical impairment of drug into the tumor due to poor absorption or increased metabolism and (ii) development of multidrug resistance (MDR) [15]. On the basis of our preliminary results, resistance to AIs mostly lies in the second category. Development of MDR is multifactorial in which the overexpression of membrane bound ATP-dependent drug effect pump (*p*-glycoprotein/MDR1) is predominant [7]. By using an established model of letrozole resistance involving LTLT-Ca cells, we have observed in our study that site-specific delivery of letrozole by PLGA nanoparticles can increase the therapeutic benefit by delivering a greater fraction of the dose to the target site, which minimizes resistance (unpublished data). Drug delivery throughout the tumor burden is crucial for the treatment to be effective, since residual cancer cell survival can promote regrowth and often imply the cause for drug resistance [28]. Brodie *et al.* have shown that in LTLT-Ca letrozole-resistant cells several other adaptive signaling pathways, including HER2 and MAPK, are circuited together to execute tumor regrowth [25]. Receptor-mediated endocytic delivery is one of the more widely accepted approaches for targeted drug delivery. CD44, the cell surface glycoprotein, is a potential molecular target for tumor-selective drug delivery. CD44 is reported to be involved in cell-to-cell interactions, adhesion, and migration. CD44 is highly expressed in a number of epithelial carcinomas, but is present in only a limited number of normal cells [9]. CD44 is the receptor for hyaluronic acid (HA), a unique glycosaminoglycan that is present in extracellular matrices and contributes to tissue hydrodynamics and repair [29]. Targeting anticancer drugs using HA nanoparticles may facilitate the selective cellular uptake by tumor cells through CD44 receptor–mediated endocytosis. In this chapter we are presenting some of our recent research for answering whether HA- bound letrozole nanoparticles would be an alternative to overcome resistance to letrozole in a preclinical mouse model of letrozole-resistant tumors.

9.3 HA-LETROZOLE NANOPARTICLES TARGET BREAST CANCER CELLS THROUGH HA-CD44 BINDING

Figure 9.1 depicts the expression of CD44 on several breast cancer cell lines.

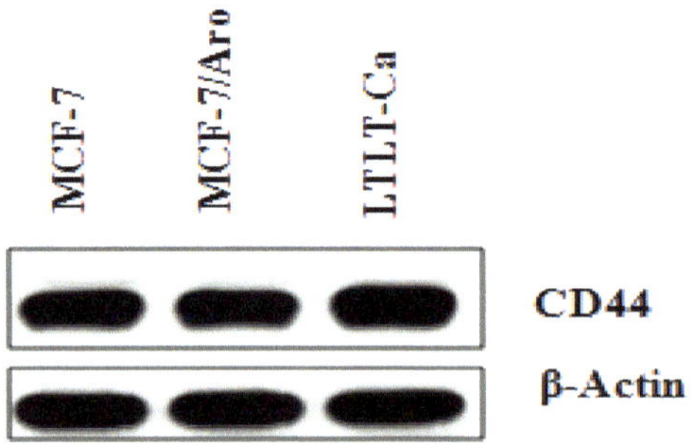

Figure 9.1 Expression of CD44 in different breast cancer cells.

Figures 9.2 and 9.3 depict size characteristics for HA-letrozole (HA-Letr) nanoparticles. MCF-7/Aro (MCF-7 breast cancer cells overexpressing aromatase gene) and LTLT-Ca (letrozole-resistant breast cancer cells) cells treated with HA-Letr nanoparticles and letrozole have been tested for cytotoxicity. The MCF-7/Aro cells treated with HA- Letr nanoparticles showed an increased IC_{50} when compared to letrozole alone. At the same time the 1 μM concentration of intact drug letrozole showed only 32% cell death in MCF-7/Aro cells in 24 hours. The letrozole-resistant LTLT-Ca cells showed an IC_{50} at 2 μM letrozole in HA-Letr nanoparticle formulation compared with that of 5 μM of intact letrozole, suggesting that acquired resistance enables the LTLT-Ca cells to overcome the dose of 2 μM letrozole. The HA-bound nanoparticles induced 55% cell death in LTLT-Ca cells compared with 17% cell death in letrozole-treated group. It is important to note that the LTLT cells were grown in 1 μM letrozole in the medium throughout the maintenance of the cells and during the experiments. The increased cytotoxicity of HA-Letr nanoparticles is directly correlated to the presence of HA on the nanoparticles, which possess drug cargo within. The hyaluronic acid present on the surface of the PLGA nanoparticles acts as an active targeting ligand for CD44 membrane cell surface receptor molecules and is preferentially expressed more on malignant cells. The HA-Letr nanoparticles treated with fluorescein isothiocynate (FITC)-labeled CD44mAb (Thermo Scientific, Fremont, CA) showed a sudden shift in fluorescence compared with letrozole nanoparticles without HA. This confirms the affinity of HA-CD44 binding (Fig 9.4). Also, the presence of viable cells after prior treatment with CD44mAb shows that removal or blocking of the receptor binding site on the cell surface may inhibit the receptor binding capability of HA. This in turn affected the binding of HA-Letr nanoparticles to the cell surface and prevented effective cell proliferation as measured by light microscopy. These results suggest that active targeting of letrozole using HA-Letr nanoparticles will require the presence of receptive CD44 cell surface receptors. Multifunctional NPs can also be used for small-interference RNA (siRNA) delivery and imaging. RNA interference (RNAi) has emerged as a powerful technology for sequence-specific silencing of genes and holds great potential as a new treatment paradigm for human diseases amenable to manipulation at the gene expression level [4, 10, 13, 24].

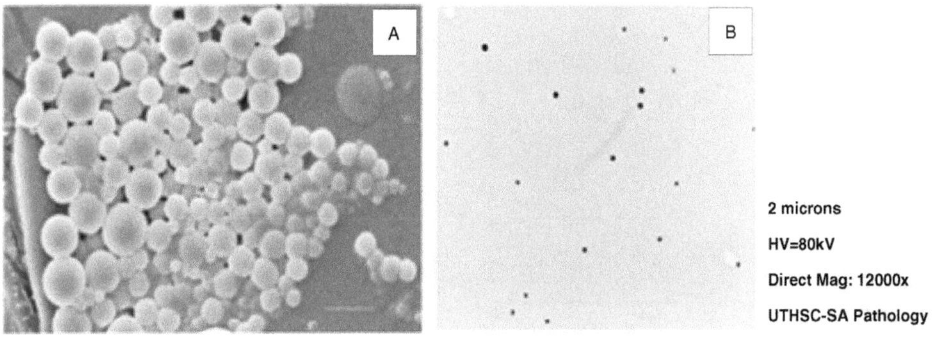

Figure 9.2 (A) SEM image of hyaluronic acid–coated letrozole nanoparticles. (B) TEM image shows that the size of the nanoparticles is below 100 nm.

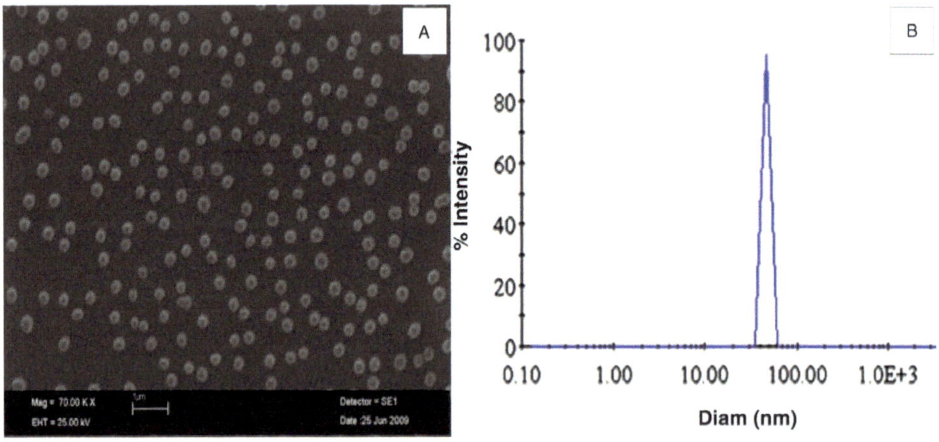

Figure 9.3 (A) Letrozole–hyaluronic acid nanoparticles were resized using membrane extruder (100 nm filter). (B) The mean diameter of particles lies below 100 nm in photon correlation spectroscopy.

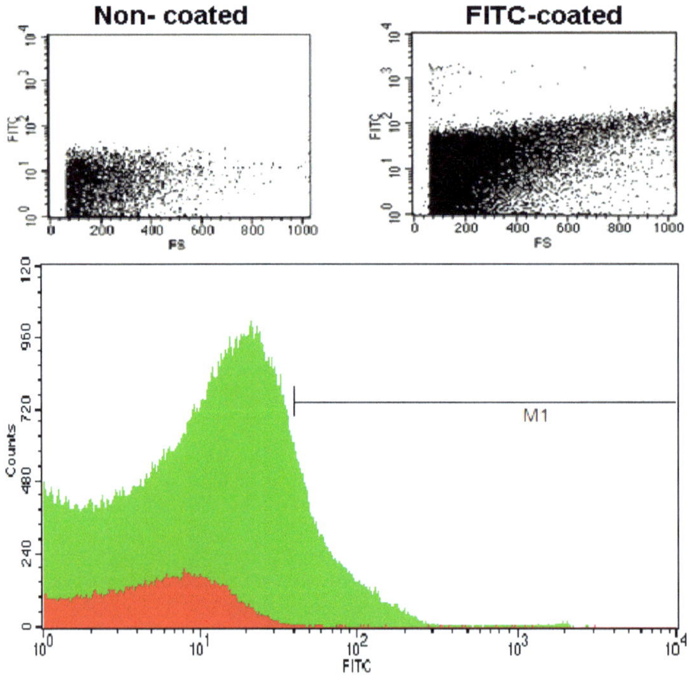

Figure 9.4 PLGA/PEG-FITC conjugation (DCC coupling) is stable: MCF-7 cells were co-incubated with control/void nanoparticles or PLGA/PEG-FITC nanoparticles for 1 hour and washed three times in PBS, and fluorescence was monitored by FACS analysis. A sudden fluorescent shift was noticed in FITC-coated nanoparticles.

9.4 *IN VIVO* ANTITUMOR ACTIVITY OF HA-LETR NANOPARTICLES IN LTLT-Ca TUMOR XENOGRAFTS

The tumor volume in mice treated with HA-Letr nanoparticles was found to be reduced to 312 mm^3 after 11 weeks compared with nontreated control (1,314 mm^3) and letrozole (400 mm^3) treated groups (Fig 9.5). The tumor volume was synchronized to ~100 mm^3 in all groups before administration of the drug. However, in the untreated control group and in empty nanoparticle–treated groups the tumor volume remains unaffected (1,314 mm^3) after 11 weeks as a result of overgrowth of tumors as per UTHSCA-IACUC regulations.

There was no significant difference in tumor volume between mice treated with letrozole or letrozole nanoparticles (Letr NPs) until the 12th week post therapy. A marked difference in tumor volume was observed between these groups after 12th week until 18th week. On the other hand, the animals treated with HA-Letr NPs showed an antitumor effect superior to Letr NPs. HA-Letr NPs expressed distinguishable activity from week 6 onward compared with all other groups of animals that undergo treatment. Even though HA-Letr NPs showed a better effect up to 10 weeks from the day of treatment, the tumors started growing back in both groups (HA-Letr NP and Letr NP), but at a very slow rate compared with untreated control animals. The mice treated with HA-empty nanoparticles also did not show any regression in tumor volume as expected. The efficacy was further evaluated in terms of body weight of the animals along with the treatment. HA-Letr NP– and Letr NP–treated mice didn't show any significant difference in body weight; however, the group receiving letrozole alone showed significant increase in body weight when compared with HA-Letr NP–treated group (data not shown). These results clearly show that encapsulation of letrozole makes a marked difference in treating letrozole-resistant tumors and even HA coating offers a better effect in therapeutic efficacy and delays the development of resistance. The inevitable acquired development of resistance to AIs is a potential problem in breast cancer treatment [27]. It is important to understand the mechanism of resistance to develop alternate strategies to circumvent or to delay the onset of resistance so as to take maximum advantage of these efficacious drugs. Moreover, the development of a successful drug to replace the current AIs might take many years to launch from bench to bedside. In this context, improvement of existing drugs has a great impact in clinics.

In this study we have shown for the first time that the potent aromatase inhibitor letrozole can be protected (from systemic release) against the onset of drug resistance in an established *in vivo* mouse model of letrozole resistance. This was achieved by coating HA on the nanoparticle surface by coprecipitation. The encapsulation of letrozole was done using the biodegradable, FDA-approved polymer poly (lactic-co-glycolic acid) (PLGA) conjugated with polyethylene glycol (PEG). The custom PEGylation of PLGA offers extended circulation time for the nanoparticles as well as reduces the potential risk of nanoparticle agglomeration and nonspecific bio-fouling, as found previously by Cheng *et al.* [5].

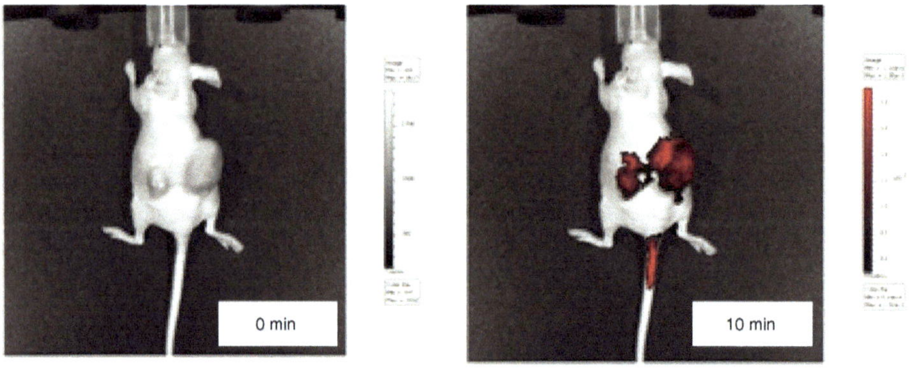

Figure 9.5 *In vivo* targeting efficacy of hyaluronic acid–coated letrozole nanoparticles to the tumor site is achieved by PLGA-PEG-HA NPs labeled with FITC. The image was taken 10 minutes after injection (i.v.) through tail vein using IVIS xenogen optical imaging station.

It has been well established that long-term letrozole-treated tumors (LTLT-Ca) overexpress HER-2 and activate the MAPK signaling pathway leading to resistance of these tumors to antiestrogens, aromatase inhibitors, androstenedione, and estradiol [25]. Interestingly, in our study, the group of mice treated with HA-Letr NPs show promising activity in reducing the tumor burden in the first 11 weeks when compared with letrozole-treated animals and untreated controls (Fig. 9.6).

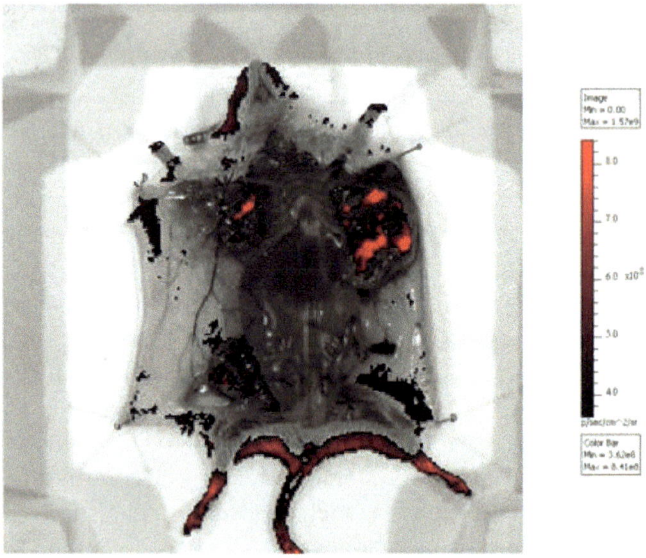

Figure 9.6 (A) Targeting HER2/neu overexpressing tumors using hyaluronic acid–coated nanoparticles. Empty PLGA-PEG nanoparticles labeled with FITC were used to show the targeting potential of HA in HER2 overexpressing mammary tumors of an HER2/neu transgenic mouse. The lower right mammary gland was used for histology. Background fluorescence in the hind limb region is due to autofluorescence from fur (IVIS xenogen imaging station).

Even though the activity was compromised after week 11, tumor growth was drastically reduced and eventually offered a significantly increased life span of 76% compared with the letrozole-treated group ($p < 0.002$). Even though the resistance to AIs is a complex phenomenon, our experimental evidence clearly shows that active targeting of AIs can alleviate or delay the onset of resistance up to an extent. Aromatase enzyme activity is a direct measure of the potency of AIs.

In our study no significant difference was noticed in the reduction of aromatase activity between Letr NP– and HA-Letr NP–treated LTLT-Ca cells. The possible reason for this may be that the extended release of encapsulated drug and active targeting of HA to CD44 in *in vivo* tumor microenvironment is more conducive than *in vitro* cell culture. Also the presence of HA on the nanoparticle enabled receptor-mediated endocytosis of letrozole that can better occur *in vivo* than during the short-term *in vitro* experiments. Since the endothelium of tumor is densely vascularized and even enhanced permeability and retention effect (EPR) exists in tumor vasculature, the interstitial pressure at tumor will be high to prevent nanoparticle uptake without an active targeting ligand [26].

Ligand-directed PEGylated nanoparticles are found to be promising in targeting HER2 overexpressed tumors using anti-HER2 immunoliposomes [22]. Recently, improved therapeutic response of folate nanoparticles charged with methotrexate demonstrated a marked decrease in toxicity and increase in antitumor capacity [19]. The reduced expression of P-glycoprotein (P-gp) levels after treatment with HA-Letr NPs on LTLT-Ca cells suggests that active targeting mediated cellular transport increases the distribution and bioavailability of letrozole by inhibiting P-gp levels. Furthermore, HA-Letr NPs can circumvent letrozole resistance up to an extent by inhibiting P-gp-mediated multidrug resistance. The encapsulated letrozole in PLGA nanospheres is believed to be internalized by endocytosis; once internalized, the nanospheres generate an intracellular sustained release of letrozole specifically inside the tumor. The maintenance of surface morphology and size distribution of the nanoparticles to ~85 nm helps minimize particle uptake and premature clearance by the mononuclear phagocytic system [6]. As noted in the results presented in section 9.4, the increased efficacy in life span or survival of HA-Letr NP–treated animals over Letr NP–treated ones may be due to the release of the drug into the extracellular space, causing systemic absorption and thereby increased toxicity and decreased efficacy. On the other hand, HA-Letr NPs are expected to get internalized into the tumor cells with subsequent intracellular release of letrozole.

9.5 TARGETING ESTROGEN RECEPTOR MODULATORS FOR ENDOCRINE THERAPY RESISTANCE

Estrogens and its receptor are critical for normal mammary development and induction of mammary tumors. A large body of evidence suggests the critical role of ERα in breast cancer initiation and progression. Hence targeted therapies to block

ERα action proved beneficial in treating the disease. Antiestrogens were reported nearly 20 years ago, and still there are concerns about increased risk of osteoporosis and heart disease and problems of drug delivery. Moreover, alkyl estradiol derivatives are difficult to synthesize, and oral availability is very low. Later on, even AIs proved to be superior agents and an alternative strategy to AE therapy, but delivery of these molecules to their site of action at the desired rate is a challenge because their transport through compartmental barriers in the body in inefficient, since they are rapidly metabolized [19]. However, the role of ERβ remains elusive. Among ~70% of estrogen-sensitive breast tumors express ERβ. Our laboratory and a few others have shown that an increase in ERβ expression decreases tumor cell proliferation. We have seen from our recent *in vivo* preclinical studies that the ratio of ERα to ERβ is critical in maintaining the normal equilibrium (unpublished data).

We have developed PLGA-PEG nanoparticles that target or modulate ERβ receptors found to be effective in reducing the tumor burden in xenograft models. These nanoparticles also show targeted accumulation of the aromatase inhibitor (letrozole) into the tumor (Fig. 9.7). We used diarylpropionitrile (DPN), a small molecule agonist for ERβ, to target these tumors.

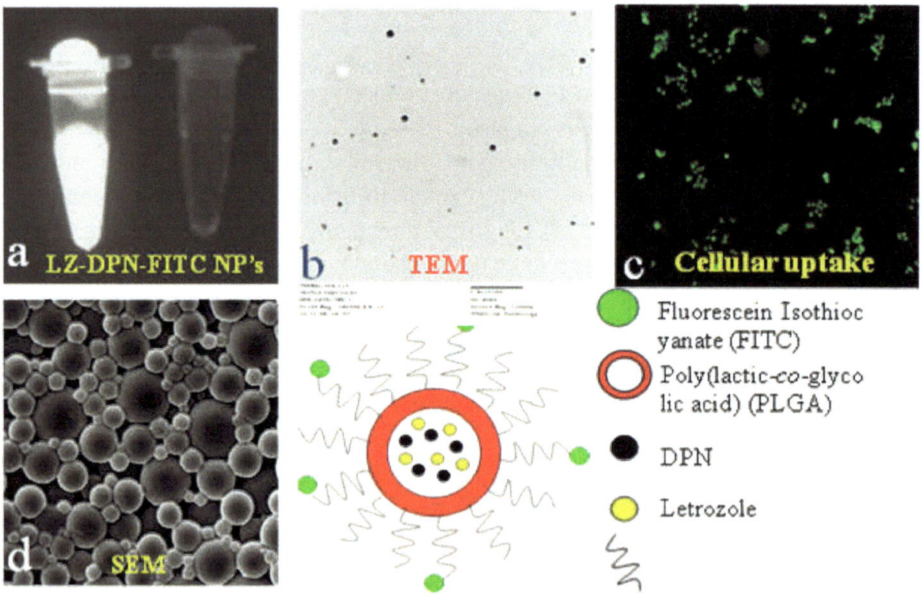

Figure 9.7 Physiochemical characterization of letrozole-DPN-FITC nanoparticles. (A) FITC-conjugated and unconjugated particles under UV light. (B) TEM image of letrozole-DPN nanoparticles. (C) Cellular uptake of nanoparticles in MCF-7 cells co-incubated for 1 hour. (D) SEM image of letrozole-DPN nanoparticles. A schematic representation of letrozole-DPN nanoparticles having a PLGA shell and decorated with FITC through PEGylation.

The nanoprecipitation technique was used to prepare letrozole-DPN NP–encapsulated nanoparticles. FITC-PLGA (100 mg) and drug (5 mg) were mixed in acetonitrile (10 mL) added dropwise to an aqueous solution rotating at 5000 rpm containing 0.1% pluronic F-68 as surfactant. The resulting dispersion of nanoparticles was vacuum-evaporated to eliminate the organic solvent. The resulting nanoparticles were centrifuged at 15,000 rpm for 15 minutes, washed with deionized water three times, and freeze-dried with 10% sucrose as a cryoprotectant. In order to track the MCF-7/Aro xenograft tumors in BALB/c nude mice (Fig 9.8), the animals were imaged 10 minutes after administration of letrozole-DPN NPs (10 mg/kg) using a fluorescent stereoscope attached to a CoolSNAP CCD camera. Nanoparticles colocalized with FITC were chemically conjugated to PEG (DCC coupling), and the stability was measured using FACS analysis. A majority of breast tumors express ERα, and about 7% of these respond to the antiestrogen tamoxifen. But prolonged use of tamoxifen leads to resistance to the drug and recurrence of the disease. Thus far, however, aromatase inhibitors have been generally safe and well tolerated. However important side effects such as musculoskeletal and bone-related problems, including risk for osteoporosis and fractures, remain a concern. In this scenario targeted depletion of ER will be an attractive alternative to reduce the side effects and increase the efficacy of AIs. Therefore a targeted drug delivery system for AEs and AIs would be of great benefit.

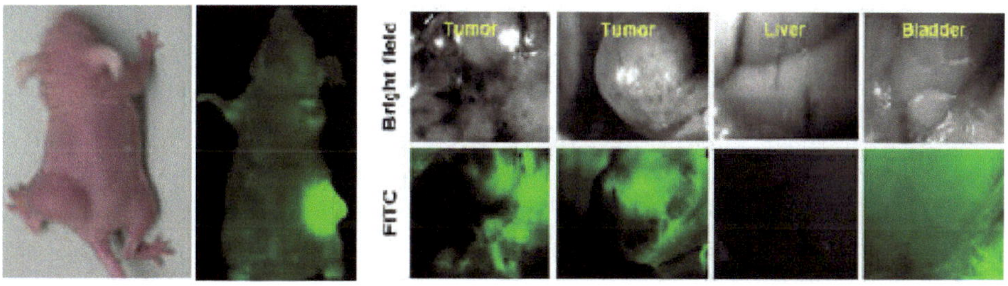

Figure 9.8 Retention of targeted nanoparticles in MCF-7/Aro (cells overexpress aromatase gene) xenograft tumor in BALB/c nude mice. The animals were imaged 30 minutes after administration of FITC-labeled letrozole-DPN nanoparticles (2.5 mg/kg) using a fluorescence stereoscope attached to a CoolSNAP CCD camera. *In vivo* distribution of DPN-letrozole nanoparticles: The animals received 2.5 mg/kg letrozole-DPN nanoparticles through tail vein injection. The tissue was removed surgically and imaged using stereo-microscope.

Nanoparticles with specific recognition ligands conjugated to or bound on the surface have a tremendous potential for site-selective drug delivery and offer a higher carrier capacity than bioconjugates. The diagnosis and treatment of cancer have been greatly improved with the recent developments in nanotechnology. PLGA-based FITC nanoparticles used in our study offer highly sensitive optical imaging of cancer at the cellular and animal level. Several functional moieties, including targeting ligands or receptors, therapeutic agents, and imaging probes, provide new potentials for clinical therapies and diagnostics and will undoubtedly play a critical role in future cancer therapy. The promising development in the area of fluorescent nanoparticles may make an impact on human cancer therapy by monitoring the efficacy of chemotherapeutic drugs. However, only limited studies are available to evaluate the safety and efficacy of PLGA-based nanoparticles. There is also concern in terms of the immunoreactivity and inflammation that these particles can generate. Hence, more studies need to be conducted to understand these possible undesirable side effects.

Acknowledgments

The authors wish to thank Dr. Angela Brodie (University of Maryland School of Medicine, Maryland) for providing LTLT-Ca cells and Dean B. Evans (Novartis Pharma, Basel, Switzerland) for providing the letrozole. We extend our thanks to Lauren Chesnut for SEM and TEM assistance and Jimmy Wewer for optical imaging. The HA-Letrozole nanoparticle study will be published separately as a research article.

References

1. Allen T. M. (2002) Ligand-targeted therapeutics in anticancer therapy. *Nat Rev Cancer* **2**, 750–763.
2. Atkins J. H., and Gershell L. J. (2002) Selective anticancer drugs. *Nat Rev Drug Discov* **1**, 491–492.
3. Brigger I., Dubernet C., and Couvreur P. (2002) Nanoparticles in cancer therapy and diagnosis. *Adv Drug Deliv Rev* **54**, 631–651.
4. Bumcrot D., Manoharan M., Koteliansky V., and Sah D. W. (2006) RNAi therapeutics: a potential new class of pharmaceutical drugs. *Nat Chem Biol* **2**, 711–719.
5. Cheng J., Teply B. A., Sherifi I., Sung J., Luther G., Gu F. X., Levy-Nissenbaum E., Radovic-Moreno A. F., Langer R., and Farokhzad O. C. (2007) Formulation of functionalized PLGA-PEG nanoparticles for *in vivo* targeted drug delivery. *Biomaterials* **28**, 869–876.
6. Colvin V. L. (2003) The potential environmental impact of engineered nanomaterials. *Nat Biotechnol* **21**, 1166–1170.
7. Dallas S., Miller D. S., and Bendayan R. (2006) Multidrug resistance-associated proteins: expression and function in the central nervous system. *Pharmacol Rev* **58**, 140–161.
8. Davis M. E., Chen Z. G., and Shin D. M. (2008) Nanoparticle therapeutics: an emerging treatment modality for cancer. *Nat Rev Drug Discov* **7**, 771–782.
9. Diaz L. K., Zhou X., Wright E. T., Cristofanilli M., Smith T., Yang Y., Sneige N., Sahin A., and Gilcrease M. Z. (2005) CD44 expression is associated with increased survival in node-negative invasive breast carcinoma. *Clin Cancer Res* **11**, 3309–3314.
10. Elbashir S. M., Harborth J., Lendeckel W., Yalcin A., Weber K., and Tuschl T. (2001) Duplexes of 21-nucleotide RNAs mediate RNA interference in cultured mammalian cells. *Nature* **411**, 494–498.
11. Farokhzad O. C., and Langer R. (2006) Nanomedicine: developing smarter therapeutic and diagnostic modalities. *Adv Drug Deliv Rev* **58**, 1456–1459.
12. Ferrari M. (2005) Cancer nanotechnology: opportunities and challenges. *Nat Rev Cancer* **5**, 161–171.

13. Fire A., Xu S., Montgomery M. K., Kostas S. A., Driver S. E., and Mello C. C. (1998) Potent and specific genetic interference by double-stranded RNA in Caenorhabditis elegans. *Nature* **391,** 806–811.

14. Fisher B., Costantino J. P., Redmond C. K., Fisher E. R., Wickerham D. L., and Cronin W. M. (1994) Endometrial cancer in tamoxifen-treated breast cancer patients: findings from the National Surgical Adjuvant Breast and Bowel Project (NSABP) B-14. *J Natl Cancer Inst* **86,** 527–537.

15. Fricker G., and Miller D. S. (2002) Relevance of multidrug resistance proteins for intestinal drug absorption *in vitro* and *in vivo*. *Pharmacol Toxicol* **90,** 5–13.

16. Gaspar MM B. D., Cruz M. E., and Alonzo M. J. (1998) Formulation of L-asparaginase loaded poly(D,L-lactide-co-glycolide) nanoparticles:influence of polymer properties on enzyme loading, activity and *in vitro* release. *J Control Release* **52,** 53–62.

17. Huang P. S., and Oliff A. (2001) Drug-targeting strategies in cancer therapy. *Curr Opin Genet Dev* **11,** 104–110.

18. Kalaria D. R., Sharma G., Beniwal V., and Ravi Kumar M. N. (2009) Design of biodegradable nanoparticles for oral delivery of doxorubicin: *in vivo* pharmacokinetics and toxicity studies in rats. *Pharm Res* **26,** 492–501.

19. Maillard S., Fattal E., Marsaud V., Sola B., and Renoir J.-M. (2006) Colloidal systems for the delivery of anticancer agents in breast cancer and multiple myeloma. *Nanomaterials for Cancer Therapy* (ed. Challa S. S. R. Kumar). Wiley-VCH Verlag, Weinheim, pp. 371–403.

20. Molineux G. (2002) Pegylation: engineering improved pharmaceuticals for enhanced therapy. *Cancer Treat Rev* **28** Suppl, **A**13–A16.

21. Moses M. A., Brem H., and Langer R. (2003) Advancing the field of drug delivery: taking aim at cancer. *Cancer Cell* **4,** 337–341.

22. Park J. W., Kirpotin D. B., Hong K., Shalaby R., Shao Y., Nielsen U. B., Marks J. D., Papahadjopoulos D., and Benz C. C. (2001) Tumor targeting using anti-her2 immunoliposomes. *J Control Release* **74,** 95–113.

23. Pritchard K. I. (2001) The role of tamoxifen and aromatase inhibitors/inactivators in postmenopausal patients. *Clin Cancer Res* **7,** 4356s–4359s; discussion 4411s–4412s.

24. Rana T. M. (2007) Illuminating the silence: understanding the structure and function of small RNAs. *Nat Rev Mol Cell Biol* **8,** 23–36.

25. Sabnis G., Schayowitz A., Goloubeva O., Macedo L., and Brodie A. (2009) Trastuzumab reverses letrozole resistance and amplifies the sensitivity of breast cancer cells to estrogen. *Cancer Res* **69,** 1416–1428.

26. Serpe L. (2006) Conventional chemotherapeutic drug nanoparticles for cancer treatment. *Nanomaterials for Cancer Therapy* (ed. Challa S. S. R. Kumar). Wiley-VCH Verlag, Weinheim, pp. 1–39.

27. Tekmal R. R., Nair H. B., Perla R. P., and Kirma N. (2007) HER-2/neu x aromatase double transgenic mice model: the effects of aromatase overexpression on mammary tumorigenesis. *J Steroid Biochem Mol Biol* **106,** 111–118.

28. van Vlerken L. E., Duan Z., Little S. R., Seiden M. V., and Amiji M. M. (2008) Biodistribution and pharmacokinetic analysis of Paclitaxel and ceramide administered in multifunctional polymer-blend nanoparticles in drug resistant breast cancer model. *Mol Pharm* **5,** 516–526.

29. Volpi N., Schiller J., Stern R., and Soltes L. (2009) Role, metabolism, chemical modifications and applications of hyaluronan. *Curr Med Chem* **16,** 1718–1745.

30. Wang Y. S., Youngster S., Grace M., Bausch J., Bordens R., and Wyss D. F. (2002) Structural and biological characterization of pegylated recombinant interferon alpha-2b and its therapeutic implications. *Adv Drug Deliv Rev* **54,** 547–570.

31. Wolf M., Wirth M., Pittner F., and Gabor F. (2003) Stabilisation and determination of the biological activity of L-asparaginase in poly(D,L-lactide-co-glycolide) nanospheres. *Int J Pharm* **256,** 141–152.

32. Zhang L., Gu F. X., Chan J. M., Wang A. Z., Langer R. S., and Farokhzad O. C. (2008) Nanoparticles in medicine: therapeutic applications and developments. *Clin Pharmacol Ther* **83,** 761–769.

Chapter 10

RADIOLABELED GOLD NANOSHELLS FOR *IN VIVO* IMAGING: EXAMPLE OF METHODOLOGY FOR INITIAL EVALUATION OF BIODISTRIBUTION OF A NOVEL NANOPARTICLE

Huan Xie,[a] Zheng Jim Wang,[b,c] Ande Bao,[b] Beth Goins,[b] and William T. Phillips[b]

[a] *Department of Pharmaceutical Sciences, College of Pharmacy and Health Sciences, Texas Southern University, Houston, Texas, USA*

[b] *Department of Radiology, University of Texas Health Science Center at San Antonio (UTHSC-San Antonio), San Antonio, Texas, USA*

[c] *MPI Research, Inc., Mattawan, Michigan, USA*

xieh@tsu.edu

10.1 INTRODUCTION

One of the most promising recent advances in cancer treatment is thermal ablation, which provides a minimally invasive or noninvasive cancer therapy technique that rapidly kills cancer cells by heat. Cancer cell death is related to both the temperature and the time period of thermal exposure. Previously investigated thermal therapies have employed a variety of heat sources, including radiofrequency, focused ultrasound, and microwaves [6]. The advantages of thermal ablation include minimal or noninvasive application, relatively simple to perform, and the potential of treating embedded tumors in vital regions where surgical resection is not feasible. However, simple heating techniques have trouble discriminating between tumors and surrounding healthy tissues, and often heat intervening tissue between the source and the target site. Nanoshell-based photothermal ablation, a laser-induced thermal therapy that utilizes special nanoshell optical properties, offers advantages over

Nanoimaging
Edited by Beth Goins and William Phillips

www.panstanford.com

traditional thermal therapies, and this approach has obtained much attention in the past several years.

Nanoparticles have been intensively studied and broadly utilized for biological applications, such as molecular imaging, molecular diagnosis, and targeted therapy. Among newly developed nanoparticles, gold nanoshells are of special interest for cancer treatment because of their unique size, composition, and physical and optical properties. Nanoshells are spherical nanoparticles consisting of a dielectric core and a metal shell, where the plasmon resonance frequency is determined by the relative size of the core and the metal shell layer [12]. By adjusting the relative core and shell thicknesses, nanoshells can be fabricated that will absorb or scatter light across the visible and near-infrared regions (NIR: 700–1300 nm) of the electromagnetic spectrum, where optical transmission through tissue is optimal. Silica core gold nanoshells are made of biocompatible materials, and their surface can be modified with "stealthing" polymers such as polyethylene glycol (PEG) to further improve their biocompatibility and prolong their circulation *in vivo*. They can also be manufactured with size ranges (60–400 nm) that can accumulate in tumors via the enhanced permeability and retention (EPR) effect, which is attributed to the leaky nature of tumor vessels [10].

Nanoshell-based photothermal ablation that utilizes special nanoshell optical properties and NIR laser has been demonstrated to be effective in the elimination of solid tumors in animal models [3, 11]. O'Neal *et al.* applied PEG-coated nanoshells (~130 nm diameter) with peak optical absorption in the NIR to treat tumor-bearing mice. Following intravenous administration, tumors were then illuminated with a diode laser (808 nm, 4 W/cm^2, 3 min). All such treated tumors were effectively abated, and treated mice appeared healthy and tumor free even after 90 days.

The fate of nanoshells in animals is critical to their application *in vivo*. In addition to these research advantages, imaging the biodistribution of gold nanoshells in patients has the potential of direct application of NIR lasers for treatment of tumors. However, there have been no prior reports on the biodistribution of nanoshells of which we are aware, whereas the *in vivo* biodistributions of other popular nanoparticles have been reported, such as carbon nanotubes and quantum dots. The biodistribution and tumor-targeting ability of radiolabeled carbon nanotubes has been investigated in rats and mice by *in vivo* positron emission tomography (PET) and single photon emission computed tomography (SPECT) [5, 7]. Cai *et al.* reported the *in vivo* targeting and imaging of tumor vasculature using RGD peptide–quantum dot conjugates [2].

Here we report a new procedure for labeling gold nanoshells with the radionuclide labels copper-64 (^{64}Cu) and indium-111 (^{111}In) through a bifunctional PEG and chelating agent. Radiolabeling the gold nanoshells with these agents permits determination of the biodistribution of the radiolabeled nanoshells in live rats bearing head and neck squamous cell carcinoma xenografts by noninvasive PET and SPECT imaging. Classical biodistribution studies that quantify radiolabeled particle deposition provide results in terms of populational averages at specific time points, but do not allow particle localization to be monitored longitudinally in individual animals. PET and SPECT are noninvasive imaging tools that allow researchers to quantitatively image the uptake of candidate nanoparticles at the target site with high sensitivity in real time. In addition to conventional *ex vivo* biodistribution techniques, the pharmacokinetic profile of new nanomaterials with potential medical applications can be obtained with PET and SPECT studies on a relatively small number of laboratory animals [16].

10.2 MATERIALS AND METHODS

10.2.1 Radiolabeling of NS and Stability Test

Gold nanoshells were synthesized as previously described [13]. Gold nanoshell (NS) formation was assessed using a UV-vis spectrophotometer (U-0080D, Hitachi) and Zetasizer (Nano-ZS, Malvern). NS manufactured in this manner have an 8–10 nm gold shell around a 110–120 nm silica sphere.

The labeling process is shown in Fig 10.1. First, a bifunctional chelating agent, *p*-NH_2-Bn-DOTA (S-2-(4-aminobenzyl)-1,4,7,10-tetraazacyclo-dodecane-tetraacetic acid, for ^{64}Cu labeling) or *p*-NH_2-Bn-DTPA (S-2(4-aminobenzyl)-diethylenetriamine pentaacetic acid, for ^{111}In labeling) (Macrocyclics), was conjugated to another bifunctional, OPSS-PEG2k-SVA (*o*-pyridyldisulfide-polyethylene glycol 2000–succinimidyl valerate) (Laysan). DOTA (or DTPA) and PEG2k were mixed with 1:1 molar ratio, and the mixture incubated overnight at RT. The resulting OPSS-PEG-DOTA (or OPSS-PEG-DTPA) was then added to NS solution (in 10 mM phosphate buffer, pH 7) with 10,000:1 molar ratio overnight at RT on a shaker, allowing the OPSS group to conjugate to the gold surface of the particles. The mixture was centrifuged, and the supernatant with unconjugated OPSS-PEG-DOTA or OPSS-PEG-DTPA was removed. The pellets of DOTA-NS and DTPA-NS were re-suspended in phosphate buffer and checked with spectrophotometer and Zetasizer to determine the NS concentration and size for further conjugation.

$^{64}CuCl_2$ (Washington University, St. Louis, MO) was diluted in 30 mM ammonium citrate buffer (pH 6.5). Then 816 μCi of ^{64}Cu was added to 150 μL of the DOTA-NS solution (~0.8 nM) prepared above and incubated at 37°C for 90 minutes. Then the blocking agent PEG5k-SH (Nektar) was added to NS solution with 300,000:1 molar ratio and incubated at RT on a shaker for 1 hour. The mixtures were then centrifuged to remove unconjugated ^{64}Cu and PEG5k. Labeling of NS with $^{111}InCl_3$ (GE Healthcare, San Antonio, TX) followed the same procedure. The supernatants and pellets were collected for radioactive counting. Labeling efficiencies were calculated by using the relation [activity in pellet/(activity in supernatant + activity in pellet)] × 100. The radiolabel stability was investigated by incubating ^{64}Cu-NS and ^{111}In-NS in PBS and serum at 37°C for up to 3 hours. Suspensions were divided into several equal parts, and at each time point 1 was taken for centrifugation. The supernatants and pellets were collected for radioactive counting. Percentages of ^{64}Cu and ^{111}In remaining on the NS were calculated by using the relation [activity in pellet/(activity in supernatant + activity in pellet)] × 100.

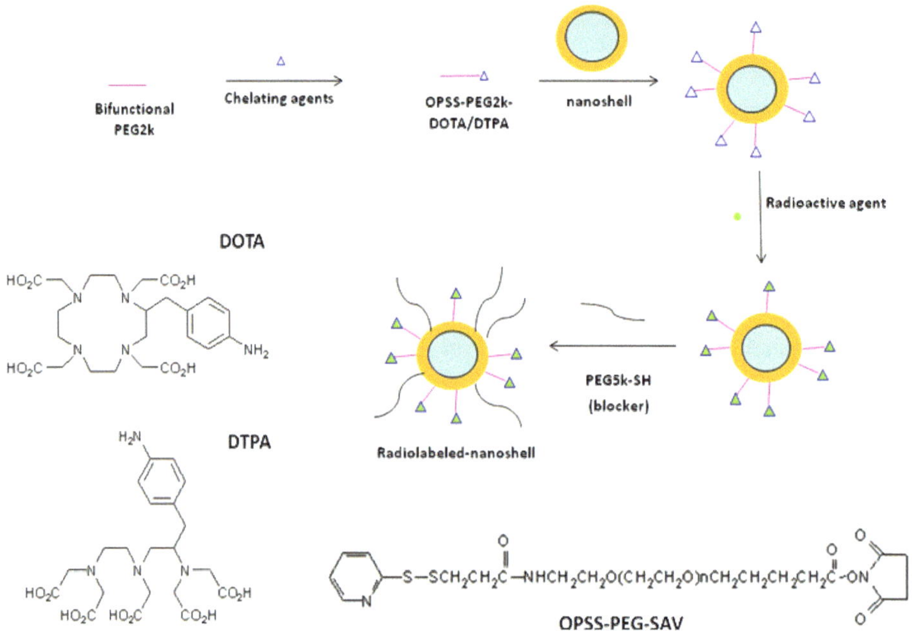

Figure 10.1 Radiolabeling process of nanoshells.

10.2.2 Intravenous Injection and PET/SPECT Imaging

A human head and neck squamous cell carcinoma (HNSCC) xenograft model in nude rats (average body weight 0.154 kg) was established via subcutaneous inoculation of an HNSCC cell line, SCC-4 [1]. On the day of imaging, the average tumor volume for the ^{64}Cu-NS group (group A, $n = 4$) and ^{111}In-NS group (group B, $n = 4$) were 1.2 cm^3 and 1.0 cm^3, respectively. During all procedures, the nude rats (~150 g) were anesthetized with Avertin (400 mg/kg) by IP injection and 0.8 mCi of ^{64}Cu-NS or ^{111}In-NS in 0.5 mL was injected manually into the rat's tail vein at about 0.5 mL/min. PET and SPECT imaging was performed at 1 hour, 4 hours, 20 hours and 44 hours after injection of radiolabeled nanoshells with the FLEX X-PET/CT/SPECT (Gamma Medica-Ideas, Inc., Northridge, CA), followed by CT image acquisition.

10.2.3 Biodistribution and Blood Circulation Studies

The post-imaging biodistribution studies were carried out 46 hours after injection of radiolabeled nanoshells. The rats were killed at that time, and the major organs were collected and wet-weighed. The radioactivity in the tissues was measured using a γ-counter (Wallac 1480, Perkin Elmer Life Sciences, Boston, MA). For each rat, the radioactivity of the tissue samples was calibrated against a known aliquot of the injectate, and normalized.

Some major organs from the biodistribution study were also subjected to neutron activation analysis (NAA). Portions of the spleen, liver, lung, kidney, and tumor tissues were collected and wet-weighed. The samples were placed into precleaned and labeled polyethylene irradiation vials. After the wet sample weight was calculated and recorded, the samples were covered and dried under a heat lamp for 48 hours before delivery into the high-temperature nuclear reactor core for NAA gold measurements. The procedure of NAA for trace gold quantification in animal tissues is described elsewhere [4]. The data were reported as gold mass vs. the original wet sample mass. Blood samples (40 μL) were collected from the opposite tail vein from the injection at 5 minutes, 1 hour, 4 hours, 20 hours, and 44 hours time points after intravenous injection. The ^{64}Cu and ^{111}In radioactivities in blood samples were measured by the γ-counter.

The data were presented as the mean value ± one standard deviation (SD). The statistical software Origin, version 7.5 (Origin Lab, Northampton, MA), was used for the calculation of the means and standard deviations and for statistical analysis. The values of blood circulation half-life were obtained by second-order exponential decay fitting of blood radioactivity with Origin.

10.3 RESULTS

The nanoshells used in this study were manufactured to be comprised of a silica core (~120 nm in diameter) and a gold shell (8~10 nm) to absorb light at the near-infrared region. The UV-vis spectrum showed the NS peak at ~760 nm. Zetasizer measurements showed the NS size of around 140 nm in diameter and zeta potential of around –50 mV. Those data met our criteria for normal NS batch manufacture.

A conjugation procedure was developed to label NS with ^{64}Cu ($t_{1/2}$ = 12.7 hours) and ^{111}In ($t_{1/2}$ = 2.8 days) using bifunctional OPSS-PEG2K-SAV and bifunctional chelating agent DOTA or DTPA (Fig. 10.1). First, OPSS-PEG2K-SAV and DOTA or DTPA were mixed and the SAV group reacted with the amine group of DOTA or DTPA to form an amide bond. The mixture was then added to an NS solution, allowing the OPSS group to attach to the gold surface of the particles. The nanoshells were then incubated with $^{64}CuCl_2$ or $^{111}InCl_3$. Finally, the longer PEG5K-thiol was added to block the remaining empty areas on the gold surface to provide better PEG coverage.

The radioactively-labeled particles were thoroughly washed twice by centrifugation and resuspension in fresh PBS buffer. The radioactive signal from both the pellet and the supernatant was measured in the dose calibrator to obtain the labeling efficiency, which is the radioactive count in the pellet vs. the total radioactive count. The average radiolabeling efficiency (n = 3) was 81.3% and 80.32% for ^{64}Cu-NS and ^{111}In-NS, respectively, after two separations.

Prior to any *in vivo* examination of ^{64}Cu-NS and ^{111}In-NS, *in vitro* stabilities in PBS and serum were tested. The labeled particles were incubated with either PBS or serum for 1 hour and 3 hours. At each time point, samples were centrifuged and the supernatant removed. The radioisotope content of both the pellet and supernatant were determined by the dose calibrator. The percentages of both ^{64}Cu and ^{111}In remaining on the particles after 3 hours' incubation with PBS and serum were larger than 90% and 80%, respectively.

Because the *in vitro* studies showed successful radiolabeling of nanoshells, we proceeded to test ^{64}Cu-NS and ^{111}In-NS in tumor-bearing rats. The rats were imaged at multiple time points post-injection by micro-PET, CT, and SPECT, and blood samples were also withdrawn at these time points for radioactive counting. The simulated dual-exponential decay circulation curves were obtained for ^{64}Cu-NS and ^{111}In-NS based on the average percentage of activity remaining in the blood samples from four rats in each group at various time points (Fig. 10.2). The retention of ^{64}Cu-NS and ^{111}In-NS in the rat blood followed a second-order exponential decay model with the following general formula:

$$y = y_0 + A_1 * \exp(-x/t_1) + A_2 * \exp(-x/t_2)$$

Here y_0 represents the Y offset (i.e., the constant retention activity other than the cleared components following the exponential decay mechanism), A_1 and A_2 represent the percentages of activity that were cleared with decay constants of t_1 and t_2, respectively, for the labeled nanoshells. Subsequently, the half clearance times correlated with the radioactivity components of A_1 and A_2 are $t_1 \times 0.693$ and $t_2 \times 0.693$ correspondingly [15].The data reveals that 61.7% of ^{64}Cu-NS have a half-life of 0.52 hours and 33.8% of ^{64}Cu-NS have 5.29 hours of half-life, while 61.53% of ^{111}In-NS have a 0.88 hour half-life and 35.9% of ^{111}In-NS have a 8.07 hour half-life in rat blood.

A

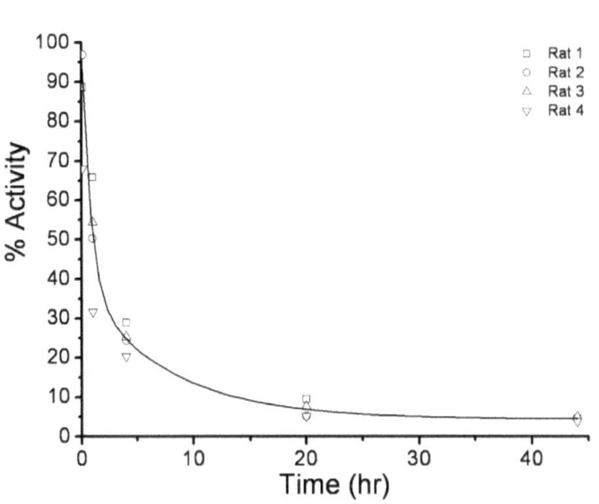

B

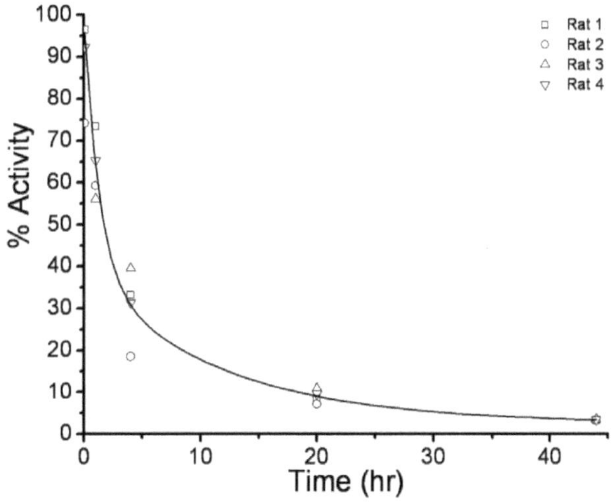

Figure 10.2 Blood circulation of ^{64}Cu-NS (A) and ^{111}In-NS (B) in nude rats (4 in each group) with head and neck squamous cell carcinoma xenograft. The following formulas were obtained from the curve fitting of Fig. 10.2:

^{64}Cu-NS: $y = 4.76 + 66.21 * \exp(-t/0.758) + 36.29 * \exp(-t/7.627)$

^{111}In-NS: $y = 2.67 + 64.20 * \exp(-t/1.269) + 37.46 * \exp(-t/11.647)$

The circulation times for both ^{64}Cu-NS and ^{111}In-NS are within the range of normal NS circulation [11, 17]. Using PET/CT and SPECT/CT imaging, we monitored the *in vivo* distribution of ^{64}Cu-NS and ^{111}In-NS in nude rats with a head and neck squamous cell carcinoma xenograft at various time points after their intravenous injection. Figure 10.3 shows the PET sagittal images of ^{64}Cu-NS of Rats A1, A2, A3, and A4 at different time points after injection and the CT images with clear locations of the tumors. As we can see, the NS accumulation at the tumor at 1 hour was not significant, but increased over time.

At 20 hours it achieved a plateau, and at 44 hours the majority of accumulated particles were still at the tumor site. Figure 10.4 are the fused SPECT/CT sagittal and the CT sagittal images of ^{111}In-NS at 20 hours after injection. They also show clear uptake of nanoshells at the tumor site, although the imaging quality is not as good as with the ^{64}Cu-radiolabeled nanoshells imaged by PET.

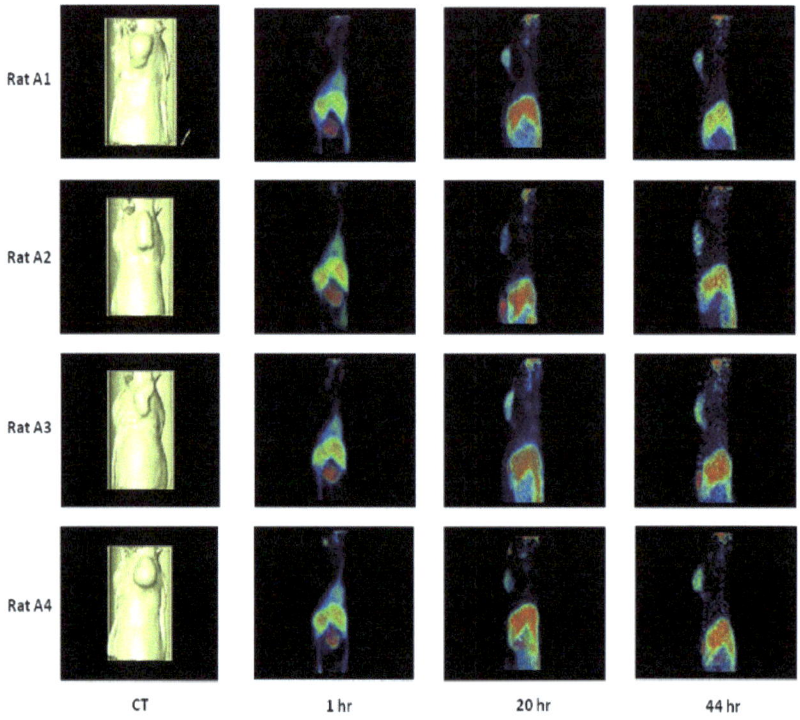

Figure 10.3 CT and PET sagittal images of four nude rats with head and neck squamous cell carcinoma xenograft at 1 hour, 20 hours, and 44 hours post tail-vein injection of ^{64}Cu-NS. CT surface images are also shown for appreciation of the average size of the tumor on the surface of the back of the neck.

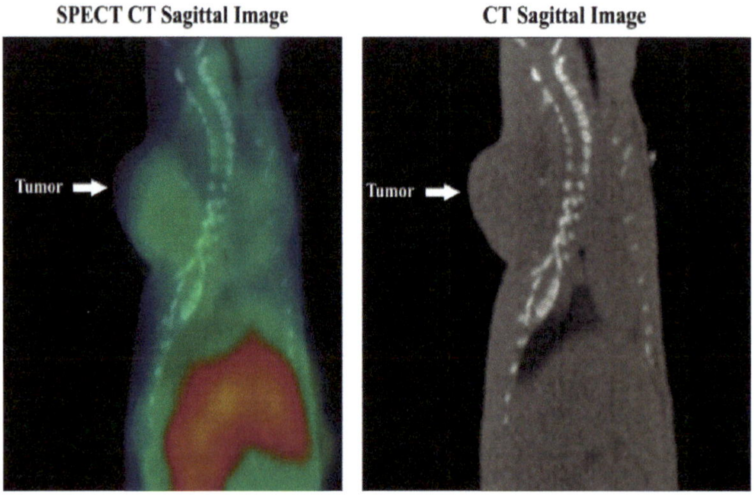

Figure 10.4 Sagittal SPECT/CT (Left) and CT image of a rat bearing head and neck tumor xenograft (arrow) acquired 20 hours after intravenous injection of the ^{111}In-NS.

The biodistribution of nanoshells is similar to other reported nanomaterials with a tendency to undergo RES uptake, including in the liver. RES uptake occurs via opsonization, i.e., antibody binding to nanomaterials in the plasma for recognition by phagocytes in the RES. Biodistribution in various organs (Fig. 10.5) after sacrificing the rats at 46 hours post-injection revealed prominent nanoshell uptake in the spleen and liver. This result was confirmed by both radioactive counting and NAA data. Figure 10.6 compared radioactivity count and gold concentration measured by NAA of several major organ tissues. The two sets of data are highly correlated (Fig. 10.6).

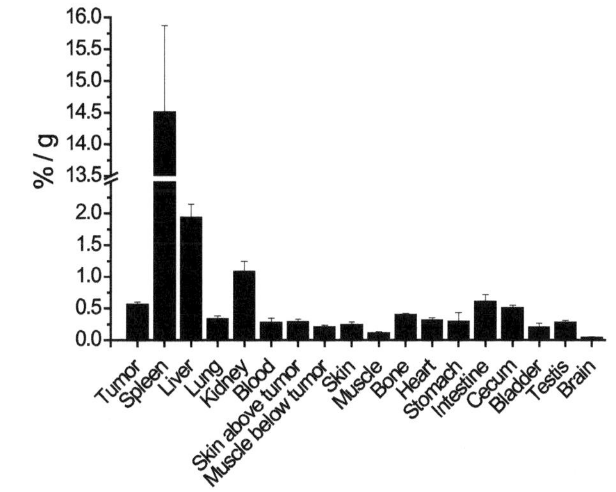

Figure 10.5 Organ distribution (% injected dose per gram tissue) at 46 hours after intravenous injection of ^{64}Cu-NS.

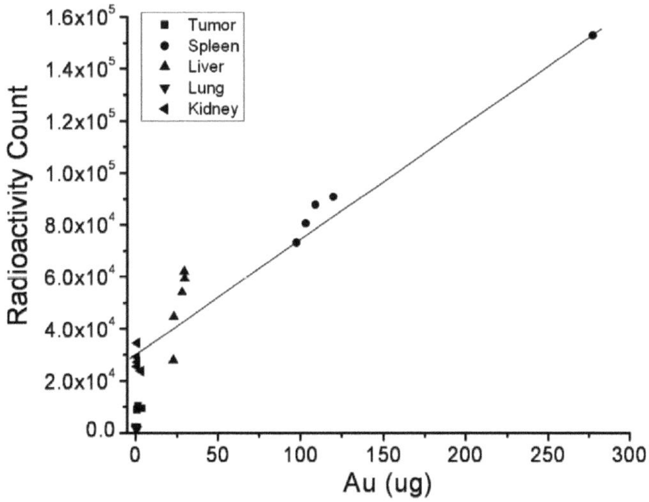

Figure 10.6 Correlation of NAA result and radioactive counting.

10.4 DISCUSSION

PET and SPECT images of the rats showed that both ^{64}Cu and ^{111}In appeared to be useful for tracking the *in vivo* distribution of nanoshells, and both ^{64}Cu-NS and ^{111}In-NS showed clear uptake of the radiolabeled nanoshells in the tumor 20 hours after injection. This uptake is likely due to the well-described EPR effect. The biodistribution data determined with ^{111}In was very close to that of ^{64}Cu, but image quality was better with the ^{64}Cu. This is likely due to the more efficient detection of the ^{64}Cu with the PET ring coincidence imaging system than the SPECT rotating camera system. Therefore, PET imaging with ^{64}Cu appears more promising for future NS imaging studies. Imaging provides the possibility not only of quantitative assessment of uptake in the tumor to ensure an adequate uptake for thermal therapy, but also of the availability of localization information to guide the application of infrared thermal therapy. Either SPECT or PET imaging of ^{111}In-labeled nanoshells or ^{64}Cu-labeled nanoshells is highly feasible in clinical patients.

Recently, small animal and clinical systems have become available that incorporate a CT scanner and a PET or SPECT camera into the same housing[14]. This permits superimposition of the PET and CT images or SPECT and CT images so that the highly sensitive tracking and quantification of a PET radiolabeled agent can be localized with the higher anatomic accuracy of CT. The very accurate localization provided by the CT imaging offers the possibility of effective image-guided application of infrared thermal ablation to specific locations of tumor in the body.

In this particular study, the percent uptake of gold nanoshells in the tumor was relatively low at <1% gram of tumor, while uptake of the nanoshells in the spleen was the highest on a per gram basis, with the liver being second highest. This high uptake in the liver and spleen is most likely due to the rapid clearance of the nanoshells from the blood by the reticuloendothelial cells of the liver and spleen. This relatively low uptake in the tumor and the high uptake in liver and spleen are consistent with the rapid clearance of radioactivity from the blood with a $T_{1/2}$ of less than 1 hour for 62% of the particles, while 38% of the particles had a $T_{1/2}$ of 5–8 hours. This rate of removal is quite rapid compared with PEGylated liposomes in rats which have an overall $T_{1/2}$ of >12 hours. The rapid removal of gold nanoshells occurred in spite of PEGylation. Although tumor accumulation of the gold nanoshells is clearly visible on the images, its relatively low accumulation is likely due to rapid removal of the nanoshells from circulation. Rapid removal from circulation interferes with the EPR mechanism of uptake in the tumor, which is dependent on the number of circulatory passes through the vasculature of the tumor. Prolonging the circulation of the nanoparticle has been shown to greatly increase nanoparticle uptake in the tumor. To achieve further prolongation in circulation for gold nanoshells, it may be necessary to use a higher percentage of PEGylation with significantly higher PEG molecular weights. One of the longest-circulating nanoparticles described had a PEG concentration of 10 mole % with a PEG molecular weight of 5000. The size of the nanoparticles is another important factor for biodistribution. Nanoparticles larger than 200 nm were trapped by splenic physical filtration, while increased renal clearance was observed for nanoparticles smaller than 20 nm. Nanoshells can passively accumulate in solid tumor due to the well-known EPR effect. However, nanoshells were too large for tumor parenchymal penetration and accumulation. Therefore, nanoshells may be linked with vascular and tumor cell markers to target tumor-specific vasculature in an attempt to further enhance their perivascular sequestration [3, 8, 9].

This study shows the feasibility of radiolabeling nanoshells for either SPECT or PET imaging using the radiolabeling procedures demonstrated in this chapter. This technique can be easily adapted for radiolabeling of other types of nanoparticles and radioactive agents. This radiolabeling procedure permits the determination of *in vivo* biodistribution and determination of tumor uptake and localization for guidance of laser thermal therapy. Further studies of the biodistribution of radiolabeled peptide-NS conjugates for the specific targeting of tumor neovasculature have been published [18] and we have already obtained promising preliminary results that also demonstrate the usefulness of PET imaging for investigation of different ^{64}Cu-labeled nanoparticle formulations [19].

Acknowledgments

This work was performed under funding from Nanospectra Biosciences, Inc., through NIST ATP Cooperative Agreement Number 70NANB4H3040.

References

1. Bao A., Phillips W. T., Goins B., McGuff H. S., Zheng X., Woolley F. R., Natarajan M., Santoyo C., Miller F. R., and Otto R. A. (2006) Setup and characterization of a human head and neck squamous cell carcinoma xenograft model in nude rats. *Otolaryngol Head Neck Surg* **135,** 853–857.

2. Cai W., Shin D. W., Chen K., Gheysens O., Cao Q., Wang S. X., Gambhir S. S., and Chen X. (2006) Peptide-labeled near-infrared quantum dots for imaging tumor vasculature in living subjects. *Nano Lett* **6,** 669–676.

3. Hirsch L. R., Stafford R. J., Bankson J. A., Sershen S. R., Rivera B., Price R. E., Hazle J. D., Halas N. J., and West J. L. (2003) Nanoshell-mediated near-infrared thermal therapy of tumors under magnetic resonance guidance. *Proc Natl Acad Sci U S A* **100,** 13549–13554.

4. James W. D., Hirsch L. R., West J. L., O'Neal P. D., and Payne J. D. (2007) Application of INAA to the build-up and clearance of gold nanoshells in clinical studies in mice. *J Radioanal Nucl Chem* **271,** 455–459.

5. Lacerda L., Soundararajan A., Singh R., Pastorin G., Al-Jamal K. T., Turton J., Frederik P., Herrero M. A., Li S., Bao A., Emfietzoglou D., Mather S., Phillips W. T., Prato M., Bianco A., Goins B., and Kostarelos K. (2008) Dynamic Imaging of Functionalized MultiWalled Carbon Nanotube Systemic Circulation and Urinary Excretion. *Adv Mater* **20,** 225–230.

6. Laeseke P. F., Sampson L. A., Haemmerich D., Brace C. L., Fine J. P., Frey T. M., Winter T. C., 3rd, and Lee F. T., Jr. (2005) Multipleelectrode radiofrequency ablation: simultaneous production of separate zones of coagulation in an *in vivo* porcine liver model. *J Vasc Interv*

Radiol **16,** 1727–1735.

7. Liu Z., Cai W., He L., Nakayama N., Chen K., Sun X., Chen X., and Dai H. (2007) *In vivo* biodistribution and highly efficient tumour targeting of carbon nanotubes in mice. *Nat Nanotechnol* **2,** 47–52.

8. Loo C., Lowery A., Halas N., West J., and Drezek R. (2005) Immunotargeted nanoshells for integrated cancer imaging and therapy. *Nano Lett* **5,** 709–711.

9. Lowery A. R., Gobin A. M., Day E. S., Halas N. J., and West J. L. (2006) Immunonanoshells for targeted photothermal ablation of tumor cells *Int J Nanomedicine* **1,** 149–154

10. Maeda H., Fang J., Inutsuka T., and Kitamoto Y. (2003) Vascular permeability enhancement in solid tumor: various factors, mechanisms involved and its implications. *Int Immunopharmacol* **3,** 319–328.

11. O'Neal D. P., Hirsch L. R., Halas N. J., Payne J. D., and West J. L. (2004) Photo-thermal tumor ablation in mice using near infrared-absorbing nanoparticles. *Cancer Lett* **209,** 171–176.

12. Oldenberg S. J., Averitt R. D., Westcott S. L., and Halas N. J. (1998) Nanoengineering of Optical Resonances. *Chem Phys Lett* **28,** 243–247.

13. Oldenburg S. J., Jackson J. B., Westcott S. L., and Halas N. J. (1999) Infrared extinction properties of gold nanoshells,. *Appl PhysLett* **111,** 2897.

14. Rowland D. J., and Cherry S. R. (2008) Small-animal preclinical nuclear medicine instrumentation and methodology. *Semin Nucl Med* **38,** 209–222.

15. Wang S. X., Bao A., Phillips W. T., Goins B., Herrera S. J., Santoyo C., Miller F. R., and Otto R. A. (2009) Intraoperative therapy with liposomal drug delivery: retention and distribution in human head and neck squamous cell carcinoma xenograft model. *Int J Pharm* **373,** 156–164.

16. Wipke B. T., Wang Z., Kim J., McCarthy T. J., and Allen P. M. (2002) Dynamic visualization of a joint-specific autoimmune response through positron emission tomography. *Nat Immunol* **3,** 366–372.

17. Xie H., Gill-Sharp K. L., and O'Neal D. P. (2007) Quantitative estimation of gold nanoshell concentrations in whole blood using dynamic light scattering. *Nanomedicine* **3,** 89–94.

18. Xie H., Parmeswaran P., Deorukhkur A. A., Goins B., Bao A., Phillips W. T., Wang Z., Schwartz J., and Krishnan S. (2011) Integrin avb3-targeted gold nanoshells augment tumor vasculature-specific imaging and therapy. *Int J Nanomedicine* **6**, 259-269.

19. Xie H., Wang Z., Bao A., Goins B., and Phillips W. T. In vivo PET imaging and biodistribution of radiolabeled gold nanoshells in rats with tumor-xenografts. *Int J Pharm* **395**, 324-330.

Chapter 11

COMBINED PHOTOACOUSTIC AND ULTRASOUND IMAGING OF METAL NANOPARTICLES *IN VIVO*

Kimberly Homan, Srivalleesha Mallidi, Erika Cooley, and Stanislav Emelianov

Department of Biomedical Engineering, Institute for Cellular and Molecular Biology, The University of Texas at Austin, Austin, TX 78712-0238, USA

emelian@mail.utexas.edu

The field of nanotechnology is poised to revolutionize medical imaging. As discussed in the previous chapters, nanoparticles can be detected *in vivo* through myriad techniques. Here we introduce another technique, photoacoustic imaging, that combines the advantages of optical and ultrasound imaging methods, resulting in the ability to detect metal nanoparticles deep in tissue in comparison with pure optical techniques. The benefits of employing combined photoacoustic and ultrasound imaging are discussed along with examples of fabrication, functionalization, and *in vivo* imaging of gold and silver nanoparticles.

11.1 INTRODUCTION

Photoacoustic imaging is an emerging technique on the precipice of reaching the clinic. The basics and history of photoacoustic imaging are presented in this section along with the benefits of combining it with the well-known and characterized ultrasound system. Furthermore, metal nanoparticles are introduced and the plasmon resonance properties of nanoparticles that enable their use as photoacoustic contrast agents are explained.

Nanoimaging
Edited by Beth Goins and William Phillips

www.panstanford.com

11.1.1 History of Photoacoustic Imaging

The photoacoustic (and, in general, thermoacoustic) effect refers to the conversion of light (electromagnetic energy) to thermal and then acoustic energy. This phenomenon was first discovered in the late 19th century by Alexander Graham Bell. In 1880, Bell patented the "photophone," a device that could produce sound when incident light striking the surface of a solid object was rapidly interrupted, typically on the order of kilohertz. Bell correctly conjectured that the resulting acoustic signal was dependent on the composition of the sample, and thus caused by the absorption of light [6]. However, it was over 100 years before Bowen *et al.* recognized the potential of applying the same phenomena for imaging of soft tissues in the body [9]. In the mid 1990s, Beard *et al.* [5] constructed an optical fiber photoacoustic-photothermal imaging probe, while Kruger *et al.* and Oraevsky *et al.* developed 3D tomographic imaging techniques for clinical applications [30, 48]. Through the work of Wang *et al.* and others, photoacoustic imaging has continued to enjoy a steady wave of advances as a noninvasive imaging modality that is capable of providing both morphological and functional information for various applications ranging from cardiovascular applications to tissue engineering and cancer [16, 27, 34, 42, 54, 58, 70].

11.1.2 Photoacoustic Imaging Basics

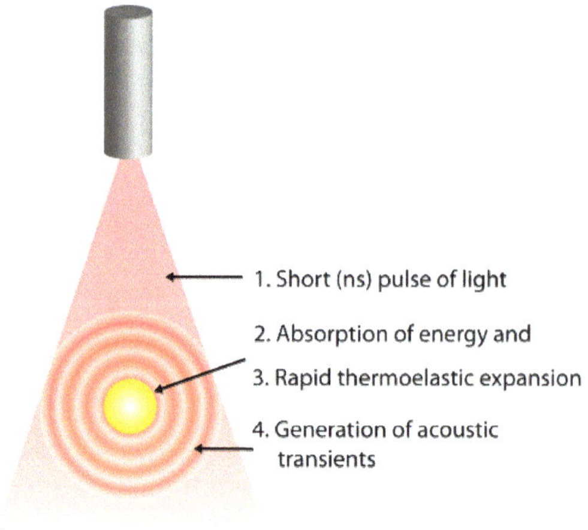

Figure 11.1 Depiction of the photoacoustic effect.

Obtaining a photoacoustic image can be simplified to a four-step process. First, the tissue of interest is illuminated with a short laser pulse, on the order of several nanoseconds. Next, the deposition of this energy in the tissue causes localized heating followed by rapid thermal expansion and the generation of acoustic pressure transients. Finally, acoustic waves, detected with an ultrasound transducer, are digitally processed to form an image of the spatial distribution of tissue optical absorption (Fig. 11.1).

11.1.2.1 Photoacoustic pressure generation

The governing equations of photoacoustics are derived from the linearized equations of fluid dynamics for an isotropic, homogeneous, inviscid fluid as well as the heat transfer equation, which relates the change of thermal energy to the change of electromagnetic radiation [28, 32]. The absorbed energy density from a laser source, E_{abs} (J/m^3), causes a localized increase in temperature, ΔT (K), giving rise to a pressure transient, ΔP (Pa), in the tissue volume given by [48]

$$\Delta P = \frac{1}{\gamma}\beta\Delta T = \frac{1}{\gamma}\beta\left(\frac{E_{\text{abs}}}{pC_v}\right) = \frac{Bc_s^2}{C_p}H\mu_a \tag{11.1}$$

where γ (Pa^{-4}) is the thermodynamic coefficient of isothermal compressibility, β (K^{-1}) is the thermal coefficient of volume expansion, ρ (kg/m^3) is the density of the medium, C_{v} and C_{p} (J/kg×K) are the heat capacities at constant volume and pressure, respectively, and c_{s} (m/s) is the speed of sound in the medium. E_{abs} is the product of the laser fluence H (J/m^2) and the absorption coefficient of the medium μ_{a} (m^{-1}). The expression ($\beta c_{\text{s}}^2/C_{\text{p}}$) represents the Grüneisen parameter Γ, which is the dimensionless, temperature-dependent factor representing the fraction of thermal energy converted into mechanical stress, yielding the simplified expression

$$\Delta P = \Gamma H\mu_a \tag{11.2}$$

Equation (11.2) is valid only in the case when the stress confinement condition is met, that is, when the heat diffusion time (τ_{D}) is significantly shorter than the stress relaxation time (τ_{SR}), i.e., the time for the acoustic waves to propagate through the irradiated volume. This condition limits the duration of the laser pulse (τ_{L}), implying that it must also be much shorter than the stress relaxation time. When the stress confinement condition is met, the input laser energy generates photoacoustic pressure with the greatest efficiency.

11.1.2.2 Optical properties of tissue

Because the strength of the photoacoustic signal is proportional to the optical absorption of photons in tissue, some consideration must be given to the optimal wavelength to be used for imaging. Figure 11.2 illustrates the absorption spectrum of four of the main tissue components that absorb in the visible (400–700 nm) and NIR (700–1100 nm) regions of the electromagnetic spectrum. These are deoxyhemoglobin (Hb), oxyhemoglobin (HbO_2), fat, and water. For clinical applications, photoacoustic imaging is likely to be performed in the region from 700 to 1100 nm, where the aggregate absorption from tissue constituents is minimized, and reasonable light penetration into tissue can be achieved.

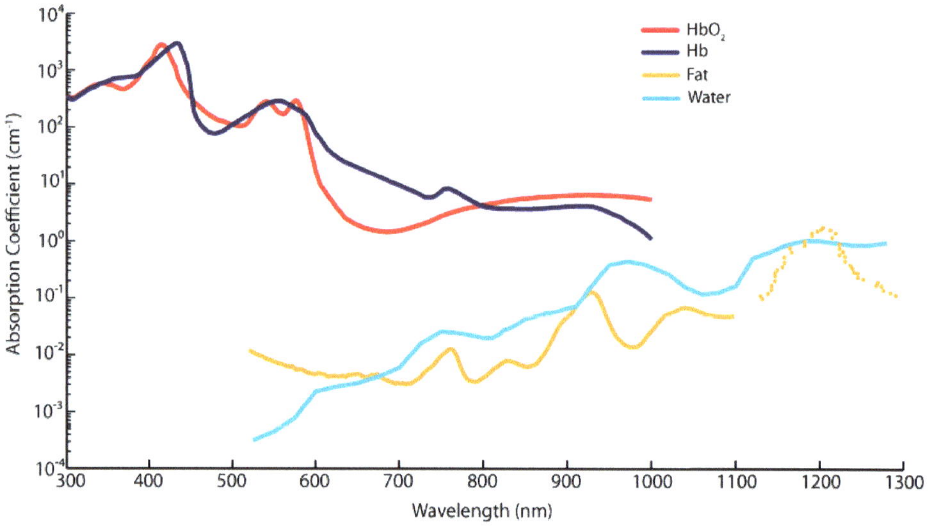

Figure 11.2 The optical absorption coefficient of compounds found in biological tissue as a function of wavelength [47].

11.1.2.3 Photoacoustic imaging resolution

Currently, photoacoustic imaging can provide high-resolution images at the microscopic scale using a high-frequency, single-element, focused ultrasound transducer. This type of imaging, called photoacoustic microscopy, provides high spatial resolution images by mechanically scanning a focused ultrasonic transducer over the tissue. Each time-resolved transient acoustic wave is converted to a 1D signal analogous to an ultrasonic A-line scan. Cross-sectional images can be formed by combining multiple A-line scans taken in the same plane [72]. The lateral resolution of photoacoustic microscopy is determined by the f-number of the ultrasonic transducer, while the axial resolution is inversely related to the bandwidth of the ultrasonic transducer [4, 16]. Lateral (d_{lat}) and axial (d_{axial}) resolution of the photoacoustic imaging system are described by the following expressions:

$$d_{lat} = \lambda \cdot f_{\#} \quad (11.3)$$

$$d_{axial} = \frac{c}{BW} \quad (11.4)$$

where λ is the wavelength of the light source, $f_{\#}$ is the f-number (ratio of focal length and aperture diameter of the transducer), c is the speed of sound, and BW is the transducer bandwidth [16]. It is clear that high spatial resolution can be achieved using a wide-bandwidth ultrasonic transducer with small $f_{\#}$. Photoacoustic microscopy uses single element transducers, while photoacoustic tomography uses linear array transducers made up of 128 or 256 elements. Photoacoustic tomography has the benefit of providing real-time B-line scan photoacoustic images, whereas in photoacoustic microscopy, single-element transducers need to be raster scanned over the tissue to obtain 2D and 3D images. By utilizing array transducers, real-time, functional information *in vivo* can be obtained. In photoacoustic tomography, 3D images are generated by combining spatial and temporal information using a variety of reconstruction algorithms [51, 72]. With the availability of high-frequency array transducers, high-resolution photoacoustic and ultrasound images could be obtained in real time.

11.1.3 Benefits of Combined Photoacoustic and Ultrasound Imaging

In a clinical setting, the photoacoustic technique provides an opportunity to augment well-established imaging modalities, such as ultrasound. Ultrasound image contrast is generated by inhomogeneities in tissue properties, such as the bulk modulus and density [16]. These variations provide structural information about the tissue that can be enhanced with the functional imaging capability of photoacoustics. Photoacoustic imaging allows for spatial detection of strongly absorbing endogenous or exogenous contrast agents, such as blood or nanoparticles. These contrast agents facilitate the detection of cellular and molecular events and can distinguish diseased from normal tissue. This provides an opportunity for image-guided clinical applications, such as cancer detection, diagnosis, and therapy guidance [16]. A schematic of a combined photoacoustic and ultrasound (PAUS) imaging system is presented in (Fig. 11.3); note that this is only one of the many ways to design and implement the PAUS imaging system. A nanosecond pulsed Nd:YAG laser is used to pump an optical parametric oscillator (OPO), which provides tunability over a broad range of wavelengths in the NIR. The desired wavelength is then focused onto a fiber-optic bundle containing 18 individual fibers. The fibers terminate in a coplanar arrangement around the surface of the ultrasound transducer, a design that permits overlap of optical and acoustic signals within the sample or tissue being analyzed (Fig. 11.4). The ultrasound transducer comprises 128 elements organized in a linear array that operates at a center frequency of 7.5 MHz, a clinically relevant frequency for deep tissue imaging. The acoustic signals, detected by an ultrasound probe and captured by an ultrasound imaging engine, are sent to a computer for image formation and display.

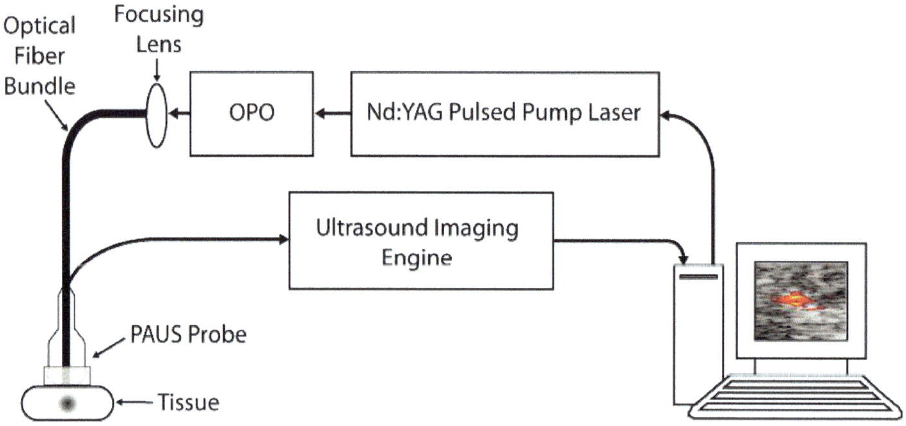

Figure 11.3 Block diagram of the combined photoacoustic and ultrasound (PAUS) imaging system.

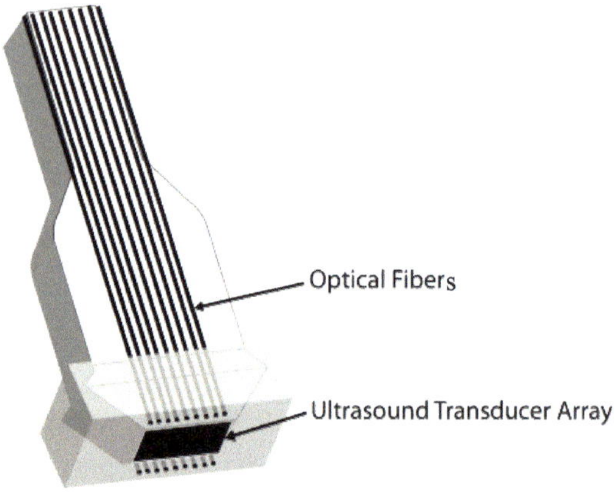

Figure 11.4 Depiction of the photoacoustic ultrasound imaging probe illustrating the coplanar arrangement of the 18-fiber-optical bundle and 128-element linear ultrasound transducer array.

For imaging, the probe (Fig. 11.4) must be coupled to the tissue via an acoustic impedance matching material, such as ultrasound gel or water. To produce ultrasound images, ultrasound pulses are generated by the transducer and transmitted into the tissue. These pressure waves interact with the tissue and produce echoes as they are reflected or scattered. The backscattered echoes are then detected by the ultrasound transducer and a gray-scale image of the tissue is produced. When the probe is operated in photoacoustic mode, light is delivered to the tissue via the fiber-optic bundle. Energy is deposited within the tissue, and the resulting pressure waves are received by the same ultrasound transducer. The signals are processed and the spatially co-registered ultrasound, and photoacoustic images are reconstructed to form a composite image of the optically absorbing components against the background anatomical structure. From the design of the PAUS system, it is apparent that the benefit of combining the two imaging techniques lies in the complementary information that is acquired, e.g., the functional properties of the tissue can be displayed in the context of the structural tissue properties.

11.1.4 Metal Nanoparticles as Photoacoustic Contrast Agents

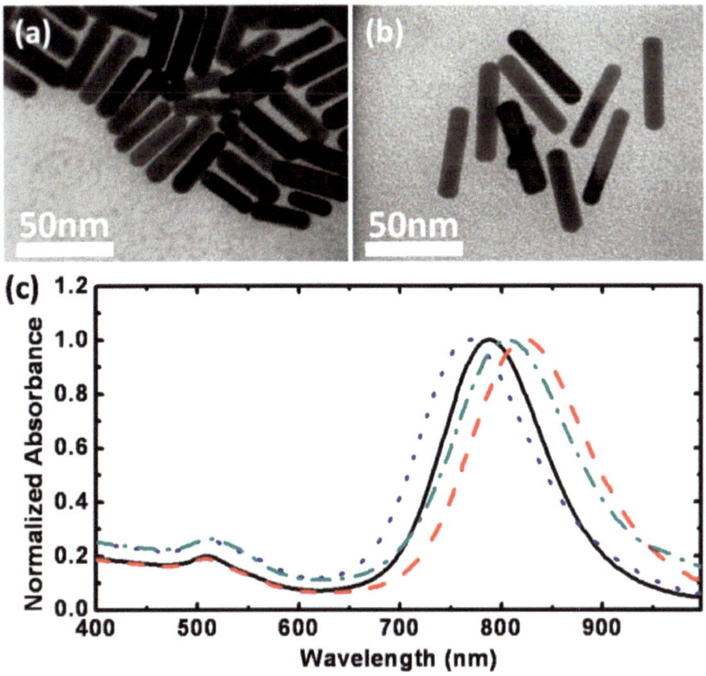

Figures 11.5 Transmission electron micrograph images of gold nanorods with aspect ratios of (a) 3.8 and (b) 4.4. (c) the normalized optical extinction spectra for samples of gold nanorods synthesized with aspect ratios between 3.8 (blue line) and 4.4 (red line), adapted from Ref. [24].

An optimal photoacoustic contrast agent should possess a large optical absorption cross section resulting in strong signal-to-noise enhancement, an acceptable biocompatibility profile, and a facile fabrication method. The utility of exogenous agents, including organic dyes and inorganic nanoparticles, has been demonstrated for use in photoacoustic imaging [26, 49]. These NIR-light-absorbing contrast agents are particularly relevant in the wavelength range of 700–1100 nm, where the optical absorption of tissue components such as hemoglobin and water is minimal (Fig. 11.2). Metal nanoparticles, specifically noble metals such as gold and silver, possess superior optical and thermal conductivity properties. These result in efficient conversion of absorbed optical energy to thermal energy that is desirable for photoacoustic imaging. At the nanometer scale, these particles have a large number of mobile surface electrons (plasmons) relative to the core. Plasmons can undergo collective, coherent oscillations when excited with a resonant laser light frequency. The resonant peak wavelength of these metal nanoparticles can be spectrally tuned by varying the size, shape, and composition of the nanoparticle, as well as the dielectric properties of the surrounding medium, as predicted by Mie theory [45]. Metal nanoparticles synthesized into anisotropic shapes with high aspect ratios (the ratio of the length to the width of the particle), such as rods, discs, shells, prisms, and nanocresents, have tunable resonance wavelength frequencies in the NIR region [65]. Gold nanorods are a common example of an anisotropic particle, as illustrated in Fig. 11.5a–c. Gold nanorods can be synthesized over a large range of aspect ratios, with a peak resonance wavelength that experiences a redshift as the aspect ratio of the rod is increased. This effect is shown in Fig. 11.5c, where the red dashed curve represents a nanorod with a higher aspect ratio than that shown by the dotted blue curve. The shift in the resonance peak results from interactions of the longitudinal dipole with the incident optical frequency, and can be calculated using the discrete dipole approximation (DDA) and experimentally verified [23, 38, 65]. For clinical applications, these anisotropic nanoparticles can be functionalized for distribution to desired tissues using targeting moieties attached to the nanoparticle surface (i.e., antibodies, peptides, aptamers, etc.). This functionalization facilitates active delivery to the tissue or cells of interest. In this text, nanoparticles functionalized with antibodies specific to upregulated receptors on the surface of cancer cells are termed "targeted nanoparticles." Nanoparticles can also accumulate passively in diseased tissue via the enhanced permeability and retention effect [3, 27]. When intravenously administered, the functionalized nanoparticles can then be used for detection and/or image guided therapy of the diseased tissue.

11.2 COMBINED PAUS IMAGING OF GOLD NANOPARTICLES *IN VIVO*

Gold nanoparticles are quickly becoming the standard for use as photoacoustic imaging contrast agents. Gold nanospheres, nanoshells, nanocages, and nanorods have high optical absorption cross sections, and therefore create strong photoacoustic signals. To date, only a handful of groups have employed metal nanoparticles as photoacoustic contrast agents *in vivo*. Wang *et al.* have performed photoacoustic imaging augmented by gold nanoshells, gold nanocages in a rat brain, and gold nanorods in sentinel lymph node mapping [64, 71]. Multiple biomarkers can be imaged using gold nanorods, as demonstrated by Li *et al* [37]. Also, Oraevsky *et al.*, Agarwal *et al.*, and Rayavarapu *et al.* have used bioconjugated gold nanorods as photoacoustic contrast agents for detection of cancer [3, 14, 37, 54]. Not only can photoacoustic imaging detect the presence of the nanoparticles in the body, it can also aid in monitoring the interactions of the gold nanoparticles at a molecular level. For example, it was shown that photoacoustic imaging of gold nanoparticles targeted to epidermal growth factor receptor (EGFR) can detect cancer cells with high sensitivity and selectivity [43, 44]. When anti-EGFR targeted gold nanoparticles bind to EGFR they tend to cluster in the same spatial distribution as EGF receptors. These receptors are known to form closely spaced assemblies upon activation with EGF, followed by endocytosis of the receptors [39]. The receptor-mediated aggregation of gold nanoparticles causes plasmon coupling of the clustered nanoparticles, leading to an optical redshift of the plasmon resonances and an increase in absorption in the red region as shown in (Fig. 11.6a) [1, 63]. An opportunity to differentiate cancer cells from surrounding benign cells using multi-wavelength photoacoustic imaging is provided by the changes in optical properties when targeted gold nanoparticles interact with cancer cells (Fig. 11.6b–g). The increase in photoacoustic contrast is attributed to the EGFR-mediated assembly of gold nanoparticles on the cytoplasmic membrane of the cancerous cells; this assembly leads to plasmon resonance coupling between adjacent gold particles and changes in their absorbance spectra shown in Fig. 11.6a. In addition to early detection of cancer, plasmon resonance coupling of gold nanoparticles has been suggested for imaging of macrophages in atherosclerotic plaques [69].

The feasibility of using the combined ultrasound and photoacoustic imaging technique to monitor the nano-molecular interactions of gold nanoparticles in nude mice with tumor xenografts *in vivo* was established. The tumors were created from A431 cells (human epithelial cancer cell line that overexpresses EGFR) injected subcutaneously in immunodeficient mice. The photoacoustic signal from the tumor region was analyzed as a function of time and optical wavelength before and after the intravenous injection of either EGFR targeted or PEGylated gold (nontargeted) nanoparticles. As can be seen in Fig. 11.7, nontargeted gold nanoparticles do not have high absorption at 720 nm. However, when targeted Au nanoparticles interact with cancer cells, there is an increased optical absorption at 720 nm. Hence molecular specific gold nanoparticles and photoacoustic imaging could be used to monitor nano-molecular interactions *in vivo*.

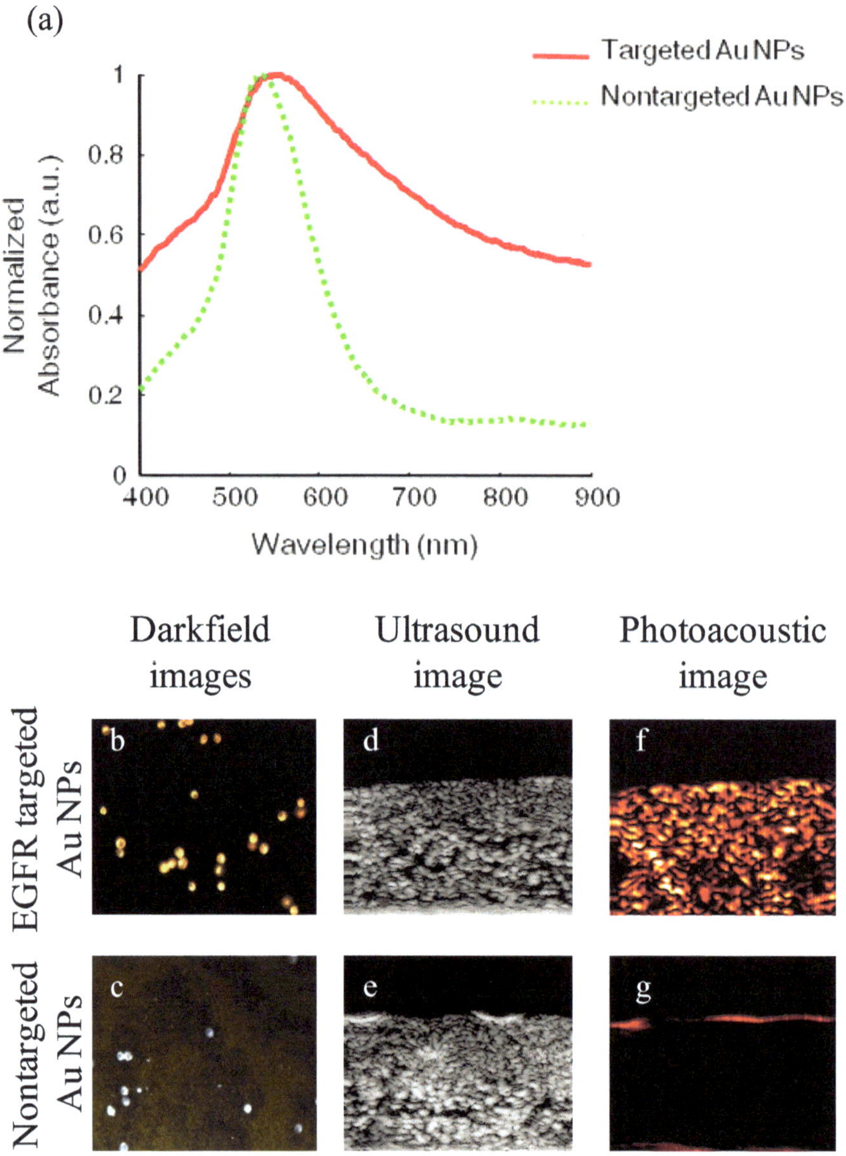

Figure 11.6 (a) Optical properties of EGFR targeted (solid red line) and nontargeted (dotted green line) gold nanoparticles when attached to cells. (b–g) Darkfield, ultrasound, and photoacoustic images (at 680 nm wavelength illumination) of A431 cells labeled with EGFR targeted Au nanoparticles and mixed with PEGylated (nontargeted) Au nanoparticles. Adapted from Ref. [43].

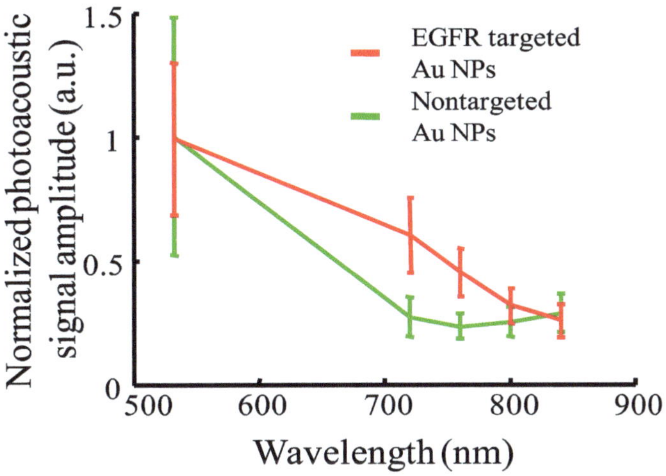

Figure 11.7 Photoacoustic signal from the tumor region as a function of wavelength 4 hours after intravenous injection of EGFR targeted ot nontargeted gold nanoparticles. The increase in photoacoustic signal in the NIR region from EGFR targeted gold nanoparticles is caused by cell induced aggregation and a redshifting of the plasmon resonance of gold nanoparticles. Thus, photoacoustic imaging can be used to specifically differentiate nanoparticles that interacted with cells.

11.3 COMBINED PAUS IMAGING OF SILVER NANOPARTICLES *IN VIVO*

Compared with gold nanoparticles, silver nanoparticles of similar shape have 10% stronger absorption cross sections [22] and therefore could be a superior contrast agent, but silver's largely disputed biocompatibility has led researchers toward adopting gold instead. Due to silver's advantage with absorbance properties and the fact that silver will eventually degrade in the body, unlike gold, use of silver nanoparticles as photoacoustic contrast agents is explored.

11.3.1 Silver Nanoplate Production and Bioconjugation

One anisotropic shape of silver that is easy to synthesize and exhibits absorbance in the NIR spectrum of light is the silver nanoplate. Silver nanoplates can consist of several shapes, including hexagons, triangles, or discs, but their common feature is their small width, making all of the shapes in a single batch appear plate-like. A colloidal suspension of silver nanoplates that absorb light at a peak wavelength of ~800 nm are generally ~100 nm in length, but only 10 nm in width. It is this large aspect ratio between length and width that enables shifting of their peak resonance wavelength into the NIR spectrum [23, 46].

A scanning transmission electron micrograph of silver nanoplates synthesized at 80°C using methods described by Ledwith *et al.* is shown in Fig. 11.8a [35]. The synthesis includes first producing small 5 nm silver seeds by reducing a solution of silver nitrate using sodium borohydride. The resulting seeds are yellow, as shown in the far left of Fig. 11.8b, and their absorbance spectrum exhibits a maximum at 400 nm, as shown in the graph in Fig. 11.8c. These seeds are then placed into a heated solution of poly(vinyl pyrrolidone) (PVP), a stabilizing agent, and trisodium citrate, a capping and reducing agent. While stirring vigorously, silver nitrate is added dropwise over a period of several minutes. Almost immediately the solution exhibits changes in color, as shown from left to right in Fig. 11.8b. The different colors are a result of the growth of silver nanoplates starting from their seeds and developing into ~100 nm plates. In theory, adding silver slowly in the second step allows for silver ions to only diffuse to the already existing seeds in solution, avoiding new nucleation of silver seeds and growth of silver spheres during this step. Unfortunately, the process of adding silver slowly does not completely thwart the production of silver spheres alongside silver nanoplates. Predominantly silver nanoplates are formed, but a population of silver spheres is clearly present, as shown in Fig. 11.8a and corroborated in Fig. 11.8c, with the peak at 400 nm still present in all of the large nanoplate solution extinction spectra. Many of the methods for rapid synthesis of silver nanoplates result in a small population of spheres also in solution [35, 78]. Other elegant ways of creating silver triangular plates with no accompanying spheres in solution were developed, but the resulting particles are so perfect that their structure is thermodynamically unstable [73]. Indeed, these silver nanotriangles are stable in water, but any change in media or addition of functional groups such as poly(ethylene glycol)-SH (PEG-SH) causes large shifts in the shape, resulting in a strong blueshift and broadening of the plasmon resonance peak. Generally, methods to avoid sphere production during synthesis or to separate the plates from the spheres post production are needed. This separation is important for *in vivo* applications because the silver spheres are not participating in enhancing photoacoustic signal, since the wavelengths used in imaging are between 700 and 1100 nm and the spheres absorb at 400 nm. Therefore, removing the sphere content will ensure that the patient is exposed to as little metal as possible when these nanoplates are used as photoacoustic contrast agents.

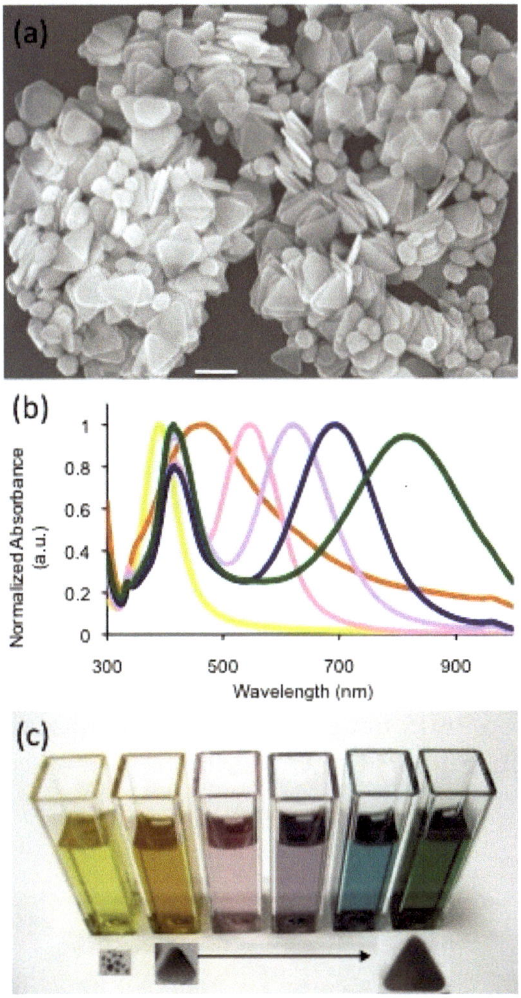

Figures 11.8 Scanning transmission electron micrograph (STEM) of silver nanoplates (a); the extinction spectra of increasing sizes of nanoplates (b) corresponding to the colors seen in solution (c) as the nanoplates are growing. Scale bar in the STEM image is 100 nm.

Using thiol chemistries, silver nanoplates can be conjugated to antibodies in the same manner as gold nanoparticles [31]. As a proof of concept, monoclonal epidermal growth factor receptor (EGFR) antibodies were bioconjugated to these silver nanoplates. The remainder of the nanoplates'surface was functionalized with PEG-SH. To test the effectiveness of bioconjugation, the functionalized nanoplates were mixed with the pancreatic cancer cell line MPanc96, which is known to overexpress EGFR. Specifically, the silver nanoplates bioconjugated with anti-EGFR were mixed in the media of MPanc96 cells and allowed to incubate for 4 hours. Clear binding of the nanoplates to the cell was observed (Fig. 11.9). However, when just PEGylated silver nanoplates (containing no antibody) were incubated with MPanc96 cells for the same amount of time, this uptake effect was not evident. Therefore, it is clear that silver nanoplates can be synthesized and conjugated to functional groups, like antibodies, that make them target and interact with cancer cells.

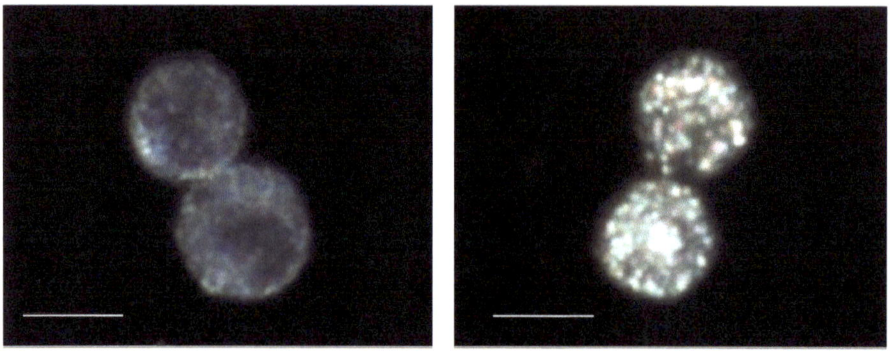

Figure 11.9 Darkfield microscopy images of MPanc96 cells mixed with PEGylated silver nanoplates (left) and antibody conjugated silver nanoplates (right). Scale bars are 20 μm.

11.3.2 Combined PAUS Imaging of Silver Nanoplates in Subcutaneous Pancreatic Tumors Grown in Mice

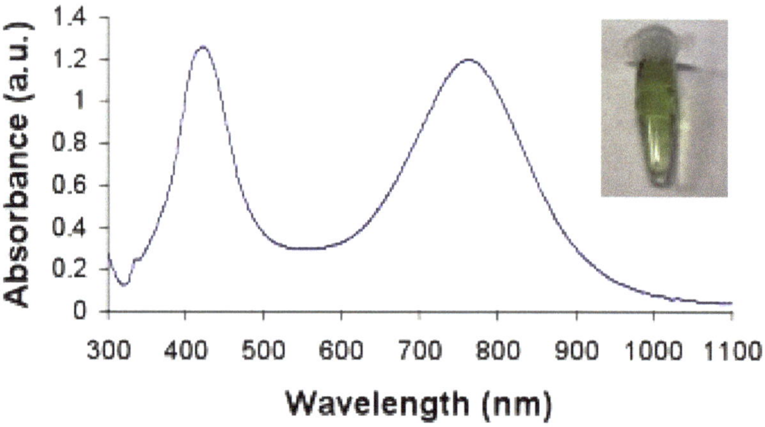

Figure 11.10 Extinction spectra and photograph of the silver nanoplates taken prior to *in vivo* intravenous injection.

The bioconjugated silver nanoplates were tested *in vivo* for use as a photoacoustic contrast agent. The animal model used was a nu/nu mouse with a xenographic pancreatic tumor grown subcutaneously on its back. Before injection of the nanoplates, photoacoustic and ultrasound images were obtained of the tumor area using an imaging system as shown in Fig. 11.3. The wavelengths of light used during photoacoustic imaging corresponded to the absorption spectra of the silver nanoplates (Fig. 11.10). The peak resonance wavelength of 760 nm was employed, along with 800 nm (85% of the absorbance at 760 nm) and 900 nm (25% of the absorbance at 760 nm).

After intravenous administration of about 10^{12} silver nanoplates in 200 μL of saline, photoacoustic, and ultrasound imaging was performed for several hours. Peak contrast enhancement was evident at 4 hours post injection and is shown in (Fig. 11.11a–k). Figures 11.11a through 11.11d are pre-injection images of the tumor area, which show the natural endogenous photoacoustic contrast, mostly from blood, in and outside the tumor area (note that photoacoustic signal is denoted in these images using a red color map and some of the photoacoustic signals are marked with a white arrow). Figures 11.11e through 11.11h were captured at 4 hours post injection and show a clear increase in photoacoustic signal in the tumor region. This increase in contrast is due to nanoplate accumulation since the photoacoustic images captured vary in intensity from highest signal at 760 nm to diminishing signal at 900 nm where the nanoplates no longer absorb as strongly. Lastly, Figs. 11.11i through 11.11k are combined PAUS images overlaying the photoacoustic signal captured at the various wavelengths on top of the ultrasound anatomical background images. These combined PAUS images are valuable since they clearly delineate where nanoparticle accumulation occurred within the tumor region. PAUS images also highlight the heterogeneous nature of the nanoplate accumulation. Indeed, nanoplates accumulated most at the base of the tumor (labeled with a white arrow in Fig. 11.11f–h), even though mild accumulation was visible inside the tumor and on a vessel in the upper left of the tumor. Upon euthanization of the mouse and during the organ harvesting process, large blood vessels were observed at the tumor base. It is understandable that this large nutrient basin for the tumor is in fact where the largest nanoplate accumulation occurred since the nanoplates were using the bloodstream to circulate and target the tumor. Furthermore, this heterogeneous accumulation of nanoplates in the tumor shows that access to all regions of the tumor via the bloodstream is limited. Therefore, strategies involving delivery of nanoparticles via the bloodstream for diagnostic imaging or therapy applications will suffer from this lack of homogeneous distribution of nanoparticles that reach and accumulate in the tumor.

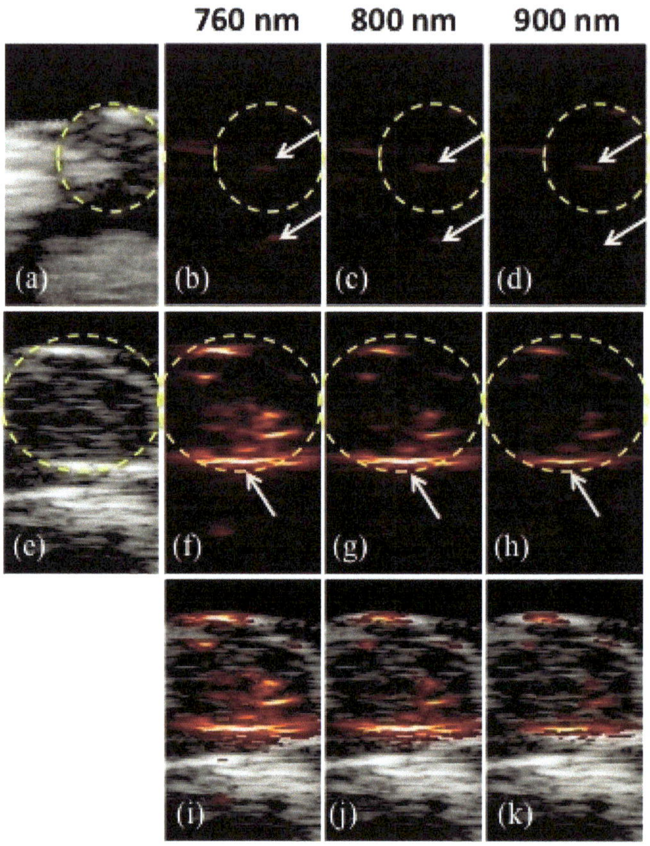

Figures 11.11 Photoacoustic and ultrasound imaging of a pancreatic cancer mass grown subcutaneously in a nu/nu mouse. (a) is an ultrasound image of the tumor prior to silver nanoplate injection, and (b)–(d) are photoacoustic images of the same region as shown in (a) acquired using 760 nm, 800 nm, and 900 nm laser light, respectively. (e) is an ultrasound image of the tumor region after intravenous injection of bioconjugated silver nanoplates, and (f) and (g) are photoacoustic images of the same region as shown in (e) acquired using 760 nm, 800 nm, and 900 nm laser light, respectively. (i) and (j) are combined photoacoustic and ultrasound images of (f) through (g), each overlayed on (e). All images represent 17.5 mm by 10.5 mm, the yellow circles outline the boundaries of the tumor area, and the white arrows point to photoacoustic signals of interest.

11.4 CYTOTOXICITY OF METAL NANOPARTICLES

In the studies presented above, gold or silver nanoparticles were targeted to the epidermal growth factor receptor, which is overexpressed in many types of epithelial cancers. These receptors are also found on normal, healthy cells, but not nearly to the extent found on cancerous cells. Therefore, there will be nonspecific uptake of anti-EGFR targeted nanoparticles by other, healthy cells. Because they are intravenously injected and will interact with healthy cells, the metal nanoparticles need to be biocompatible. Gold and silver are both noble metals, and therefore share some intrinsic reactivity properties, but their biocompatibility profiles are different and are discussed separately below. Physicians have been using colloidal gold to treat rheumatoid arthritis for almost a century [2]. The widespread use of gold in that treatment was our first indication of gold's biocompatibility. Researchers now employ gold for biomedical applications because once in colloidal form, gold is stable, nonreactive, and easy to functionalize. In the last two decades, researchers have functionalized the surface of gold with polymers, antibodies, drugs, aptamers, and so on. These different coatings add various intended functionalities to the gold nanoparticles. For example, the coatings help enhance nanoparticle interaction with cell membranes, endosomes, or other specific intracellular targets. Generally, these coatings largely determine the cytotoxicity and biocompatibility of the nanoparticle, since it is the coating that interacts most with the cellular environment. A comprehensive review of the cytotoxicity of gold nanoparticles, including gold nanoshells and nanorods, with various coatings is outlined by Lewinski *et al.* [36]. As stated in their review, gold is known to be biocompatible up to concentrations of 200 μg/mL, at which point it lowered cell viability between 15% and 20% in a few select studies [17, 57, 61, 66].

Silver has been used throughout history even more extensively than gold. The antimicrobial properties of silver have been known and exploited for centuries [19]. Recent increases in bacterial resistance to antibiotics have led researchers to explore alternative biocides, and thus today we see a renewed interest in this age-old remedy. Today silver is being researched or is already employed as a germicide in a variety of products to include catheter coatings, external fixation pins, and dental amalgams [7, 56, 72, 74]. Silver is even used in swimming pools. Silver has also been used in the spin cycle of laundry machines and embedded in toilets to reduce bacterial proliferation [18, 62]. The most prominent use of silver in biomedical applications is silver sulphadiazine, which has been the gold standard for use in burn treatments to limit infection since World War II [29]. Silver's efficacy in killing bacteria is directly related to the amount of silver ions (Ag^{+1}) present in the solution. Unfortunately, at high Ag^{+1} concentrations (>1 μg/mL), in addition to killing bacteria, silver cations will also bind with proteins and/or chloride ions and possibly disrupt cellular processes [52]. Drinking excessive colloidal silver is known to cause a cosmetic condition called argyria, which is a bluing of the skin [8, 13, 55]. The mode of action of the silver cation in relation to cytotoxicity was elucidated by incorporating silver salts into biological fluids *in vitro*, but similar studies using silver nanoparticles have yet to be extensively researched. Nanoparticles of silver will not release silver cations into solution as readily as silver salts and therefore might possess a more biocompatible profile. In one preliminary study by our group, pancreatic cancer and breast cancer cells remained viable up to a concentration of 2 mg/mL of silver nanoparticles incorporated in the media. Interestingly, silver nanoparticles have also recently been shown to inhibit action of the HIV virus [15, 67]. Thus, silver is an attractive material for several of its unique properties to include optical absorption and scattering as well as antibacterial and antiviral, but we should study at length the specific formulation of silver employed to properly access biocompatibility.

Since gold and silver nanoparticles are excellent photoabsorbers, cells labeled with these nanoparticles may have to endure photodamage when interacting with laser pulses. Though the photothermal effects associated with nanoparticles are required to destroy cancer cells, it is an undesired effect during the photoacoustic imaging regime. Therefore to confirm the safety of photoacoustic imaging using metal nanoparticles, viability experiments were conducted on a monolayer of cancer cells labeled with gold nanoparticles. Cell death was observed at laser fluencies of approximately 500 mJ/cm^2. Studies indicating photothermal destruction of cancer cells using gold nanoparticles also indicate cell death at these fluencies [53, 75–77]. Photoacoustic imaging utilizes a diffused laser beam with fluencies lower than 25-30 mJ/cm^2. At these fluencies no cell death was observed, and that indicates the nondestructive nature of photoacoustic imaging with metal nanoparticles.

11.5 DISCUSSION AND FUTURE DIRECTIONS

As mentioned above, high concentrations of contrast agents could be toxic to the body. Therefore, highly sensitive imaging techniques are required to detect small concentrations of nanoparticles. Nuclear and optical imaging modalities use nanocontrast agents at picomolar concentrations. Magnetic resonance imaging can detect superparamagnetic iron oxide nanoparticles at nanomolar concentrations [11, 50]. Photoacoustic imaging has the potential to detect gold nanoparticles at picomolar concentrations, indicating that highly sensitive detection of cancer cells, on par with existing techniques, is possible using photoacoustics [14, 44].

The unique optical absorption properties of metal nanoparticles have been exploited by scientists for both imaging and therapeutic applications. Photothermal therapy of cancer cells *in vitro* and *in vivo* has recently been demonstrated with gold nanoparticles of different shapes especially gold nanorods, nanoshells, and nanocages [10, 12, 20, 40, 68]. Either continuous or pulsed laser irradiation was used for localized photothermal destruction of cancer cells labeled with targeted nanoparticle bioconjugates [20, 41, 53, 75]. Moreover, photodamage to the surrounding healthy tissue is minimized with the use of targeted gold nanoparticles. Studies have demonstrated that metal nanoparticles can cause thermal or mechanical damage to the adjacent cells upon laser irradiation either by generating excessive heat or by formation of microbubbles [25]. For example, a 2 minute exposure to continuous laser irradiation (7 mW power) killed most of the human oral squamous carcinoma cells labeled with EGFR conjugated gold nanospheres (30 nm diameter) [21]. Zharov *et al.* have shown 10% viability of bacteria loaded with 40 nm diameter gold nanospheres by using 12 ns laser pulses [77]. The size of the nanoparticles and the number of nanoparticles present in a cell also play a major role in determining the extent of photothermal damage induced [53, 77].

Photothermal therapy with metal nanoparticles has numerous advantages over other phototherapeutic applications. First, the number of metal particles required for cell destruction is a few orders lower than immunotargeted photodynamic methods (approximately 10^6 molecules per cell) [53]. Secondly, the therapeutic procedure has no additional unpalatable side effects accompanying current therapeutic procedures, thereby sparing the surrounding benign tissue. Finally, metal nanoparticles can be used simultaneously as photoacoustic imaging and therapeutic contrast agents; i.e, the nanoparticles can initially be used as photoacoustic contrast agents to demarcate the tumor location followed by inducing localized photothermal damage to the tumor region by using continuous or pulsed laser illumination. In addition, photoacoustic imaging could also aid in monitoring photothermal therapy. Shah *et al.* have also proposed a combined ultrasound and photoacoustic imaging system to plan, guide, and monitor photothermal therapy of cancer [60]. In the study presented by Shah *et al.*, laser therapy experiments were carried out in tissue-mimicking phantoms and in *ex vivo* tissue [59]. Ultrasound imaging was first used to image tissue structure and identify tissue composition. Photoacoustic imaging detected the presence and location of nanoparticles. Both ultrasound and photoacoustic imaging were then used to monitor temperature during photothermal therapy. Quantitative analysis of the data indicated that temperature monitoring using photoacoustic imaging has higher signal-to-noise ratio in the presence of exogenous optical absorbers [16, 33].

In summary, a noninvasive imaging tool—combined ultrasound and photoacoustic imaging—that can detect and diagnose pathologies such as cancer using molecularly specific plasmonic metal nanoparticles was demonstrated in this chapter. The advantages and limitations of the molecular specific photoacoustic imaging technique were also enumerated. In addition, the possibility of using complex nanostructures such as nanorods and the extension of the imaging tool to a combined diagnostic and therapeutic procedure was presented. Overall, the combination of molecular specific nanoparticles and the ultrasound-guided photoacoustic imaging technique has the potential to be translated to the clinic for highly sensitive detection of cancer and monitoring of molecular specific therapeutic procedures.

Acknowledgments

The authors thank Dr. Andrei Karpiouk for his assistance in the mechanics of the photoacoustic setup, Pratixa Joshi and Yun-Sheng Chen for their assistance in gold nanoparticle synthesis, and Mike Souza and Ryan Truby for their assistance in silver nanoplate synthesis and characterization. The personnel at the Animal Resource Center (ARC) at the University of Texas at Austin also provided valuable assistance and expertise. Lastly, funding from the National Institute of Health is recognized.

References

1. Aaron J., Nitin N., Travis K., Kumar S., Collier T., Park S. Y., Jose-Yacaman M., Coghlan L., Follen M., Richards-Kortum R., and Sokolov K. (2007) Plasmon resonance coupling of metal nanoparticles for molecular imaging of carcinogenesis *in vivo*. *J Biomed Opt* **12,** 034007.
2. Aaseth J., Haugen M., and F¯rre ÿ. (1998) Rheumatoid arthritis and metal compounds: perspectives on the role of oxygen radical detoxification. *The Analyst* **123,** 3–6.
3. Agarwal A., Huang S. W., O'Donnell M., Day K. C., Day M., Kotov N., and Ashkenazi S. (2007) Targeted gold nanorod contrast agent for prostate cancer detection by photoacoustic imaging. *J Appl Phys* **102,** 064701.
4. Andreev V. G., Karabutov A. A., and Oraevsky A. A. (2003) Detection of ultrawide-band ultrasound pulses in optoacoustic tomography. *IEEE Trans Ultrason Ferroelectr Frequency Control* **50,** 1383–1390.
5. Beard P. C., and Mills T. N. (1995) Evaluation of an optical fiber probe for *in vivo* measurement of the photoacoustic response of tissues. *Proc SPIE,* **2388,** 446.
6. Bell A. G. (1880) Apparatus for Signalling and Communicating Called "Photophone". American Bell Telephone Company.
7. Bosetti M., Masse A., Tobin E., and Cannas M. (2002) Silver coated materials for external fixation devices: *in vitro* biocompatibility and genotoxicity. *Biomaterials* **23,** 887–892.
8. Bouts B. A. (1999) Images in clinical medicine. Argyria. *N Engl J Med* **340,** 1554.
9. Bowen T. (1981) Radiation-induced thermoacoustic soft tissue imaging. 1981 IEEE Ultrasonics Symposium Chicago, IL, pp. 817–822.
10. Chen J., Wang D., Xi J., Au L., Siekkinen A., Warsen A., Li Z.-Y., Zhang H., Xia Y., and Li X. (2007) Immuno gold nanocages with tailored optical properties for targeted photothermal destruction of cancer cells. *Nano Lett* **7,** 1318–1322.
11. Cheon J., and Lee J. H. (2008) Synergistically integrated nanoparticles as multimodal probes for nanobiotechnology. *Acc Chem Res* **41,** 1630–1640.
12. Ding A.-A., Chen Y.-Y., C. W. C., Li P.-C., and Shieh D.-B. (2008) HER-2 antibody conjugated gold nano rod for in vivo photothermal therapy. NANO '08. 8th IEEE Conference on Nanotechnology, 2008, pp. 882–885.
13. Drake P. L., and Hazelwood K. J. (2005) Exposure-related health effects of silver and silver compounds: a review. *Ann Occup Hyg* **49,** 575–585.
14. Eghtedari M., Oraevsky A., Copland J., Kotov N., Conjusteau A., and Motamedi M. (2007) High sensitivity of *in vivo* detection of gold nanorods using a laser optoacoustic imaging system. *Nano Lett* **7,** 1914–1918.
15. Elechiguerra J. L., Burt J. L., Morones J. R., Camacho-Bragado A., Gao X., Lara H. H., and Yacaman M. J. (2005) Interaction of silver nanoparticles with HIV-1. *J Nanobiotechnol* **3,** 6.
16. Emelianov S. Y., Aglyamov S. R., Karpiouk A. B., Mallidi S., Park S., Sethuraman S., Shah J., Smalling R. W., Rubin J. M., and Scott W. G. (2006) Synergy and applications of combined

ultrasound, elasticity, and photoacoustic imaging. *Proceedings of IEEE Ultrasonics Symposium Vancouver*, British Columbia, pp. 405–415.

17. Fu W., Shenoy D., Li J., Crasto C., Jones G., Dimarzio C., Sridhar S., and Amiji M. (2005) Biomedical applications of gold nanoparticles functionalized using hetero-bifunctional poly(ethylene glycol) spacer. Vol 845, Materials Research Society, Warrendale, PA, 1999, p. 223.

18. Hill J. (2004) *Colloidal Silver: A Literature Review: Medical Uses, Toxicology, and Manufacture*. Clear Springs Press, LLC.

19. Hippocrates (1994–2000) *On Ulcers*. Written 400 B.C.E., translated by Francis Adams.

20. Hirsch L. R., Stafford R. J., Bankson J. A., Sershen S. R., Rivera B., Price R. E., Hazle J. D., Halas N. J., and West J. L. (2003) Nanoshell-mediated near-infrared thermal therapy of tumors under magnetic resonance guidance. *Proc Natl Acad Sci U S A* **100**, 13549–13554.

21. Huang X., Qian W., El-Sayed I. H., and El-Sayed M. A. (2007) The potential use of the enhanced nonlinear properties of gold nanospheres in photothermal cancer therapy. *Lasers Surg Med* **39**, 747–753.

22. Jackson J. B., and Halas N. J. (2001) Silver nanoshells: variations in morphologies and optical properties. *J Phys Chem B* **105**, 2743–2746.

23. Jain P. K., Lee K. S., El-Sayed I. H., and El-Sayed M. A. (2006) Calculated absorption and scattering properties of gold nanoparticles of different size, shape, and composition: applications in biological imaging and biomedicine. *J Phys Chem B* **110**, 7238.

24. Joshi P., Chen Y.-S., Kim S., Shah J., and Emelianov S. (2009) Molecular therapeutic agents for noninvasive photoacoustic image-guided photothermal therapy. *31st Annual International IEEE The Engineering in Medicine and Biology Conference (EMBC) Proceedings*, **1158**, 4106–4109.

25. Khlebtsov B., Zharov V., Melnikov A., Tuchin V., and Khlebtsov N. (2006) Optical amplification of photothermal therapy with gold nanoparticles and nanoclusters. *Nanotechnology* **17**, 5167–5179.

26. Kim G., Huang S. W., Day K. C., O'Donnell M., Agayan R. R., Day M. A., Kopelman R., and Ashkenazi S. (2007) Indocyanine-green-embedded PEBBLEs as a contrast agent for photoacoustic imaging. *J Biomed Opt* **12**, 044020.

27. Kim K., Huang S.-W., Ashkenazi S., O'Donnell M., Agarwal A., Kotov N. A., Denny M. F., and Kaplan M. J. (2007) Photoacoustic imaging of early inflammatory response using gold nanorods. *Appl Phys Lett* **90**, 223901.

28. Kinsler L. E., Frey A. R., Coppens A. B., and Sanders J. V. (2000) *Fundamentals of Acoustics*. John Wiley and Sons, Inc., New York.

29. Klasen H. J. (2000) Historical review of the use of silver in the treatment of burns. I. Early uses. *Burns* **26**, 117–130.

30. Kruger R. A., Liu P., and Appledorn C. R. (1995) Photoacoustic ultrasound (PAUS)—reconstruction tomography. *Med Phys* **22**, 1605.

31. Kumar S., Aaron J., and Sokolov K. (2008) Directional conjugation of antibodies to nanoparticles for synthesis of multiplexed optical contrast agents with both delivery and targeting moieties. *Nat Protoc* **3**, 314–320.

32. Landau L. D., and Lifshitz E. M. (1987) *Fluid Mechanics.* Pergamon Press.

33. Larina I. V., Larin K. V., and Esenaliev R. O. (2005) Real-time optoacoustic monitoring of temperature in tissues. *J Phys D* **38**, 2633.

34. Laufer J., Delpy D., Elwell C., and Beard P. (2007) Quantitative spatially resolved measurement of tissue chromophore concentrations using photoacoustic spectroscopy: application to the measurement of blood oxygenation and haemoglobin concentration. *Phys Med Biol* **52**, 141–168.

35. Ledwith D. M., Whelan A. M., and Kelly J. M. (2007) A rapid, straight-forward method for controlling the morphology of stable silver nanoparticles. *J Mater Chem* **17**, 2459–2464.

36. Lewinski N., Colvin V., and Drezek R. (2008) Cytotoxicity of nanoparticles. *Small* **4**, 26–49.

37. Li P., Wang C., Shieh D., Wei C., Liao C., Poe C., Jhan S., Ding A., and Wu Y. (2008) *In vivo* photoacoustic molecular imaging with simultaneous multiple selective targeting using antibody-conjugated gold nanorods. *Opt Express* **16**, 18605–18615.

38. Li P. C., Wei C. W., Liao C. K., Chen C. D., Pao K. C., Wang C. R. C., Wu Y. N., and Shieh D. B. (2006) Multiple targeting in photoacoustic imaging using bioconjugated gold nanorods. *Proc SPIE* **6086**, 60860M.

39. Lidke D. S., Nagy P., Heintzmann R., Arndt-Jovin D. J., Post J. N., Grecco H. E., Jares-Erijman E. A., and Jovin T. M. (2004) Quantum dot ligands provide new insights into erbB/HER receptor-mediated signal transduction. *Nat Biotechnol* **22**, 198–203.

40. Loo C., Lin A., Hirsch L., Lee M. H., Barton J., Halas N., West J., and Drezek R. (2004) Nanoshell-enabled photonics-based imaging and therapy of cancer. *Technol Cancer Res Treat* **3**, 33–40.

41. Loo C., Lowery A., Halas N., West J., and Drezek R. (2005) Immunotargeted nanoshells for integrated cancer imaging and therapy. *Nano Lett* **5**, 709–711.

42. Mallidi S., Aglyamov S. R., Karpiouk A. B., Park S., and Emelianov S. Y. (2006) Functional and morphological ultrasonic biomicroscopy for tissue engineers. *Proc SPIE Med Imaging* **6147**, 61470Y61471-61477.

43. Mallidi S., Larson T., Aaron J., Sokolov K., and Emelianov S. (2007) Molecular specific optoacoustic imaging with plasmonic nanoparticles. *Opt Express* **15**, 6583–6588.

44. Mallidi S., Larson T., Tam J., Joshi P. P., Karpiouk A., Sokolov K., and Emelianov S. (2009) Multiwavelength photoacoustic imaging and plasmon resonance coupling of gold nanoparticles for selective detection of cancer. *Nano Lett* **9**, 2825–2831.

45. Mie G. (1908) Contributions to the optics of turbid media, particularly of colloidal metal solutions. *Ann Phys (Leipzig)* **25**, 377–445.

46. Murphy C. J., Sau T. K., Gole A. M., Orendorff C. J., Gao J., Gou L., Hunyadi S. E., and Li T. (2005) Anisotropic metal nanoparticles: synthesis, assembly, and optical applications. *J Phys Chem B* **109**, 13857–13870.

47. OMLC (2001) Optical Properties Spectra, vol. 2009.

48. Oraevsky A. A., Jacques S. L., and Tittel F. K. (1997) Measurement of tissue optical properties by time-resolved detection of laser-induced transient stress. *Appl Opt* **36**, 402–415.

49. Oraevsky A. A., Karabutov A. A., and Savateeva E. V. (2001) Enhancement of optoacoustic tissue contrast with absorbing nanoparticles. *Proc SPIE* **4434**, 60–69.

50. Pardoe H., Chua-anusorn W., St Pierre T. G., and Dobson J. (2003) Detection limits for ferrimagnetic particle concentrations using magnetic resonance imaging based proton transverse relaxation rate measurements. *Phys Med Biol* **48**, N89–N95.

51. Park S., Karpiouk A. B., Aglyamov S. R., and Emelianov S. Y. (2008) Adaptive beamforming for photoacoustic imaging. *Opt Lett* **33**, 1291–1293.

52. Percival S. L., Bowler P. G., and Russell D. (2005) Bacterial resistance to silver in wound care. *J Hosp Infect* **60**, 1–7.

53. Pitsillides C. M., Joe E. K., Wei X., Anderson R. R., and Lin C. P. (2003) Selective cell targeting with light-absorbing microparticles and nanoparticles. *Biophys J* **84**, 4023–4032.

54. Rayavarapu R. G., Petersen W., Ungureanu C., Post J. N., van Leeuwen T. G., and Manohar S. (2007) Synthesis and bioconjugation of gold nanoparticles as potential molecular probes for light-based imaging techniques. *Int J Biomed Imaging* **2007**, 29817.

55. Rosenman K. D., Moss A., and Kon S. (1979) Argyria: clinical implications of exposure to silver nitrate and silver oxide. *J Occup Med* **21**, 430–435.

56. Rupp M. E., Fitzgerald T., Marion N., Helget V., Puumala S., Anderson J. R., and Fey P. D. (2004) Effect of silver-coated urinary catheters: efficacy, cost-effectiveness, and antimicrobial resistance. *Am J Infect Control* **32**, 445–450.

57. Salem A., Searson P., and Leong K. (2003) Multifunctional nanorods for gene delivery. *Nat Mater* **2**, 668–671.

58. Sethuraman S., Amirian J. H., Litovsky S. H., Smalling R. W., and Emelianov S. Y. (2008) Spectroscopic intravascular photoacoustic imaging to differentiate atherosclerotic plaques. *Opt Express* **16**, 3362–3367.

59. Shah J., Aglyamov S. R., Sokolov K., Milner T. E., and Emelianov S. Y. (2008) Ultrasound imaging to monitor photothermal therapy — feasibility study. *Opt Express* **16**, 3776–3785.

60. Shah J., Park S., Aglyamov S., Larson T., Ma L., Sokolov K., Johnston K., Milner T., and Emelianov S. Y. (2008) Photoacoustic imaging and temperature measurement for photothermal cancer therapy. *J Biomed Opt* **13**, 034024.

61. Shenoy D., Fu W., Li J., Crasto C., Jones G., DiMarzio C., Sridhar S., and Amiji M. (2006) Surface functionalization of gold nanoparticles using hetero-bifunctional poly (ethylene glycol) spacer for intracellular tracking and delivery. *Int J Nanomed* **1**, 51.

62. Silver S., Phung le T., and Silver G. (2006) Silver as biocides in burn and wound dressings and bacterial resistance to silver compounds. *J Ind Microbiol Biotechnol* **33**, 627–634.

63. Sokolov K., Follen M., Aaron J., Pavlova I., Malpica A., Lotan R., and Richards-Kortum R. (2003) Real-time vital optical imaging of precancer using anti-epidermal growth factor receptor antibodies conjugated to gold nanoparticles. *Cancer Res* **63**, 1999–2004.

64. Song K., Kim C., Maslov K., and Wang L. (2009) Noninvasive *in vivo* spectroscopic nanorod-contrast photoacoustic mapping of sentinel lymph nodes. *Eur J Radiol* **70**, 227–231.

65. Sosa I., Noguez C., and Barrera R. G. (2003) Optical properties of metal nanoparticles with arbitrary shapes. *J Phys Chem B* **107**, 6269–6275.

66. Su C., Sheu H., Lin C., Huang C., Lo Y., Pu Y., Weng J., Shieh D., Chen J., and Yeh C. (2007) Nanoshell magnetic resonance imaging contrast agents. *J Am Chem Soc* **129**, 2139–2146.

67. Sun R. W., Chen R., Chung N. P., Ho C. M., Lin C. L., and Che C. M. (2005) Silver nanoparticles fabricated in Hepes buffer exhibit cytoprotective activities toward HIV-1 infected cells. *Chem Commun (Camb)* 5059–5061.

68. Tong L., Wei Q., Wei A., and Cheng J. X. (2009) Gold nanorods as contrast agents for biological imaging: optical properties, surface conjugation and photothermal effects. *Photochem Photobiol* **85**, 21–32.

69. Wang B., Yantsen E., Larson T., Karpiouk A. B., Sethuraman S., Su J. L., Sokolov K., and Emelianov S. Y. (2009) Plasmonic intravascular photoacoustic imaging for detection of macrophages in atherosclerotic plaques. *Nanoletters* **9**, 2212–2217.

70. Wang X., Pang Y., Ku G., Xie X., Stoica G., and Wang L. V. (2003) Noninvasive laser-induced photoacoustic tomography for structural and functional *in vivo* imaging of the brain. *Nat Biotechnol* **21**, 803–806.

71. Wang Y., Xie X., Wang X., Ku G., Gill K. L., O'Neal D. P., Stoica G., and Wang L. V. (2004) Photoacoustic tomography of a nanoshell contrast agent in the *in vivo* rat brain. *Nano Lett* **4**, 1689–1692.

72. Xu M., and Wang L. V. (2006) Photoacoustic imaging in biomedicine. *Rev Sci Instrum* **77**, 041101.

73. Xue C., Metraux G. S., Millstone J. E., and Mirkin C. A. (2008) Mechanistic study of photomediated triangular silver nanoprism growth. *J Am Chem Soc* **130**, 8337–8344.

74. Yoshida K., Tanagawa M., and Atsuta M. (1999) Characterization and inhibitory effect of antibacterial dental resin composites incorporating silver-supported materials. *J Biomed Mater Res* **47**, 516–522.

75. Zharov V. P., Galitovskaya E. N., Johnson C., and Kelly T. (2005) Synergistic enhancement of selective nanophotothermolysis with gold nanoclusters: Potential for cancer therapy. *Lasers Surg Med* **37**, 219–226.

76. Zharov V. P., Kim J. W., Curiel D. T., and Everts M. (2005) Self-assembling nanoclusters in living systems: application for integrated photothermal nanodiagnostics and nanotherapy. *Nanomedicine* **1**, 326–345.

77. Zharov V. P., Mercer K. E., Galitovskaya E. N., and Smeltzer M. S. (2006) Photothermal nanotherapeutics and nanodiagnostics for selective killing of bacteria targeted with gold nanoparticles. *Biophys J* **90,** 619–627.

78. Zou X., Ying E., Chen H., and Dong S. (2007) An approach for synthesizing nanometer-to micrometer-sized silver nanoplates. *Colloids Surf A* **303,** 226–234.

Chapter 12

IMAGING CARBON NANOTUBES *IN VIVO*: A VIGNETTE OF IMAGING MODALITIES AT THE NANOSCALE

Khuloud T. Al-Jamal[a] and Kostas Kostarelos[b]
[a] *Institute of Pharmaceutical Science, King's College London, Franklin-Wilkins Building, London SE1 9NH, UK*
[b] *Nanomedicine Lab, Centre for Drug Delivery Research, The School of Pharmacy, University of London, 29–39 Brunswick Square, London WC1N 1AX, UK*
kostas.kostarelos@pharmacy.ac.uk

12.1 INTRODUCTION

Nanotechnology is an interdisciplinary research effort bridging many scientific fields from physics and chemistry to engineering, biology, and medicine. The result of such interconnections is holding great potential for the early detection, diagnosis, and personalized treatment of disease. The nanoscale range at which nanosystems operate, i.e., one-thousandth smaller than a human cell, can offer facile transport across the human body and intracellular interactions with many cell components that would otherwise be inaccessible. Imaging for early detection and diagnosis of diseases using newly emerged nanoparticles such as quantum dots (QDs) [21], carbon nanotubes (CNTs) [19], nanoshells [10], paramagnetic nanoparticles [34], and others [6, 9] have been an area of interest over the last few years.

Carbon nanotubes are a type of nanomaterial that offers unique intrinsic properties that make them very interesting candidates as imaging contrast agents. Their identification in the early 1990s [12] and their further biomedical development [18] opened a new era in the development of novel delivery systems for therapeutics and diagnostics. Carbon nanotubes are mainly classified as single-walled (SWNTs)

Nanoimaging
Edited by Beth Goins and William Phillips

www.panstanford.com

or multiwalled (MWNTs) according to the number of concentric layers of graphitic sheets rolled into cylindrical structures. The high aspect ratio of CNT offers great advantages over other nanoparticle types, since the high surface area provides multiple attachment sites for drugs, targeting ligands, and imaging probes. Advancements achieved in filling CNT with small molecules or other nanomaterials also indicate an opportunity to use them as imaging probes *in vivo* [11, 30, 35]. One of the main disadvantages of CNT is their hydrophobicity, so advances in solubilization and dispersion methodologies, including chemical functionalization of the CNT surface [4] or coating with amphiphilic molecules such as PEGylated phospholipids or polymers [13], have been an essential breakthrough to allow exploitation of the biomedical applications of CNT. Recently, successful studies in our laboratories as well as others have reported that CNT can translocate into cells using several analytical techniques, including optical microscopy, micro-Raman spectroscopy, single-particle tracking (SPT), transmission electron microscopy (TEM), flow cytometry, and fluorescence microscopy. Moreover, proof-of-concept studies have established that CNT can act as delivery systems for drugs (methotrexate, amphotericin) [26, 37], antigens, and genes (plasmid DNA, siRNA) [14, 32] into prokaryotic and mammalian cells with minimal cytotoxicity both *in vitro* [3, 5] and *in vivo* [2, 27, 40]. This indeed allows for using CNTs not only as multimodal imaging probes but also as theranostic devices.

In this chapter, we will specifically focus on the applications of CNTs as imaging contrast agents *in vivo*, with an overview on the different imaging techniques available for CNT imaging, and also discuss their limitations. Although the studies discussed here all fall within the preclinical settings, the unique properties of CNTs and the ongoing efforts on improving their biocompatibility, in addition to advances in the imaging field, hold a great promise for clinical applications in the future. Imaging modalities that will be discussed here and that have successfully tested CNTs as imaging contrast agents preclinically *in vivo* include single photon emission computed tomography (SPECT), positron emission tomography (PET), magnetic resonance imaging (MRI), optical fluorescence, and ultrasound. The list of the techniques and the comprehensive list of relevant studies are shown in Table 12.1.

12.2 MODALITIES FOR *IN VIVO* IMAGING

CNTs developed as imaging contrast agents can be classified on the basis of their surface properties into three main types: (i) pristine CNTs (pCNTs), (ii) coated (noncovalently surface modified) CNTs, and (iii) covalently modified CNTs. A schematic showing the type of modification and the modalities used to image CNT-based contrast agents is shown in Fig. 12.1.

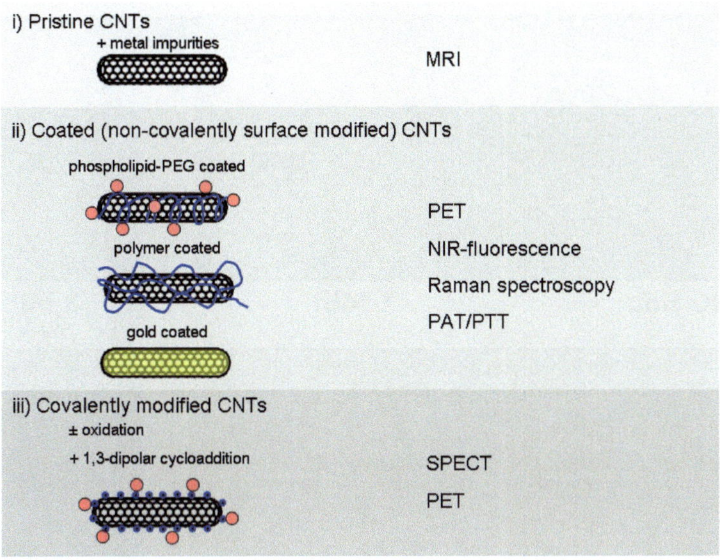

Figure 12.1 Types of CNT studied as imaging contrast agents *in vivo* and the imaging techniques involved.

12.2.1 Single Photon Emission Computed Tomography

SPECT allows whole-body dynamic imaging of the same animal repeatedly, and therefore it is suitable to monitor and study changes in tissue distribution following administration within a timeframe determined by the half-life of the radioisotope used. Conventional gamma counting can be used in parallel with imaging studies to allow more accurate quantification of the biodistribution profile. Our laboratory has collaborated with others to report the first study of CNTs used in dynamic imaging of living small animals using SPECT [20]. The CNT surface was chemically functionalized by 1,3-dipolar cyclo-addition reaction without prior oxidation (MWNT-NH_3^+), where the ammonium functional groups were covalently conjugated to the chelating molecule diethylentriaminepentaacetic (DTPA) and labeled with the gamma emitter indium-111 (^{111}In) for imaging purposes ($t_{1/2}$= 2.81 days). The radiolabeled CNTs were injected intravenously via tail vein and showed kidney clearance with no radioactivity left in the body within the first 24 hours. SPECT showed that the nanotubes enter the systemic blood circulation and within 5 minutes begin to permeate through the renal glomerular filtration system into the bladder (Fig. 12.2). Clearance of imaging agents can have an impact on their toxicological profile, and fast clearance makes their use greatly favorable. We have also previously reported the *in vivo* biodistribution profile of SWNTs chemically functionalized by the same method as reported with *f*-MWNT [31]. The same biodistribution profile of MWNT-NH_3^+ was observed suggesting the suitability of both functionalized SWNTs and MWNTs for SPECT imaging.

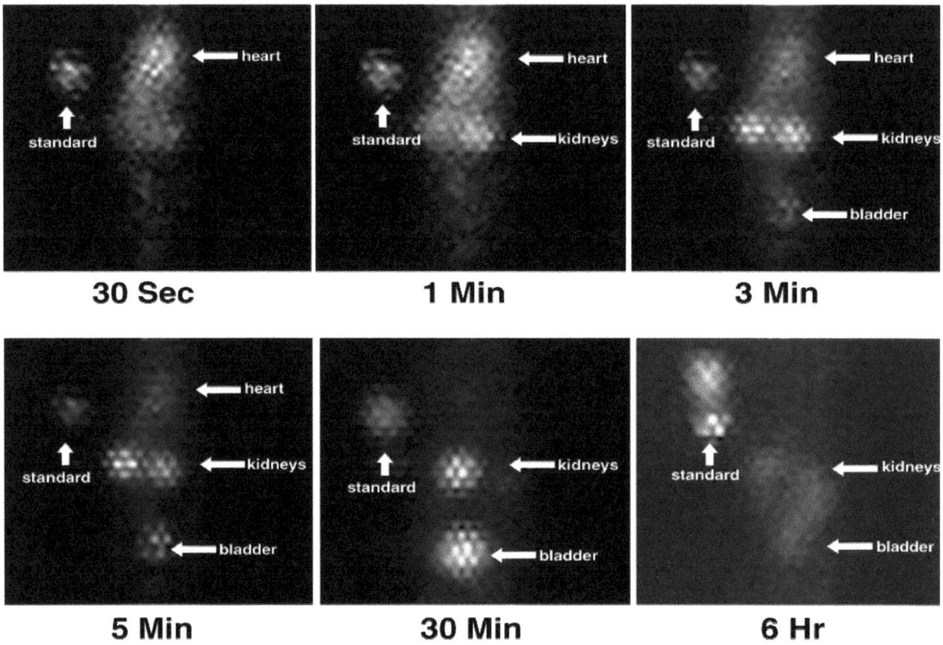

Figure 12.2 Serial single-photon planar images of ^{111}In radiolabeled functionalized MWNTs administered intravenously to a normal rat. Radiolabeled nanotubes are visualized in the blood pool of the heart at 30 seconds, but very rapidly at 1 minute the CNTs begin to accumulate in the kidneys and move into the bladder by 3 minutes. By 6 hours, the majority of the activity has cleared the body, as can be observed by comparing the rat image with the image standard. The pixilated appearance of the image is due to the large size of the camera collimator holes required for imaging high-energy ^{111}In photons. (Images obtained during collaboration with Drs. B. Goins and W. Phillips, University of Texas Health Science Center, San Antonio. [20])

12.2.2 Positron Emission Tomography

The first PET study, described by McDevitt *et al.* [24], used shortened (carboxylated) ammonium functionalized SWNT conjugated with the chelating molecule 1,4,7,10-tetraazacyclo-dodecane-1,4,7,10-tetraacetic acid (DOTA) for labeling with the short half-lived positron-emitting isotope yttrium-86 (^{86}Y) ($t_{1/2}$= 14.7 hours) for PET imaging. The labeled CNTs were imaged, and the biodistribution profile was concluded. The administration was via the intraperitoneal (IP) or via the intravenous (IV) route through the retro-orbital sinus in athymic nude mice. PET imaging showed clear signals in the liver, spleen, and kidney, primarily the renal cortex. The spleen and the liver showed less uptake after IP administration compared with IV administration. Within 24 hours, kidney clearance was observed, while liver and spleen signals remained almost unchanged. The blood clearance profile indicated rapid clearance from the blood compartment with no radioactivity detected at 3 hours. PET imaging was conclusive on the biodistribution profile of the contrast agent at early time points after administration restricted by the half-life of yttrium. However, deep tissue imaging was possible with excellent penetration capacity of the radionuclide signal. Although no active targeting was employed in this PET imaging study, a previous study performed by the same group had shown that the same CNT construct used in PET imaging was able to target a disseminated (bone marrow) human lymphoma murine xenograft model after chemical conjugation of the monoclonal antibody Rituximab onto the CNT surface. Rituximab targeted the CD20 epitope on human Burkitt lymphoma cells for radionuclide targeting to the cells. In conclusion, the work by McDevitt *et al.* [24, 25] using shortened ammonium functionalized SWNTs (<100 nm in length) conjugated to DOTA showed rapid clearance of both targeted and nontargeted radiolabeled CNT constructs from the blood stream. However, conjugation of Rituximab antibody to the DOTA-SWNT dramatically altered the biodistribution, with lower kidney and higher liver accumulation. The differential organ accumulation observed here could be due to the different hydrophobicity/bundling of the DOPA-functionalized SWNTs, as has been reported by these authors [24]. This means that by appropriately designing CNT-based diagnostics and therapeutic systems, one may be able to establish platforms with altered lung, liver, and spleen accumulation versus kidney clearance. This is again a criterion necessary for selective organ imaging.

Another *in vivo* PET study was performed by Dai and colleagues [22], who imaged SWNTs coated with phospholipids bearing polyethylene glycol chains on their polar headgroups. The radiolabel used was copper-64 (^{64}Cu) with a $t_{1/2}$ = 12.7 hours, attached to the PEG chain chemically conjugated to DOTA at their distal ends. RGD peptides were also conjugated to the radiolabeled SWNTs to serve as targeting ligands specific for integrin-positive tumors via RGD- $\alpha_v\beta_3$ integrin binding. The length of PEG was changed, and two molecular weights were used: 2000 Da and 5400 Da. The results reported showed a prolonged blood circulation profile and low uptake by the reticuloendothelial system (RES) at early time points following administration. Tumor targeting was also achieved by conjugation of RGD to the PEG termini coating the CNTs after tail vein administration in mice bearing $\alpha_v\beta_3$ integrin-positive U87MG subcutaneous tumors. SWNT coated with both lipopolymer types eventually deposited prominently in the liver and the spleen, with lower uptake in these organs of the SWNT-PEG5400 compared with SWNT-PEG2000. SWNT-PL-PEG5400 also exhibited longer blood circulation time ($t_{1/2}$ = 0.5 hour). Other organs such as solid tumor, muscle, bone, and skin showed low uptake. RGD-functionalized SWNT-PL-PEG2000 showed only slight increase in tumor uptake compared with nontargeted SWNT-PL-PEG2000. However, SWNT-PL-PEG5400-RGD exhibited higher tumor uptake (>10% ID/g tissue) compared with nontargeted SWNT-PL-PEG5400 (3–4% ID/g tissue), with a peak at 6 hours and leveling off at 20 hours post-injection. The PET imaging results reported in this study were confirmed semiquantitatively using *ex vivo* Raman spectroscopy.

12.2.3 Near-Infrared Fluorescence

One of the major drawbacks of optical imaging in living animals is the poor tissue penetration and intense scattering of light. The near-infrared (NIR) photoluminescence intrinsic to only individualized semiconducting SWNTs, mostly in the infrared-A (IR-A, 1–1.4 μm) region, has made them promising as novel NIR fluorescence contrast agents. In preclinical small animal imaging studies, low autofluorescence and deep tissue penetration in the NIR region beyond 1 mm can be achieved using such probes.

NIR imaging of SWNTs in live animals, however, remains challenging. The first study that employed the intrinsic NIR fluorescence of SWNTs to measure the blood elimination kinetics in rabbits and to identify the organs in which they concentrate was performed by Cherukuri *et al.* [7]. That study demonstrated that pristine SWNTs, dispersed by coating of the block copolymer Pluronic F108, can be detected through their intrinsic NIR after jugular vein administration. NIR signals were detected in blood samples and excised tissues. Blood circulation half-life was 0.1 hour, and after 2 hours most of the dose was detected in the liver. Although this technique does not require complex modification of the SWNTs as in linking or conjugation chemistries of fluorophores and radioisotopes, preparation of SWNTs to render them water dispersible and biocompatible have been shown to affect their quantum yield, thereby making the use of this technique limited to *in vitro* imaging or superficial *in vivo* imaging of pristine SWNTs [36]. The main limitation of this technique is the low quantum efficiency of SWNTs and the diffraction to resolution. Also, quantitative data are difficult to obtain using this technique. Recently, Welsher *et al.* reported that the method of SWNT coating is essential and dictates the sensitivity levels of this imaging technique [36]. The study has shown nude mice injected with two types of coated SWNTs: (i) direct SWNTs coated with phospholipid–PEG and (ii) SWNTs solubilized with sodium cholate first, followed by surfactant exchange to form phospholipid–PEG-coated nanotubes. The exchange procedure was found to improve the sensitivity of detection of signals by 15 times, allowing for deep tissue penetration and low autofluorescence background for imaging of tumor vessels beneath thick skin using high-resolution intravital microscopy, with an InGaAs camera in the 1–1.7 mm spectral range. Authors have attributed this to less damage to the nanotubes that improved the quantum yield during sonication.

12.2.4 Raman Spectroscopy

The first preclinical study to analyze CNTs deposited in tissue samples by Raman spectroscopy *ex vivo* was performed by Liu *et al.* [22] using SWNTs coated with PEGylated lipids. Such analytical tool was employed to confirm the tissue distribution results obtained by PET imaging of physically adsorbed PEG (2000 and 5400 Da) onto the surface of SWNTs targeted with RGD peptide. Confirmation of the presence of SWNTs (for an injected dose of 0.5 mg/kg) in liver and tumor tissues by using the intrinsic optical properties of SWNTs was offered on the basis of the strong resonance Raman bands, including the G band at ~1580 cm^{-1}, which is characteristic of graphitic carbon. In a subsequent study [23] the intrinsic Raman scattering intensity of SWNTs was used to quantify the concentration of nanotubes in blood and mice organs *ex vivo* (tissue lysates). It was shown that the longer and more branched the PEG chains, the longer the blood circulation time and lower the liver and spleen uptake. Also, slow but persistent decrease in the SWNT Raman signal suggested clearance of SWNTs through the biliary route and kidney. SWNT Raman signals were found in the intestine, feces, kidney, and bladder.

Kang *et al.* [16] have studied the biodistribution of shortened SWNTs (50–200 nm in length and 1–3 nm in diameter) coated with chitosan or chitosan-Alexa Fluor 488, injected via the tail vein of nude mice at a dose of 5 or 20 μg/animal. Tissue uptake was monitored at 0.5 hour to 24 hours after administration by fluorescence microscopy and Raman spectroscopy. The liver was found to have the highest nanotube uptake, with a maximum CNT accumulation at 3 hours followed by relatively high levels in the spleen and kidney. Raman signals in other organs were low (<5% ID/g). The sensitivity threshold was 3–4.5% ID/g tissue at 20 μg dose. Blood circulation half-life was 3–4 hours. No reduction in liver uptake was observed up to 24 hours. Similarly, rapid uptake and slow clearance was found in the spleen, whereas the kidney showed faster clearance of SWNT compared with the liver and spleen, where almost 50% dose detected in the kidney at 3 hours was excreted out 24 hours after injection. No Raman signals were detected in urine and feces, and this was attributed to the low sensitivity of the technique.

Raman spectroscopy offers the advantage of long-term tracking of SWNT in organs up to several months and offers high sensitivity in the case of noncovalent coating of SWNTs. On the other hand, disadvantages include (i) limited detection sensitivity, with minimum detection levels reported at 0.04 μg/ml in blood (0.2%ID/g) and 0.2 μg/ml in other tissues (1%ID/g) [23]; (ii) unsuitability for chemically functionalized SWNTs that generally induces defects on the nanotube backbone, thereby dramatically compromising signal intensity; (iii) and restriction to *ex vivo* (blood and dissected tissue) sample analyses. However, in a recent study performed by Zavaleta *et al.* [39], SWNTs coated with lipid-PEG5000-RGD and injected intravenously in mice bearing U87MG $\alpha_v\beta_3$ integrin-positive subcutaneous tumors, it was reported that Raman spectroscopy can also be used successfully for noninvasive tissue imaging of the superficial tumors. Optimization of the technique is still required to improve the depth and accuracy of the imaging (maximum depth achieved was 2 mm), with more quantification data still needed to correlate with *ex vivo* analyses.

Raman spectroscopy has been used to study the tissue distribution and *in vivo* stability of covalently PEGylated SWNT (PEG_{1500}-SWNT) after systemic administration in correlation with photo-bioluminescence measurements. De-PEGylation from the SWNTs was reported taking place in the liver at 4–8 weeks after injection, but not in the spleen [38]. Generally, pristine and acid-treated SWNTs exhibit a characteristic Raman spectrum with a distinct G-band peak at around 1590 cm^{-1}. However, this peak is less intense upon covalent functionalization, as in the case of PEGylated SWNT studied in the report by Yang *et al.*, leading to significant interference and the G-band signals becoming barely detectible. Therefore, detection sensitivity is a serious limitation of this imaging technique, particularly when it comes to chemically functionalized CNTs.

12.2.5 Magnetic Resonance Imaging

MRI detects the interaction of water protons with each other and with the surrounding media. Different tissues have different relaxation times, and MRI contrast agents can enhance the relaxation time further, which is represented by the shortening of either T_1 (longitudinal) or T_2 (transverse) relaxation time. Super-paramagnetic iron oxides (SPIOs) affect the T_2 relaxation time more significantly than T_1, whereas gadolinium chelates predominantly affect T_1. Although MRI offers great resolution to reveal the spatial anatomical structure, it suffers from inherently low sensitivity. This can be improved by the use of contrast agents, prolonging the scanning time or increasing the strength of the magnetic field. CNTs have shown potential as MRI contrast agents, where Gd^{+3}-grafted SWNTs (gadonanotubes) have been reported to have an MRI efficacy 40 times greater than other Gd^{+3}-based contrast agents [33]. In addition, the confinement of Gd^{+3} ions onto ultrashort SWNT has made them super-paramagnetic. They have also shown pH-dependent response, which can be useful in imaging the acidic environment of tumors. However, no *in vivo* studies have been performed using such gadonanotubes to illustrate their imaging efficacy in the biological environment. A noninvasive follow-up study would be beneficial to evaluate the biodistribution and effect of nanotube tissue deposition after exposure directly *in vivo*.

The only study that has shown *in vivo* MRI capacity of SWNTs performed today [1] combined He-3 and H-1 magnetic resonance in a rat model to evaluate the biodistribution and biological impact of raw single-wall (raw SWNT) and superpurified (SP SWNT) nanotubes. The effects induced by metal impurities in the intrapulmonary instilled raw SWNT samples induced a significant decrease in magnetic field homogeneity detected in He-3 MRI acquired under spontaneous breathing conditions using a multi-echo radial sequence. H-1 MRI allowed detection of intravenously injected raw SWNTs in the spleen and kidneys using a gradient echo sequence sensitive to changes of relaxation time values. Histological analysis confirmed the absence of SWNTs in organs other than the lung following intrapulmonary instillation of the material. This study concluded that hyperpolarized He-3 can be used for the investigation of CNT pulmonary biodistribution, while standard H-1 MRI may best be utilized for imaging nanotube dispersions following systemic administration.

12.2.6 Photo-acoustic Tomography

PAT is a cross-sectional imaging technique based on the photo-acoustic effect of nanomaterials. PAT offers higher spatial resolution and better tissue penetration than conventional optical imaging techniques. In a recent study, SWNTs coated with lipid-PEG-RGD molecules were used as a contrast agent for PAT in U87MG tumor models in living mice [8]. Intravenous administration of this nanotube conjugate to tumor-bearing animals showed eight times photo-acoustic signal enhancement in the tumor compared with mice injected with nontargeted (no RGD) SWNTs. Results were confirmed by *ex vivo* Raman spectroscopy; however, the SWNTs studied lacked sensitivity, since at an SWNT concentration of 50 nM the photo-acoustic signal produced was equivalent to background tissue, and only at higher nanotube concentrations a signal could be detected *in vivo*.

CNTs have also been shown to be promising photothermal contrast agents [15, 29], even though *in vivo* applications have been limited by the relatively low absorption displayed by nanotubes at NIR wavelengths and the concerns over toxicity. Kim *et al.* [17] have recently deposited a thin layer of gold (4–8 nm thick) around shortened CNTs (100 nm in length and 1.5–2 nm in diameter) forming what they termed *golden carbon nanotubes* (GNTs). The gold layer acted as an NIR absorption enhancer and could modulate their toxicity. GNTs have shown the capacity to act as photoacoustic and photothermal contrast agents, exhibiting enhanced near-infrared contrast (approximately 10^2-fold) for targeting lymphatic vessels in mice using extremely low laser fluence levels. Antibody-conjugated GNTs were also used to bind to the lymphatic endothelial receptors in order to map the lymphatic endothelium, and preliminary *in vitro* viability tests showed that GNTs have minimal toxicity. More such approaches of combinatory imaging modalities are expected in the near future illustrating some of the advantages offered by CNTs.

Table 12.1 Imaging modalities using CNT as contrast agents

Imaging modality	*Type of CNT*		*Source of signals*	*Preclinical animal model*	*In vivo/ ex vivo*	*Conclusion*	*Ref.*
SPECT	Covalently modified CNTs by 1,3-dipolar cycloaddition	$MWNT\text{-}NH_3^+$ chemically conjugated to DTPA	^{111}In	Nude rats	*In vivo*	*Pros* • Clinically relevant • High sensitivity (picomolar levels)/ medium resolution • Quantitative • Results in 3D image • Good tissue penetration *Cons* • Chemical modification of CNTs is required • It detects the radiolabel, not the CNT, so stability of radiolabeling is critical	[20]
PET	Covalently modified CNTs by oxidation, 1,3-dipolar cycloaddition	$SWNT\text{-}NH_3^+$ chemically conjugated to DOTA	^{86}Y	Athymic nude mice	*In vivo*		[24]
	Coated (noncovalently surface-modified) CNTs	pSWNT coated with PL-PEG-DOTA: SWNT-PEG2000-RGD SWNT-PEG2000 SWNT-PEG5400-RGD SWNT-PEG5400	^{64}Cu	U87MG or HT-29 tumor-bearing nude mice	*In vivo*		[22]
Near IR-fluorescence	Coated (noncovalently surface-modified) CNTs	pSWNT coated with Pluronic F108	Intrinsic NIR fluorescence signals of SWNT	New Zealand rabbits	*Ex vivo*	*Pros* • No chemical modification of SWNTs is required *Cons* • Restricted to individualized semimetallic SWNTs • Low quantum efficiency of SWNT and the diffraction to resolution (medium sensitivity/ medium resolution) • Poor tissue penetration • Quantitative data are difficult to obtain using this technique.	[7]
		pSWNT coated with PL-PEG after lipid exchange with sodium cholate: exchange SWNT-PEG5000 direct SWNT-PEG5000		LS174T tumor-bearing mouse	*In vivo*		[36]

Imaging modality	*Type of CNT*		*Source of signals*	*Preclinical animal model*	*In vivo/ ex vivo*	*Conclusion*	*Ref.*
Raman spectroscopy	Coated (noncovalently surface-modified) CNTs	pSWNT coated with PL-PEG-DOTA: SWNT-PEG2000-RGD SWNT-PEG2000 SWNT-PEG5400-RGD SWNT-PEG5400	Intrinsic Raman signals of SWNT	U87MG tumor-bearing nude mice	*Ex vivo*	*Pros* • Differentiates the spectral fingerprint of different particles allowing for multiplexing imaging • Lack of background signals • No labeling or encapsulation of Raman probes is required due to the inherent Raman signature of SWNT • Raman signals are stable against photobleaching so suitable for long-term and repeated imaging weak signal intensity, which limits its sensitivity *Cons* • Medium sensitivity/ poor resolution • Poor signal (light) penetration (limited to below few centimeters of tissues) • Laser source used to generate Raman signals can damage tissues	[22]
		pSWNT coated with PL-PEG: SWNT-linear-PEG2000 SWNT-linear-PEG5000 SWNT-branched-PEG7000		Balb/c mice	*Ex vivo*		[23]
		pSWNT coated with chitosan or chitosan Alexa Fluor 488		Nude mice	*Ex vivo*		[16]
		pSWNT coated with PL-PEG: SWNT-PEG5000-RGD SWNT-PEG5000		U87MG tumor-bearing nude mice	*In vivo*		[39]
MRI	Pristine CNTs	Raw SWNT Superpurified SWNT	Metal impurities inducing a drop in magnetic field homogeneity	Sprague-Dawley rats	*In vivo*	*Pros* • High resolution • Good tissue penetration *Cons* • Low sensitivity	[1]

Imaging modality	*Type of CNT*		*Source of signals*	*Preclinical animal model*	*In vivo/ ex vivo*	*Conclusion*	*Ref.*
Photoacoustic tomography (PAT)	Coated (noncovalently surface-modified) CNTs	pSWNT coated with PL-PEG: SWNT-PEG5000-RGD SWNT-PEG5000	Intrinsic PA signals of SWNT	U87MG tumor-bearing nude mice	*In vivo*	*Pros* • Offers high spatial resolution (submillimeter)/ medium sensitivity • Better tissue penetration than optical imaging techniques but still lower than nuclear imaging *Cons* • Long acquisition time (20 min for single image of 100 mm^3 sized tumor)	[8]
		pSWNT dispersed in Pluronic F127	Intrinsic PA signals of SWNT	Sprague-Dawley rats	*In vivo*		[28]
	CNT coated with gold (GNT)	GNT with or without anti-LYVE-1 antibody	Intrinsic PA/ PT signals of GNT	Murine mesentry with lymph node of Nu/ nu mice	*In vivo*		[17]

12.3 CONCLUSIONS

The exploitation of CNTs in the biomedical arena has become an increasingly interesting area of research both at the basic biological level and also in the context of various therapeutic or diagnostic applications. The field has experienced an exponential increase in the number of studies and laboratories using CNTs in contact with various cells and tissues that will surely increase in the next few years. Several preclinical studies have been carried out to establish the suitability of CNT-based nanodevices as imaging contrast agents for small animal imaging *in vivo*. The sensitivity and resolution of these techniques varies; however, CNTs have also been shown to be successful candidates for multimodal imaging, i.e., using a combination of two or more imaging modalities. Such findings open a new era for the use of CNTs in multimodal imaging *in vivo*. Furthermore, CNTs have been shown to be able to carry therapeutic drug molecules in addition to contrast agents, thus allowing their use as theranostic devices. More work is required to optimize the sensitivity and to improve the biocompatibility of CNT-based devices to allow for smooth translation of these studies for clinical applications.

References

1. Al-Faraj A., Cieslar K., Lacroix G., Gaillard S., Canet-Soulas E., and Cremillieux Y. (2009) *In vivo* imaging of carbon nanotube biodistribution using magnetic resonance imaging. *Nano Lett* **9**, 1023–1027.

2. Bhirde A. A., Patel V., Gavard J., Zhang G., Sousa A. A., Masedunskas A., Leapman R. D., Weigert R., Gutkind J. S., and Rusling J. F. (2009) Targeted killing of cancer cells *in vivo* and *in vitro* with EGF-directed carbon nanotube-based drug delivery. *ACS Nano* **3**, 307–316.

3. Bianco A. (2004) Carbon nanotubes for the delivery of therapeutic molecules. *Expert Opin Drug Deliv* **1**, 57–65.

4. Bianco A., Kostarelos K., Partidos C. D., and Prato M. (2005a) Biomedical applications of functionalised carbon nanotubes. *Chem Commun* 571–577.

5. Bianco A., Kostarelos K., and Prato M. (2005b) Applications of carbon nanotubes in drug delivery. *Curr Opin Chem Biol* **9**, 674–679.

6. Cai W., and Chen X. (2007) Nanoplatforms for targeted molecular imaging in living subjects. *Small* **3**, 1840–1854.

7. Cherukuri P., Gannon C. J., Leeuw T. K., Schmidt H. K., Smalley R. E., Curley S. A., and Weisman R. B. (2006) Mammalian pharmacokinetics of carbon nanotubes using intrinsic near-infrared fluorescence. *Proc Natl Acad Sci US A* **103**, 18882–18886.

8. De la Z. A., Zavaleta C., Keren S., Vaithilingam S., Bodapati S., Liu Z., Levi J., Smith B. R., Ma T. J., Oralkan O., Cheng Z., Chen X., Dai H., Khuri-Yakub B. T., and Gambhir S. S. (2008) Carbon nanotubes as photoacoustic molecular imaging agents in living mice. *Nat Nanotechnol* **3**, 557–562.

9. Ferrari M. (2005) Cancer nanotechnology: opportunities and challenges. *Nat Rev Cancer* **5**, 161–171.

10. Hirsch L. R., Gobin A. M., Lowery A. R., Tam F., Drezek R. A., Halas N. J., and West J. L. (2006) Metal nanoshells. *Ann Biomed Eng* **34**, 15–22.

11. Huh Y., Shao L., Tobias G., Green M. L. (2006) Opening and reversible filling of single-walled carbon nanotubes with various materials. *J Nanosci Nanotechnol* **6**, 3360–3363.

12. Iijima S. (1996) Structural flexibility of carbon nanotubes. *J Jap Soc Tribol* **41**, 724–729.

13. Kam N. W., Liu Z., and Dai H. (2005a) Functionalization of carbon nanotubes via cleavable disulfide bonds for efficient intracellular delivery of siRNA and potent gene silencing. *J Am Chem Soc* **127**, 12492–12493.

14. Kam N. W., Liu Z., Dai H. (2006) Carbon nanotubes as intracellular transporters for proteins and DNA: an investigation of the uptake mechanism and pathway. *Angew Chem Int Ed Engl* **45**, 577–581.

15. Kam N. W., O'Connell M., Wisdom J. A., and Dai H. (2005b) Carbon nanotubes as multifunctional biological transporters and near-infrared agents for selective cancer cell destruction. *Proc Natl Acad Sci USA* **102**, 11600–11605.

16. Kang B., Yu D., Dai Y., Chang S., Chen D., and Ding Y. (2009) Biodistribution and

accumulation of intravenously administered carbon nanotubes in mice probed by Raman spectroscopy and fluorescent labeling. *Carbon* **47**, 1189–1206.

17. Kim J. W., Galanzha E. I., Shashkov E. V., Moon H. M., and Zharov V. P. (2009) Golden carbon nanotubes as multimodal photoacoustic and photothermal high-contrast molecular agents. *Nat Nanotechnol* **4**, 688–694.
18. Kostarelos K., Bianco A., and Prato M. (2009) Promises, facts and challenges for carbon nanotubes in imaging and therapeutics. *Nat Nanotechnol* **4**, 627–633.
19. Lacerda L., Bianco A., Prato M., and Kostarelos K. (2006) Carbon nanotubes as nanomedicines: From toxicology to pharmacology. *Adv Drug Delivery Rev* **58**, 1460–1470.
20. Lacerda L., Soundararajan A., Singh R., Pastorin G., Al-Jamal K. T., Turton J., Frederik P., Herrero M. A., Bao S. L. A., Emfietzoglou D., Mather S., Phillips W. T., Prato M., Bianco A., Goins B., and Kostarelos K. (2008) Dynamic Imaging of functionalized multi-walled carbon nanotube systemic circulation and urinary excretion. *Adv Mater* **20**, 225–230.
21. Li Z. B., Cai W., and Chen X. (2007) Semiconductor quantum dots for *in vivo* imaging. *J Nanosci Nanotechnol* **7**, 2567–2581.
22. Liu Z., Cai W., He L., Nakayama N., Chen K., Sun X., Chen X., and Dai H. (2007) *In vivo* biodistribution and highly efficient tumour targeting of carbon nanotubes in mice. *Nat Nanotechnol* **2**, 47–52.
23. Liu Z., Davis C., Cai W. B., He L., Chen X. Y., and Dai H. J. (2008) Circulation and long-term fate of functionalized, biocompatible single-walled carbon nanotubes in mice probed by Raman spectroscopy. *Proc Natl Acad Sci US A* **105**, 1410–1415.
24. McDevitt M. R., Chattopadhyay D., Jaggi J. S., Finn R. D., Zanzonico P. B., Villa C., Rey D., Mendenhall J., Batt C. A., Njardarson J. T., Scheinberg D. A. (2007a) PET imaging of soluble yttrium-86-labeled carbon nanotubes in mice. *PLoS ONE* **2**, e907.
25. McDevitt M. R., Chattopadhyay D., Kappel B. J., Jaggi J. S., Schiffman S. R., Antczak C., Njardarson J. T., Brentjens R., and Scheinberg D. A. (2007b) Tumor targeting with antibody-functionalized, radiolabeled carbon nanotubes. *J Nucl Med* **48**, 1180–1189.
26. Pastorin G., Wu W., Wieckowski S., Briand J. P., Kostarelos K., Prato M., and Bianco A. (2006) Double functionalization of carbon nanotubes for multimodal drug delivery. *Chem Commun (Camb)* 1182–1184.
27. Podesta J. E., Al-Jamal K. T., Herrero M. A., Tian B., li-Boucetta H., Hegde V., Bianco A., Prato M., and Kostarelos K. (2009) Antitumor activity and prolonged survival by carbon-nanotube-mediated therapeutic siRNA silencing in a human lung xenograft model. *Small* **5**, 1176–1185.
28. Pramanik M., Song K. H., Swierczewska M., Green D., Sitharaman B., and Wang L. V. (2009a) *In vivo* carbon nanotube-enhanced noninvasive photoacoustic mapping of the sentinel lymph node. *Phys Med Biol* **54**, 3291–3301.
29. Pramanik M., Swierczewska M., Green D., Sitharaman B., and Wang L. V. (2009b) Single-walled carbon nanotubes as a multimodal-thermoacoustic and photoacoustic-contrast agent. *J Biomed Opt* **14**, 034018.

30. Shao L., Lin T. W., Tobias G., and Green M. L. (2008) A simple method for the containment and purification of filled open-ended single wall carbon nanotubes using C60 molecules. *Chem Commun (Camb)* 2164–2166.

31. Singh R., Pantarotto D., Lacerda L., Pastorin G., Klumpp C., Prato M., Bianco A., and Kostarelos K. (2006) Tissue biodistribution and blood clearance rates of intravenously administered carbon nanotube radiotracers. *Proc Natl Acad Sci USA* **103**, 3357–3362.

32. Singh R., Pantarotto D., McCarthy D., Chaloin O., Hoebeke J., Partidos C. D., Briand J. P., Prato M., Bianco A., and Kostarelos K. (2005) Binding and condensation of plasmid DNA onto functionalized carbon nanotubes: toward the construction of nanotube-based gene delivery vectors. *J Am Chem Soc* **127**, 4388–4396.

33. Sitharaman B., Kissell K. R., Hartman K. B., Tran L. A., Baikalov A., Rusakova I., Sun Y., Khant H. A., Ludtke S. J., Chiu W., Laus S., Toth E., Helm L., Merbach A. E., and Wilson L. J. (2005) Superparamagnetic gadonanotubes are high-performance MRI contrast agents. *Chem Commun (Camb)*, 3915–3917.

34. Thorek D. L., Chen A. K., Czupryna J., and Tsourkas A. (2006) Superparamagnetic iron oxide nanoparticle probes for molecular imaging. *Ann Biomed Eng* **34**, 23–38.

35. Tilmaciu C. M., Soula B., Galibert A. M., Lukanov P., Datas L., Gonzalez J., Barquin L. F., Rodriguez F. J., Gonzalez-Jimenez F., Jorge J., and Flahaut E. (2009) Synthesis of superparamagnetic iron(III) oxide nanowires in double-walled carbon nanotubes. *Chem Commun (Camb)* 6664–6666.

36. Welsher K., Liu Z., Sherlock S. P., Robinson J. T., Chen Z., Daranciang D., and Dai H. (2009) A route to brightly fluorescent carbon nanotubes for near-infrared imaging in mice. *Nat Nanotechnol* **4**, 773–780.

37. Wu W., Wieckowski S., Pastorin G., Benincasa M., Klumpp C., Briand J. P., Gennaro R., Prato M., and Bianco A. (2005) Targeted delivery of amphotericin B to cells by using functionalized carbon nanotubes. *Angew Chem Int Ed Engl* **44**, 6358–6362.

38. Yang S. T., Wang H., Meziani M. J., Liu Y., Wang X., and Sun Y. P. (2009) Biodefunctionalization of functionalized single-walled carbon nanotubes in mice. *Biomacromolecules* **10**, 2009–2012.

39. Zavaleta C., De la Z. A., Liu Z., Keren S., Cheng Z., Schipper M., Chen X., Dai H., and Gambhir S. S. (2008) Noninvasive Raman spectroscopy in living mice for evaluation of tumor targeting with carbon nanotubes. *Nano Lett* **8**, 2800–2805.

40. Zhang Z., Yang X., Zhang Y., Zeng B., Wang S., Zhu T., Roden R. B., Chen Y., and Yang R. (2006) Delivery of telomerase reverse transcriptase small interfering RNA in complex with positively charged single-walled carbon nanotubes suppresses tumor growth. *Clin Cancer Res* **12**, 4933–4939.

Chapter 13

ULTRASOUND CONTRAST MICROBUBBLES: *IN VIVO* IMAGING AND POTENTIAL THERAPEUTIC APPLICATIONS

Amanda Caissie,[a] Raffi Karshafian,[b] Kullervo Hynynen,[c,d] and Gregory J. Czarnota[a,c,d,e]

[a] *Department of Radiation Oncology, University of Toronto, Toronto, Ontario, Canada*
[b] *Department of Physics, Ryerson University, Toronto, Ontario, Canada*
[c] *Department of Medical Biophysics, University of Toronto, Toronto, Ontario, Canada*
[d] *Imaging Research, Sunnybrook Health Sciences Centre, Toronto, Ontario, Canada*
[e] *Department of Radiation Oncology, Sunnybrook Health Sciences Centre, Toronto, Ontario, Canada*

Gregory.Czarnota@sunnybrook.ca

13.1 INTRODUCTION

In contrast to nanoparticles, this chapter will discuss microbubbles, which are ultrasound contrast agents with a diameter on the order of micrometers. Frequently, microbubbles are made of the same kinds of molecules used to make nanoparticles. In addition, hybrid microbubble nanoparticle systems are under investigation for applications such as gene and drug delivery [22, 105, 140]. Beginning with a brief overview of the physics behind microbubbles and their uses, this chapter will focus on the various current and emerging imaging and therapeutic applications of microbubbles. In terms of imaging, cardiology has been the main field of medicine using microbubbles, specifically in echocardiography [61]. With advances in technology, microbubbles are now being used in various other fields, including neurology for the assessment of brain microcirculation in stroke [92, 123] and oncology for the assessment of tumor microcirculation [49]. Microbubble research has broadened the range of potential biomedical therapeutic applications of ultrasound to include

Nanoimaging
Edited by Beth Goins and William Phillips

www.panstanford.com

drug and gene delivery [133], sonothrombolysis [93], sonoporation [58], and tumor radiosensitization [16, 24, 25].

Contrast enhancement with ultrasound imaging was first noted in a 1970 study of the left ventricle reported by Feigenbam *et al.* [33]. While contrast was initially thought to be originating from the intracardiac injection of indocyanine green dye, other work identified bubbles at the catheter tip as the true source of contrast [42, 74]. Studies later showing improved contrast when saline was mixed with patient blood [34] led to the development of one of the first commercially available microbubbles Albunex® (Mallinckrodt Inc. Hazelwood, MO, USA). Comprising air microbubbles coated in a serum albumin shell [61, 62, 83], Albunex® microbubbles are no longer clinically available. Subsequent generations of microbubbles have been developed with improvements in design that make them more stable and more amenable to ultrasound scanning.

In terms of general composition, microbubbles are generally described by properties such as physical size, core gas, and shell material. The majority of microbubbles fall within a 1 to 5 μm diameter range [61]. The gas core usually consists of air, perfluorocarbon, nitrogen, or sulfur hexafluoride. The shell consists of albumin protein, lipid, or polymer material, and its thickness ranges from 10 to 200 nm [13]. Many different types of microbubbles exist, including coated gas bubbles, echogenic liposomes, and phase shift emulsions. Similar to phospholipid-coated gas microbubbles such as SonoVue® (Bracco International BV, Amsterdam, The Netherlands) [4], are echogenic liposomes, composed of a gas-liquid mixture encapsulated by phospholipid bi-layers. As a new emerging bubble type, phase shift emulsions are unique, being composed of a stabilized emulsion of volatile liquid nanoparticles, which vaporize into microbubbles upon injection or exposure to ultrasound. The advantages and disadvantages of the various developments with respect to microbubble composition are beyond the scope of this chapter. Nevertheless, details on microbubbles may be found in various reviews such as those by Stride [130] and Qin *et al.* [113], which thoroughly discuss the physics behind microbubble insonification, a topic that will only be summarized in the following section to give context to subsequent discussions of microbubble applications.

13.2 MICROBUBBLE PHYSICS

Acoustic cavitation refers to a process caused by the interaction of acoustic waves with a gas bubble or gas nucleus in which bubbles collapse. The bubbles can be naturally present (as in lung and intestine), created through ultrasound exposure (a process known as cavitation inception), or intentionally introduced into the body through injection of microbubble contrast agents [80, 101]. With insonification by ultrasound, microbubble contrast agents can oscillate or be disrupted upon exposure to a pressure wave above a particular amplitude threshold. The acoustic response of a microbubble depends on a number of factors, including (i) its physical properties (e.g., size, gas, and shell characteristics), (ii) ultrasound exposure parameters (e.g., peak negative pressure, frequency, and pulse repetition frequency), and (iii) environmental conditions (e.g., boundary condition, ambient pressure, and temperature) [17, 28, 32].

A microbubble placed in an ultrasound field above an acoustic threshold will oscillate about an equilibrium radius and emit an acoustic pressure wave. In a sound field of sufficient amplitude, microbubbles exhibit a highly nonlinear oscillation and emit harmonic, ultraharmonic, and subharmonic signals. These nonlinearities have been exploited to detect microbubbles, forming the basis of contrast-enhanced ultrasound imaging methods [14]. The magnitude of echoes from microbubbles depends on the pulse center frequency, in addition to the amplitude and phase of the ultrasound pulse. Microbubble detection strategies utilize pulse sequences, in which two or more transmitted pulses interact with the microbubble agent and the resulting echoes are combined to improve microbubble detection. In general, both the phase and amplitude of the transmitted pulses are modulated in order to detect the nonlinear signals from the microbubbles [108, 127]. More recently, it was demonstrated that microbubbles can be detected by exposing them to low- and high-frequency pulses [18, 19].

A unique characteristic of microbubbles is that despite serving as contrast agents they can be disrupted by ultrasound. This phenomenon depends on ultrasound exposure parameters. Generally, microbubble disruption efficiency increases with acoustic pressure and decreases with acoustic frequency. As well, the disruption threshold of encapsulated-microbubbles may be higher in smaller vessels ($< \sim 200$ μm) [119] rather than larger. Microbubbles can be disrupted through various physical mechanisms. The shell of encapsulated microbubbles can be cracked with acoustic energy releasing gas and forming a non-encapsulated free gas bubble [9, 112]. These free gas bubbles respond to ultrasound more vigorously than their encapsulated parent microbubbles [8, 146]. Free-gas microbubbles may disappear through passive and ultrasound-driven diffusion. Furthermore, optical observations at high speeds ($\sim$10 million frames per second) have shown bubbles undergoing fragmentation [111]. A single microbubble can be fragmented into multiple microbubbles and further respond to ultrasound. Bubbles can also be disrupted in the presence of high enough acoustic pressures through the mechanism of inertial cavitation, in which significant radial expansion of a bubble is followed by rapid and violent collapse [36].

Physical mechanisms of microbubbles may occur through both inertial and stable cavitation. Inertial cavitation refers to bubble collapse and subsequent release of energy. Stable cavitation involves bubble oscillation and growth via rectified diffusion within the ultrasound field, at low acoustical power levels [48]. Associated physical effects, including microstreaming, microjet formation, and acoustic radiation force, will be discussed later in the section on blood–brain barrier disruption. In addition to physical effects, microbubbles may also cause thermal effects. Heating occurs as a result of energy absorption as ultrasound propagates through the medium. Microbubbles enhance a frequency-dependent heating effect as they generate higher-frequency harmonics. Microbubbles may also dissipate heat via conduction during compression or due to viscous friction between bubble coating and surroundings [47]. Great rises in temperature are beneficial in therapeutic applications such as HIFU or drug and gene delivery relying on thermal activation [110]. However, heating effects will have to be controlled in therapeutic applications with temperature-sensitive therapy agents. Microbubble chemical effects include the potential for production of free radicals or chemicals such as hydrogen peroxide, accounting for cell damage seen outside of ultrasound exposure field [21, 114]. Microbubbles may also cause chemical effects in the cell membrane, including effects on ion channels [55]. While the majority of microbubble mechanistic research has focused on physical effects, future work will be particularly important to investigate the various potential thermal and chemical effects that current research is just now uncovering.

13.3 IMAGING APPLICATIONS

13.3.1 Functional Imaging

The introduction of microbubbles has significantly changed the field of ultrasound imaging, which previously lacked effective contrast agents compared with other fields of imaging such as computed tomography (CT) or magnetic resonance imaging (MRI), which use iodinated and gadolinium contrast agents, respectively [12, 46]. Microbubbles have been used for a few nonvascular imaging applications, including the assessment of fallopian tube patency in women [121], and vesico-ureteric reflux in children [56]. The majority of diagnostic applications nevertheless involve imaging of the vasculature, be it macro- or microcirculation.

Injected intravenously, microbubbles circulate and remain for the most part in the blood pool [54]. Certain microbubble agents have, however, been shown to exhibit tropism by mechanisms that are yet unclear. For example, SonoVue® microbubbles have been shown to exhibit splenic tropism [82], leading to investigations of contrast-enhanced sonography for the imaging of splenic pathologies [106]. The liver tissue-specific phase of microbubbles has been more widely investigated, with Levovist® (Schering AG, Berlin, Germany) being well known as a liver-tropic microbubble [11]. While its strong liver specific phase is an advantage, Levovist is a poor real-time imaging agent, requiring high power output settings that disrupt the microbubbles. Levovist use is therefore being replaced by real-time methods using subsequent generation microbubbles such as SonoVue [4, 136], which are lower mechanical index (MI) agents. Whereas once inferior to CT and MRI in the study of benign and malignant hepatic lesions, the accuracy of ultrasound has now improved significantly with the use of contrast agents [136]. In hepatic malignancies, microbubbles have proven useful not only in diagnostic imaging but also in treatment monitoring [78, 126].

While radiology has certainly found a role for ultrasound contrast microbubbles, cardiology is the field of medicine that uses microbubbles most often, with their primary application in echocardiography for ventricular opacification and delineation of endocardial borders [2, 129]. Other uses have included investigations of carotid artery stenosis [15] and myocardial infarction [61, 84]. As further research explores the potential role of microbubbles in functional imaging of other disease sites such as pleural based lesions [41], the research of microbubbles is also expanding significantly in the field of molecular imaging.

13.3.2 Molecular Imaging and Targeted Microbubbles

In order to enhance localization in imaging, targeting ligands may be attached to the microbubble lipid shell [27, 85], typically using a biotin-streptavidin technique. Depending on the size of ligand, it may be attached before or after microbubble formation. Larger ligands are generally applied after microbubble generation, which involves agitation and increased temperatures [69, 77]. Initial *in vitro* studies [67, 68] used microbubbles attached to avidin-coated petri dishes. *In vivo*, following intravenous injection of contrast, imaging is performed over time as microbubbles accumulate in the field prior to an applied destructive ultrasound pulse. Images taken prior to the destructive pulse represent circulating and bound microbubbles, whereas those taken after the destructive pulse represent only circulating agents [31, 85, 127].

Various imaging applications of microbubbles are now being investigated from imaging areas of tumor angiogenesis to inflammation [109]. Angiogenesis is an important process in cerebrovascular ischemia, and studies of chronic limb ischemia have looked at microbubbles targeted to integrins expressed on neovascular endothelium. The majority of angiogenesis research with respect to targeted microbubbles has, however, focused on tumor growth and metastasis [127, 145]. Perfusion and microvascular blood volume information obtained from standard contrast ultrasound imaging may be enhanced by information obtained by molecular imaging with microbubbles targeting components of vascular remodeling such as integrins (e.g., α_V-integrin) [31, 81] or growth factor receptors (e.g., vascular endothelial growth factor receptor 2 [VEGFR-2]). Applications of such targeted microbubbles have included not only the assessment of tumor prognosis and metastatic potential but also response to anti-angiogenic treatment [30, 102].

In order to assess inflammation [86] using microbubble contrast imaging, two different strategies may be employed. The first relies on the interaction of microbubbles with activated leukocytes, with applications including the assessment of acute ischemia-reperfusion injury in the heart [20] and cardiac transplant rejection [72, 144]. In a second, more specific strategy, microbubbles may be made to target endothelial cell adhesion molecules, such as intercellular adhesion molecule (ICAM), which are upregulated in inflammation and promote leukocyte recruitment [87]. Microbubbles targeted to the vascular cell adhesion molecule (VCAM-1) have been used to investigate atherosclerosis, another process in which an immune response plays a key role [60, 97]. In a similar manner, the imaging of intravascular thrombi may be enhanced in diseases such as stroke by using microbubbles targeted to platelets [122, 138]. By targeting thrombi, these microbubbles may also have therapeutic applications since ultrasound exposure of the microbubbles can increase endogenous and exogenous thrombolysis [93, 147].

13.4 THERAPEUTIC APPLICATIONS

While microbubbles are clinically approved for various imaging applications, research is critical to explore the potential therapeutic applications ranging from sonothrombolysis to the enhancement of high-intensity focused ultrasound (HIFU) or tumor-targeted radiation. The following subsections will discuss in short some of these therapeutic applications.

13.4.1 Sonothrombolysis

Sonothrombolysis represents one of the therapeutic applications of ultrasound, with or without microbubbles, that has been moved extensively into clinical study. In the 1960s [3], sonothrombolysis was initially proposed as an alternative treatment to thrombotic diseases such as atherosclerosis and peripheral vascular disease. In the 1970s, the first patient was treated with sonothrombolysis [137]. In the 1980s, investigations revealed that ultrasound could be used to enhance the effect of systemically administered thrombolytics [76]. In 2004, the Combined Lysis of Thrombus in Brain Ischemia with Transcranial Ultrasound and Systemic t-PA (CLOTBUST) randomized trial reported a significant benefit with transcranial Doppler ultrasound when added to tissue plasminogen activator (t-PA) for induced recanalization in acute ischemic stroke [1]. Unfortunately, the Transcranial Low-Frequency Ultrasound-Mediated Thrombolysis in Brain Ischemia (TRUMBI) study terminated early due to increased hemorrhagic complications in the treatment group, while also revealing lack of benefit with respect to recanalization or clinical improvement [26]. These findings obviously raised significant concerns regarding the safety and efficacy of low-frequency transcranial isonation in acute ischemic stroke.

Another stream of research has investigated the use of ultrasound microbubbles to enhance thrombolysis. In 1995, Tachibana *et al.* reported that ultrasound in conjunction with microbubbles accelerated thrombolysis, the breakdown of blood clots, with urokinase (a thrombolytic drug) [131]. While early experiments used low-frequency ultrasound [10], further studies used sonications at standard diagnostic frequencies. In 2006, a clinical trial with combined microbubbles, transcranial Doppler ultrasound and t-PA showed increased rates of recanalization without significantly higher rates of symptomatic intracranial hemorrhage in patients with acute ischemic stroke [96].

13.4.2 High-Intensity Focused Ultrasound

HIFU is another form of ultrasound that has been investigated for its use in sonothrombolysis with [115] or without t-PA [128]. Originally conceived as a treatment in the 1940s [89], the technique was first used by the Fry brothers [38] in Parkinson patients in the 1950s. The technique has been investigated and refined since. As described above, microbubbles may be formed *in vivo* by a process called nucleating cavitation, in which gas is drawn out of solution in tissue as pressure falls within an ultrasound focus region during the rarefaction phase. Cavitation of *de novo* microbubbles *in vivo* can only be achieved with high-intensity ultrasound levels as in HIFU, which may cause undesired surrounding tissue damage in addition to target volume ablation. Ultrasound contrast agent microbubbles are, therefore, now being used with lower-intensity levels of ultrasound for therapeutic applications [113, 130]. Microbubbles have, however, also been used to overcome various technical limitations of HIFU, thereby enhancing the speed and therapeutic efficacy of HIFU in the treatment of tumors located in the kidney and the liver [57, 148]. Such microbubbles have been used not only therapeutically with HIFU but also for imaging of HIFU response [63].

13.4.3 Drug and Gene Delivery

Microbubble contrast agents have also found applications in improving the therapeutic efficacy of biologically active molecules [133]. Several possible mechanisms include the enhancement of (i) the concentration of therapeutic biomolecules in the vascular compartment of the target area, (ii) increased therapeutic agent delivery by extravasation through blood vessels, and (iii) potentially enhanced intracellular delivery. Molecules of the therapeutic agent can be attached to the outer shell of bubbles, incorporated within the bubble shell or loaded into the interior of microbubbles, and released in the vascular compartment through ultrasound-induced microbubble disruption [35, 141]. The extravasation of a therapeutic agent is achieved through the permeabilization of blood vessels with ultrasound and microbubbles [39, 73]. Radiation forces exerted on bubbles also displace them in the direction of the traveling sound wave. This force can thus direct microbubbles loaded with a therapeutic agent toward a vessel wall to enhance targeting through receptor-ligand contact [88] or to improve extravasation through the release of the therapeutic agent close to the vessel wall [116]. The application of ultrasound in the presence of microbubbles has been demonstrated to improve intracellular delivery of therapeutic molecules by increasing the permeability of plasma cell membranes as discussed below in the sonoporation section [53, 58, 64, 117, 143].

The ultrasound-microbubble-based delivery of therapeutic agents has one main advantage over other techniques using colloidal drug carriers such as nanoparticles or liposomes. The microbubble-based technique may be targeted through the external control of the ultrasound. It is hoped that this localized approach may then improve the therapeutic efficacy of drugs, such as routinely used chemotherapeutic agents like paclitaxel, by lowering the dose of agent to normal tissue as it is more targeted, with a consequent minimization of unwanted drug effects away from the treatment site [132]. Ultrasound-based delivery of gene therapy [139, 140] is also a current area of interest in microbubble research, with target diseases investigated to date ranging from hepatocellular carcinoma treated with IFN_{β} pDNA [118] to myocardial ischemia treated with VEGF pDNA [150]. While ultrasound-based drug and gene delivery must deal with the issue of the endothelium that acts as a biological barrier, it is not surprising that many studies have investigated these therapeutic applications of microbubbles in the context of disrupting the specialized endothelial cells of the blood–brain barrier (BBB).

13.4.4 Blood–Brain Barrier Disruption

The BBB is a specialized endothelial-based structure, protecting sensitive central nervous system (CNS) structures under normal conditions, but consequently acting as a barrier to therapeutics in treatments of CNS disorders. Improved drug delivery and gene therapy, via BBB disruption (BBBD), is therefore particularly important in diseases of the CNS. The various strategies for drug delivery into the CNS include trans-cranial drug delivery via an invasive catheter or the use of small molecule drugs, which often do not cross the BBB or are not effective with a particular CNS disease [103]. BBBD has therefore been another strategy; however, traditional non-ultrasound techniques have caused reversible BBBD with nonlocalized drug delivery, using intracarotid arterial infusion of noxious substances, including hyper-osmolar solutions like mannitol and many other agents [29, 98, 103, 142]. BBBD with noninvasive ultrasound methods in the absence of contrast bubbles was first reported in 1956 [6] and has continued over the decades from the 1950s to 1990s. It was not until 2001 that the first study of ultrasound, using contrast microbubbles, showed reproducible, noninvasive evidence of BBBD [50]. That original study showed magnetic-resonance-detected BBBD using ultrasound with contrast microbubbles applied through an opened window of rabbit skull. With the majority of subsequent studies focusing on BBBD in rats with intact skull, ultrasound parameters have been refined to produce consistent barrier disruption with minimal normal tissue effects [51, 91]. Nanoparticles have more recently been investigated for potential use in BBBD [61]; however, most studies of BBBD with ultrasound have focused on microbubbles up to this time [142].

Various applications of this technique have since been investigated with proof of principle that the technique may enhance the delivery of therapeutic agents in the CNS. Studies have shown successful CNS delivery of antibodies, including antibodies against the dopamine D4 receptor [66] as well as Herceptin, a humanized antibody now critical in the treatment of breast cancer [65]. Herceptin is currently not effective in treating breast cancer brain metastases in the presence of an intact BBB. Other studies using ultrasound and microbubbles have shown CNS delivery of other cancer treatments, including the widely used chemotherapeutic agent doxorubicin [135]. This technique of BBBD therefore holds great promise in the field of oncology.

Despite the refinement of method to produce BBBD and the various investigated therapeutic applications, the mechanisms behind the microbubble effect remain relatively unknown. Proposed physical mechanisms include both inertial and stable cavitation as mentioned above in the physics section. Since McDannold *et al.* have, however, shown that BBBD is possible in the absence of inertial cavitation [90], possible physical BBBD mechanisms may include bubble oscillation via stable cavitation and associated microstreaming of fluid around the bubbles, which causes sheer stress on endothelial cells [75, 94, 99]. Acoustic radiation force may also play a role, causing bubbles to move in the direction of wave propagation and subsequently stimulate endothelial cells [79, 142].

While the majority of studies have focused on the physics behind the ultrasound microbubble technique of BBBD, more recent work has begun to explore the underlying biological mechanisms. In a paper just published, immunoelectron microscopy has shown that ultrasound-microbubble-induced leakage of molecules such as horseradish peroxidase across BBB is associated temporally with the transient disruption of BBB tight junction proteins such as occludin, zonula occludin, and claudin-5 [124]. Future studies will be needed to further explore the underlying mechanisms behind the BBB disrupting effect of ultrasound microbubbles. The results of technique efficacy have, however, already led to other studies now looking at novel therapeutic applications of the technique, including the radiosensitization of tumors.

13.4.5 Sonoporation

In oncology, as well as other medical fields, the therapeutic potential of microbubbles in sonoporation is promising, as increased cell permeability may enhance delivery of drugs including chemotherapeutic agents. Sonoporation is a transient and reversible increase in the permeability of plasma cell membranes by ultrasound. Through the facilitation of molecule transport across the cell membrane, sonoporation has been shown to enhance the intracellular delivery of small compounds [43], macromolecules [44, 95], chemotherapeutic drugs [149], and genetic materials [5, 64] in a variety of cell types. Sonoporation may enhance drug delivery of nanoparticle carried drugs. While sonoporation-mediated drug and gene delivery is currently being developed for therapeutic applications in cancer, cardiovascular, and blood–brain-barrier-limited treatments [7, 52, 118], consistent and controllable delivery is challenging. The successful implementation of sonoporation-mediated therapeutic applications depends on understanding its mechanism, which is unclear at present, along with the effects of ultrasound and microbubble exposure conditions.

Karshafian *et al.* [58] have recently demonstrated an optimized set of ultrasound parameters to facilitate cell permeability. Specifically, *in vitro* KHT-C cells in suspension were exposed to ultrasound and microbubbles using near 100 exposure conditions (Fig. 13.1).

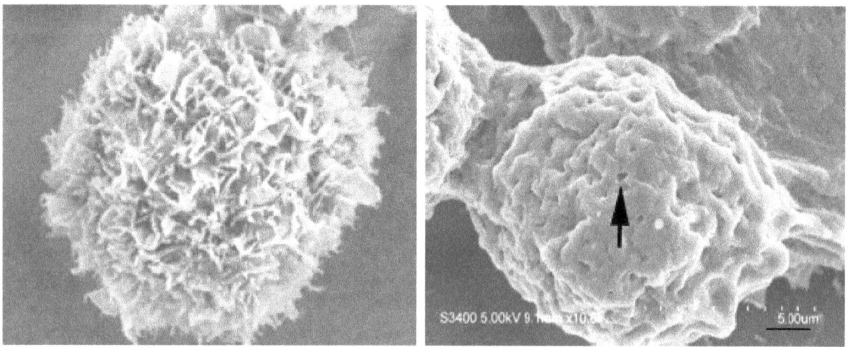

Figure 13.1 (Left Panel) Unexposed KHT fibrosarcoma cells. (Right Panel) KHT cells exposed to ultrasound and microbubbles under sonporating conditions. Note the destruction of the cell's normal ruffled appearance. The arrow indicates the formation of porelike structures in the cell membrane consistent with enhanced permeability of cells. The scale bar indicates 5 μm. From Karshafian, Ph.D. thesis, University of Toronto, 2009.

Effects on cells were assessed through colony assay, the uptake of cell-impermeable molecules (10 kDa to 2 MDa FITC-dextran), cell viability, and observations of plasma membrane integrity using flow cytometry and electron microscopy techniques (Fig. 13.2).

In that study, sonoporation was a result of the interaction of ultrasound and microbubbles with the cell membrane. Disruptions ranging in size from 30 nm to 100 nm were generated on the cell membrane, allowing otherwise cell-impermeable molecules to cross the membrane. Under these conditions molecules up to 2 MDa in size were delivered at high efficiency (~70% permeabilization). Sonoporation was short lived; cells re-established their barrier function within 1 minute, which allowed compounds introduced into cells to remain inside the cells. Following uptake, cells remained viable, with ~50% of sonoporated cells proliferating afterwards. The exposure conditions could be tailored to achieve a desired effect, ranging from cell permeability of nearly 70% of cells with 25% cell death versus a permeability of nearly 35% with only 2% cell death. Microbubble disruption was a necessary but insufficient indicator of ultrasound-induced permeabilization.

Other researchers have developed liposomal bubbles that contain perfluoropropane gas to use as a novel gene delivery system linked to bubble collapse induced by ultrasound exposure [100]. In that study, colon-26 xenograft tumors demonstrated complete remission when treated with dendritic cells injected intratumorally when tumors were *a priori* treated with liposomal bubbles in the presence of ultrasound. Ultrasound presumably broke up cells enhancing antitumor immunity when dendritic cells were injected. Those authors were able to induce complete tumor remission in such animals with the combination therapy and consequent complete cohort survival. Other groups are using microbubble-based sonoporation to selectively treat infected cervical cancer tumors by targeting the HPV viral E6 oncoprotein with antibodies [23].

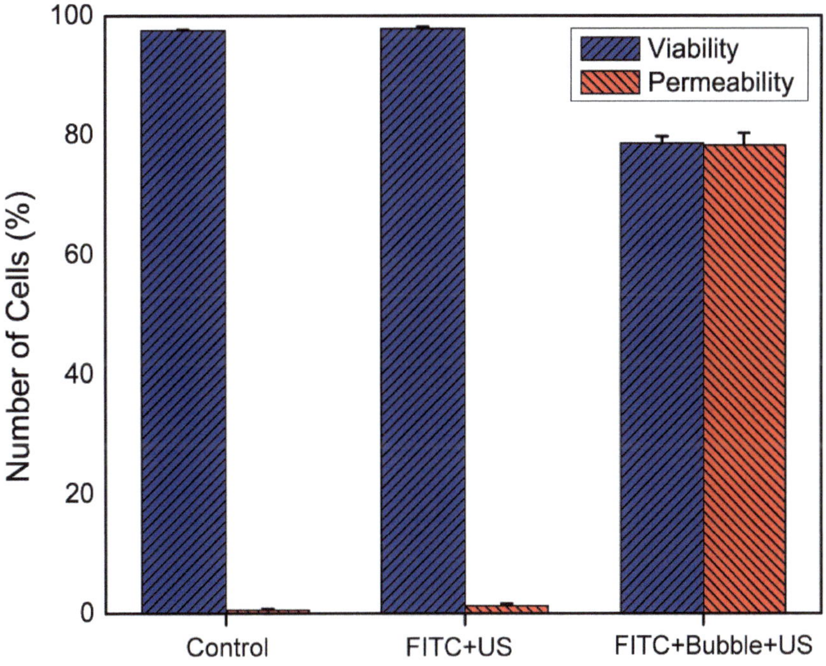

Figure 13.2 KHT-C fibrosarcoma cell uptake of fluorescein isothiocyanate (FITC)-dextran during sonoporation and 1 minute following termination of the treatment. Cell permeability (PR), the percentage of cells stained with FITC-dextran and unstained with PI, and cell viability (VPI), the percentage of cells unstained with PI, are shown for four conditions: untreated, ultrasound alone, ultrasound and microbubbles with FITC-dextran added 60 seconds before treatment, and ultrasound and microbubbles but with FITC-dextran added 60 seconds after terminating the ultrasound treatment. Intracellular delivery of FITC-dextran is achieved only when present during treatment with ultrasound and microbubbles. VPI was reduced to 80% by ultrasound and microbubbles.

Other groups are combining magnetic resonance guided high intensity focused ultrasound and low temperature sensitive liposomal nanoparticles in order to deliver therapeutics to targeted tissues with accuracy and precision not currently available. Such particles have just recently been developed and soon will be evaluated *in vivo* [105]. Other particle types that are truly nanoparticles and that can be converted into microdroplets with ultrasound have also been recently generated and tested for sonoporation effects [22]. These particles are being designed for tissue tattooing and can be formulated as double emulsions. Couture *et al.* [22] have recently developed particles that have a primary emulsion consisting of nanoparticles of fluorescein-saturated water encapsulated in perfluorohexane. This emulsion is then encapsulated in aqueous Pluronic (1%) using a flow-focusing microfluidic system. The combined effect of the vaporization of these particles and sonoporation is now being tested in the vascular beds of fertilized chicken eggs and mice.

13.4.6 Radiosensitization

Radiation is a standard anticancer therapy, currently used to treat the majority of oncology patients, with or without the addition of chemotherapy. Whereas it has long been thought that radiation acts primarily by damaging cancer cell DNA leading to tumor cell death [45], recent data indicate that radiation-induced apoptotic effects on blood-vessel cells can lead to vascular destruction and subsequent secondary tumor cell death [37, 40, 70, 71, 104, 107, 120]. It is therefore crucial to exploit the relative contribution of this radiation-induced vascular effect in order to optimize radiation treatment of tumors. The above evidence of microbubble effects on the endothelium, including the BBB, has led to recent investigations demonstrating for the first time that ultrasound-induced targeting of tumor vasculature may be used to enhance radiation response [24, 25].

A prostate cancer xenograft model was used to demonstrate this microbubble radiosensitization effect *in vivo*. Results indicated a synergistic effect with combined treatments, shown to enhance the vascular response that radiation is recognized to induce [37, 40, 70, 71, 104, 107, 120]. Histology revealed a viable peripheral rim and central tumor cell kill (Fig. 13.3), consistent with histology seen with other classical vascular disrupting agents [134].

Endothelial cell apoptosis was induced by ultrasound and microbubble treatments, and power Doppler ultrasound analyses revealed significant reduction of blood flow in tumors treated with the combined modalities. Single treatments of ultrasound-activated microbubbles and radiation produced over 40% tumor cell death when ultrasound activation of bubbles was used in combination with only 2 Gy of radiation. With ultrasound-microbubble-induced radiosensitization, this combined treatment modality shows great promise in the field of oncology, with potential to conformally enhance the effects of radiation in a tumor, thus minimizing radiation effects on neighboring normal tissue.

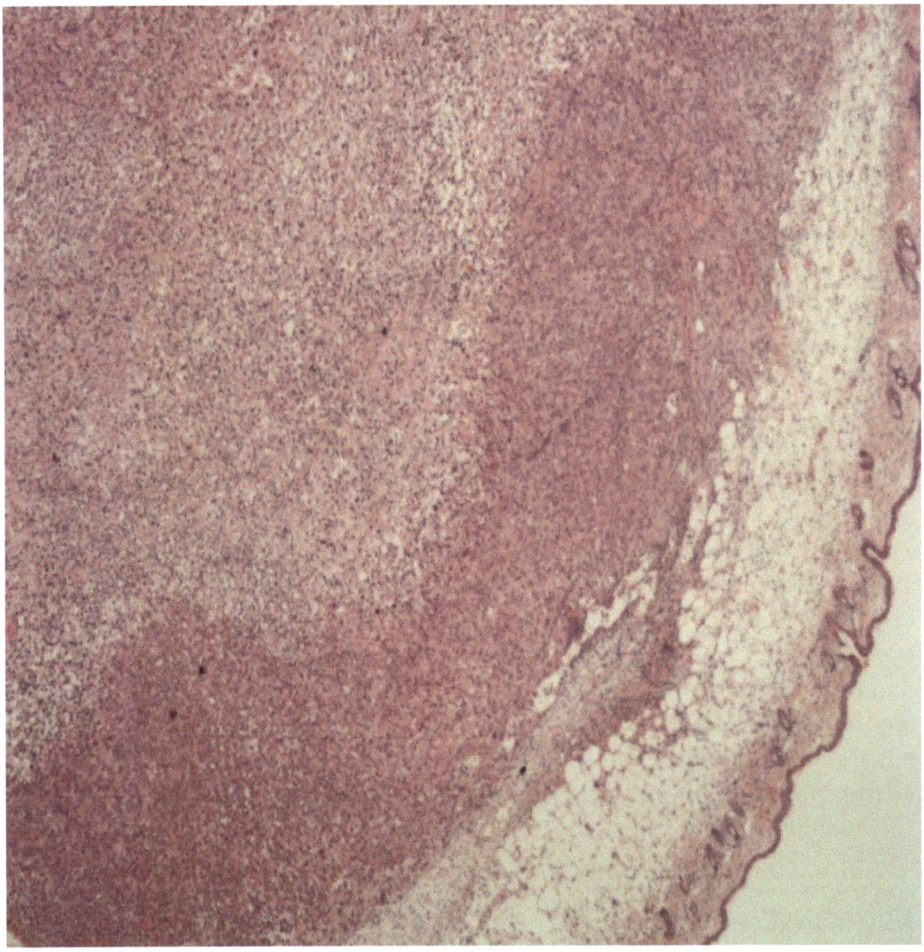

Figure 13.3 Hematoxylin- and eosin-stained tumor from a prostate cancer (PC3) mouse xenograft treated with ultrasound microbubbles (2.5 μm field of view obtained at 25× objective magnification). Note the rim of viable tumor surrounding a necrotic core as seen with other well-known vascular disrupting agents.

13.5 ADVERSE EFFECTS

As microbubbles are used routinely in diagnostics and new therapeutic applications are constantly being explored, potential adverse effects of microbubbles are particularly relevant. For diagnostic imaging applications, several ultrasound microbubble contrast agents are clinically approved worldwide, such as those currently approved in Canada, including Optison (Nycomed/Amersham, Buckinghamshire, UK), Definity (Lantheus Medical Imaging, Billerica, MA, USA), and Sonovue (Bracco, Moneglia, Italy). Extensive safety studies have therefore been performed. These agents are very safe with complication rates comparable to conventional intravenously administered contrast agents used with other imaging modalities such as CT. The rates of adverse events are comparable to those of iodinated contrast agents (1/40,000 major and 1/160,000 life-threatening events). Unlike iodinated contrast agents, however, ultrasound microbubbles have not been associated with a risk of nephrotoxicity [59, 125].

Rarely observed side effects include headache (2% of patients), altered sensation at the injection site (2%), as well as nausea, paresthesias, and altered taste. As with any other imaging contrast or therapeutic agent, the frequency of very rare side effects may be documented once the agent is used in a large number of patients. In terms of microbubbles, the list of rare side effects (<1% of patients) documented includes "abdominal pain, nonspecific pain, respiratory disorder, pharyngitis, pruritus, rash, abnormal vision, dry mouth, dizziness, personality disorder, insomnia, nervousness, hyperglycemia, peripheral edema, ecchymosis and sensory-motor paresis." Allergy/hypersensitivity (skin erythema, bradycardia, hypotension, or anaphylactic shock) and vasovagal reactions have been rare. Although no direct cardiac toxicity has been shown, the microbubble contrast-induced bradycardia and hypotension have been associated with myocardial ischemia and very rare cases of death in patients with underlying coronary artery disease. [12]. The adverse effects discussed above have been identified in the clinically relevant imaging applications of microbubbles. The potential adverse effects of microbubbles in the various therapeutic applications are, however, unknown. Some unusual and unexpected side effects may occur, but we may infer that the majority of adverse effects will arise from the various effects microbubbles have, including physical, thermal, and chemical effects on surrounding tissues.

13.6 CONCLUSIONS

In conclusion, there is a broad range of applications for ultrasound contrast microbubbles, many of which are just recently emerging. Many more uses will inevitably be uncovered as the research field expands. The identification of microbubbles as ultrasound contrast agents dates back four decades, with microbubbles now routinely used in their clinical application of diagnostic imaging. Innovation first came with the development of subsequent microbubble generations, each with improved design that strives for greater bubble stability and enhanced ability to perform real-time imaging. In terms of functional imaging, research has identified new areas of clinical applicability that extends to several nonvascular imaging applications such as fallopian tube patency. The field of molecular imaging with targeted microbubbles has certainly broadened the range of imaging applications even further. Microbubble molecular imaging has required knowledge of ligands such as growth factor receptors, thereby moving the field of microbubble research into the realm of biology.

Biology will certainly be the foundation for research delving into the therapeutic applications of microbubbles such as drug and gene delivery, as well as enhanced delivery of nanoparticles. Medical fields such as oncology benefit from the use of microbubble-based imaging in diagnostics and treatment monitoring, but may also benefit in the future from therapeutic microbubble-based strategies such as drug/gene delivery and sonoporation to improve chemotherapy efficacy, as well as radiosensitization, a novel characteristic of ultrasound-activated microbubbles. The concept of microbubbles as potential radiosensitizers with vascular disrupting effects came about from review of BBBD literature, which has begun to explore the biological effects of microbubbles on BBB endothelium and associated tight junctions. Like other areas of microbubble research such as HIFU, BBBD research was founded in the field of physics. Of course this physics foundation remains strong and will inevitably continue to contribute to the research field that is constantly expanding and identifying novel imaging and therapeutic applications of microbubbles.

References

1. Alexandrov A. V., Molina C. A., Grotta J. C., Garami Z., Ford S. R., and Alvarez-Sabin J., et al. (2004)CLOTBUST Investigators: ultrasound-enhanced systemic thrombolysis for acute ischemic stroke, *N Engl J Med* **351**, 2170–2178.
2. Al-Mansour H. A., Mulvagh S. L., Pumper G. M., Klarich K. W., and Foley D. A. (2000) Usefulness of harmonic imaging for left ventricular opacification and endocardial border delineation by Optison, *Am J Cardiol* **85**, 795–799.
3. Anschuetz R., and Bernard H. R. (1965) Ultrasonic irradiation and atherosclerosis, *Surgery* **57**, 549–553.
4. Arditi M., Brenier T., and Schneider M. (1997) Preliminary study in differential contrast echography, *Ultrasound Med Biol* **23(8)**, 1185–1194.
5. Bao S., Thrall B. D., and Miller D. L. (1997) Transfection of a reporter plasmid into cultured cells by sonoporation *in vitro*. *Ultrasound Med Biol* **23**, 953–959.
6. Bakay L., Hueter T. F., Ballantine H. T. Jr., and Sosa D. (1956) Ultrasonically produced changes in the blood-brain barrier, *Arch Neurol Psychat* 457–467.
7. Bekeredjian R., Grayburn P. A., and Shohet R. V. (2005) Use of ultrasound contrast agents for gene or drug delivery in cardiovascular medicine, *J Am Coll Cardiol* **45**, 329–335.
8. Bevan P. D., Karshafian R., Tickner E. G., and Burns P. N. (2007) Quantitative measurement of ultrasound disruption of polymer-shelled microbubbles, *Ultrasound Med Biol* **33**(11), 1777–1786.
9. Bevan P., Karshafian R., and Burns P. N. (2008) The influence of fragmentation on the acoustic response from shrinking bubbles, *Ultrasound Med Biol* **34(7)**, 1152–1162.
10. Birnbaum Y., Luo H., Nagai T., Fishbein M. C., Peterson T. M., and Li S., et al. (1998) Noninvasive *in vivo* clot dissolution without a thrombolytic drug : recanalization of thrombosed iliofemoral arteries by transcutaneous ultrasound combined with intravenous infusion of microbubbles, *Circulation* **97**, 130–134.

11. Blomley M., Albrecht T., Cosgrove D., Jayaram V., Butler-Barnes J., and Eckersley R. (1998) Stimulated acoustic emission in liver parenchyma with Levovist, *Lancet* **351**(9102), p. 568.

12. Blomley M., Claudon M., and Cosgrove D. (2007) WFUMB Safety Symposium on Ultrasound Contrast Agents: clinical applications and safety concerns, *Ultrasound Med Biol* **33**(2), 180–186.

13. Bouakaz A., and de Jong N. (2007) WFUMB Safety Symposium on Echo-Contrast Agents: nature and types of ultrasound contrast agents, *Ultrasound Med Biol* **33**(2), 187–196.

14. Burns P. N. (2002) Instrumentation for contrast echocardiography, *Echocardiography* **19**(3), 241–258.

15. Caiati C., Aragona P., Iliceto S., and Rizzon P. (1996) Improved Doppler detection of proximal left anterior descending coronary artery stenosis after intravenous injection of a lung-crossing contrast agent: a transesophageal Doppler echocardiographic study, *J Am Coll Cardiol* **27**, 1413–1421.

16. Caissie A. L., Furukawa M., Karshafian R., Giles A., Li Y., Wong S., and Czarnota G. J. (2009) Ultrasound-activated microbubbles as novel radiosensitizers with vascular disrupting effects, Proceedings Annual Meeting of the American Association of Cancer Research.

17. Chen W. S., Matula T. J., Brayman A. A., and Crum L. A. (2003) A comparison of the fragmentation thresholds and inertial cavitation doses of different ultrasound contrast agents, *J Acoust Soc Am* **113**, 643–651.

18. Chen J., and Zhu Z. (2006) Ultrasound attenuation in encapsulated microbubble suspensions: the multiple scattering effects, *Ultrasound Med Biol* **32**(6), 961–969.

19. Cherin E., Brown J., Masoy S. E., Shariff H., Karshafian R., and Williams R., et al. (2008) Radial modulation imaging of microbubble contrast agents at high frequency, *Ultrasound Med Biol* **34**(6), 949–962.

20. Christiansen J. P., Leong-Poi H., Klibanov A. L., Kaul S., and Lindner J. R. (2002) Noninvasive imaging of myocardial reperfusion injury using leukocyte-targeted contrast echocardiography, *Circulation* **105**(15), 1764–1767.

21. Clarke P. R., and Hill C. R. (1970) Physical and chemical aspects of ultrasonic disruption of cells, *J Acoust Soc Am* **47**, 649–653.

22. Couture O., Pannacci N., Tabeling P., Fink M., Servois V., and Tanter M. (2009) Ultrasound-inducible fluorescent particles for internal tattooing, Proceeedings 2009 International Society of Therapeutic Ultrasound Meeting, S3B-5.

23. Curiel L., Pichardo S., Lee K., and Zehbe I. (2009) Sonoporation of cervical carcinoma cells affected with E6-oncoprotein for the treatment of uterine cancer, Proceeedings 2009 International Society of Therapeutic Ultrasound Meeting.

24. Czarnota G. J., Karshafian R., Giles A., Banihashemi B., Lee J., and Burns P. (2008) Microbubble and ultrasound enhancement of radiation-induced tumor cell death *in vivo*, Proceedings Annual Meeting of the American Association of Cancer Research.

25. Czarnota G. J., Karshafian R., Burns P. N., Wong S., Lee J., Caissie A. L., Furukawa M., Al Mahrouki A., and Giles A. (2009) Novel ultrasound induced microbubble enhancement

of radiation response (submitted to *Science-Translational Medicine*).

26. Daffertshofer M., Gass A., Ringleb P., Sitzer M., Sliwka U., and Els T., et al. (2005)Transcranial low-frequency ultrasoundmediated thrombolysis in brain ischemia: increased risk of hemorrhage with combined ultrasound and tissue plasminogen activator: results of a phase II clinical trial, *Stroke.*, ***36***, 1441–1446.
27. Dayton P. A., and Ferrara K. W. (2002) Targeted imaging using ultrasound, *J Magn Reson Imaging* **16**(4), 362–377.
28. de Jong N., Emmer M., Chin C. T., Bouakaz A., Mastik F., Lohse D., and Versluis M. (2007) "Compression-only"behavior of phospholipid-coated contrast bubbles, *Ultrasound Med Biol* **33**, 653–656.
29. Doolittle N. D., Miner M. E., Hall W. A., Siegal T., Jerome E., and Osztie E., et al. (2000) Safety and efficacy of a multicenter study using intraarterial chemotherapy in conjunction with osmotic opening of the blood-brain barrier for the treatment of patients with malignant brain tumors, *Cancer* **88**(3), 637–647.
30. Eckersley R. J., Sedelaar J. P., Blomley M. J., Wijkstra H., deSouza N. M., and Cosgrove D. O., et al. (2002) Quantitative microbubble enhanced transrectal ultrasound as a tool for monitoring hormonal treatment of prostate carcinoma, *Prostate* **51**(4), 256–267.
31. Ellegala D. B., Leong-Poi H., Carpenter J. E., Klibanov A. L., Kaul S., and Shaffrey M. E., et al. (2003) Imaging tumor angiogenesis with contrast ultrasound and microbubbles targeted to alpha(v)beta3, *Circulation* **108**(3), 336–341.
32. Emmer M., van Wamel A., Goertz D. E., and de Jong N. (2007) The onset of microbubble vibration, *Ultrasound Med Biol* **33**(6), 941–949.
33. Feigenbaum H., Stone J. M., Lee D. A., Nasser W. K., and Chang S. (1970) Identification of ultrasound echoes from the left ventricle by use of intracardiac injections of indocyanine green, *Circulation* **41**(4), 615–621.
34. Feinstein S. B., Ten Cate F. J., Zwehl W., Ong K., Maurer G., and Tei C., et al. (1984) Two-dimensional contrast echocardiography. I. *In vitro* development and quantitative analysis of echo contrast agents, *J Am Coll Cardiol* **3**(1), 14–20.
35. Ferrara K., Pollard R., and Borden M. (2007) Ultrasound microbubble contrast agents: fundamentals and application to gene and drug delivery, *Annu Rev Biomed Eng* **9**, 415–447.
36. Flynn H. G. (1964) Physics of acoustic cavitation in liquids, in *Physical Acoustics, Principles and Methods* (ed. Mason W. P.), Academic Press, New York.
37. Folkman J., and Camphausen K. (2001) Cancer. What does radiotherapy do to endothelial cells? *Science* **293**(5528), 227–228.
38. Fry W. J., Mosberg W. H. Jr., Barnard J. W., and Fry F. J. (1954) Production of focal destructive lesions in the central nervous system with ultrasound. *J Neurosurg* **11**, 471–479.
39. Gaber M. H., Wu N. Z., Hong K., Huang S. K., Dewhirst M. W., and Papahadjopoulos D. (1996) Thermosensitive liposomes: extravasation and release of contents in tumor microvascular networks, *Int J Radiat Oncol Biol Phys* **36**(5), 1177–1187.

40. Garcia-Barros M., Paris F., Cordon-Cardo C., Lyden D., Rafii S., and Haimovitz-Friedman A., et al. (2003) Tumor response to radiotherapy regulated by endothelial cell apoptosis, *Science* **300**(5622), 1155–1159.

41. Görg C., Bert T., and Gorg K. (2005) Contrast-enhanced sonography for differential diagnosis of pleurisy and focal pleural lesions of unknown cause, *Chest* **128**(6), 3894–3899.

42. Gramiak R., and Shah P. M. (1968) Echocardiography of the aortic root, *Invest Radiol* **3**(5), 356–366.

43. Guzmán H. R., Nguyen D. X., Khan S., and Prausnitz M. R., (2001) Ultrasound-mediated disruption of cell membranes. I. Quantification of molecular uptake and cell viability, *J Acoust Soc Am* **110**, 588–596.

44. Guzmán H. R., Nguyen D. X., McNamara A. J., and Prausnitz M. R., (2002) Equilibrium loading of cells with macromolecules by ultrasound: effects of molecular size and acoustic energy, *J Pharm Sci* **91,** 1693–1701.

45. Hall E. J. (2000) *Radiobiology for the Radiobiologist*, Lipincott, Williams and Wilkins.

46. Halvorsen R. A. (2008) Which study when? Iodinated contrast-enhanced CT versus gadolinium-enhanced MR imaging, *Radiology* **249(1)**, 9–15.

47. Hilgenfeldt S., Lohse D., and Zomack M. (2000) Sound scattering and localized heat deposition of pulse-driven microbubbles, *J Acoust Soc Am.,* **107**(6), 3530–3539.

48. Hsieh D. Y., and Plesset M. S. (1961) Theory of rectified diffusion of mass into gas bubbles, *J Acoust Soc Am* **33**, 206–211.

49. Huang S. F., Chang R. F., Moon W. K., Lee Y. H., Chen D. R., and Suri J. S. (2008)Analysis of tumor vascularity using three-dimensional power Doppler ultrasound images, *IEEE Trans Med Imaging* **27**, 320–330.

50. Hynynen K., McDannold N., Vykhodtseva N., and Jolesz F. A. (2001) Noninvasive MR imaging-guided focal opening of the blood-brain barrier in rabbits, *Radiology* **220**(3), 640–646.

51. Hynynen K., McDannold N., Vykhodtseva N., Raymond S., Weissleder R., and Jolesz F. A., et al. (2006) Focal disruption of the blood-brain barrier due to 260-kHz ultrasound bursts: a method for molecular imaging and targeted drug delivery, *J Neurosurg* **105**(3), 445–454.

52. Hynynen K. (2008) Ultrasound for drug and gene delivery to the brain, *Adv Drug Deliv Rev* **60**, 1209–1217.

53. Iwanaga K., Tominaga K., Yamamoto K., Habu M., Maeda H., and Akifusa S., et al. (2007) Local delivery system of cytotoxic agents to tumors by focused sonoporation, *Cancer Gene Ther* **14**(4), 354–363.

54. Jayaweera A. R., Edwards N., Glasheen W. P., Villanueva F. S., Abbott R. D., and Kaul S. (1994) *In vivo* myocardial kinetics of air-filled albumin microbubbles during myocardial contrast echocardiography. Comparison with radiolabeled red blood cells, *Circ Res* **74**(6), 1157–1165.

55. Juffermans L. J. M., Kamp O., Dijkmans P. A., Visser C. A., and Musters R. J. P. (2008) Low-

intensity ultrasound-exposed microbubbles provoke local hyperpolarization of the cell membrane via activation of BKCa channels, *Ultrasound Med Biol* **34**, 502–508.

56. Kaneko K., Kuwatsuru R., Fukuda Y., Yamataka A., Yabuta K., and Katayama H., et al. (1994) Contrast sonography for detection of vesicoureteral reflux, *Lancet* **344**(8923), 687.

57. Kaneko Y., Maruyama T., Takegami K., Watanabe T., Mitsui H., and Hanajiri K., et al. (2005) Use of a microbubble agent to increase the effects of high intensity focused ultrasound on liver tissue, *Eur Radiol* **15**(7), 1415–1420.

58. Karshafian R., Bevan P., Williams R., Samac S., and Burns P. N. (2009) Sonoporation by ultrasound-activated microbubble contrast agents: effect of acoustic exposure parameters on cell membrane permeability and cell viability, *Ultrasound Med Biol* **35**(5), 847–860.

59. Katayama H., Yamaguchi K., Kozuka T., Takashima T., Matsuura K., and Nakata H., et al. (1991) Full-scale investigation into adverse reaction in Japan. Risk factor analysis, The Japanese Committee on the Safety of Contrast Media, *Invest Radiol* **26**(Suppl 1), S33–S36; discussion S40–31.

60. Kaufmann B. A., Sanders J. M., Davis C., Xie A., Aldred P. and Sarembock, I. J., et al. (2007) Molecular imaging of inflammation in atherosclerosis with targeted ultrasound detection of vascular cell adhesion molecule-1, *Circulation* **116**(3), 276–284.

61. Kaul S. (2008) Myocardial contrast echocardiography: a 25-year retrospective, *Circulation* **118**(3), 291–308.

62. Keller M. W., Glasheen W., and Kaul S. (1989) Albunex: a safe and effective commercially produced agent for myocardial contrast echocardiography, *J Am Soc Echocardiogr* **2**(1), 48–52.

63. Kennedy J. E., ter Haar G. R., Wu F., Gleeson F. V., Roberts I. S., and Middleton M. R., et al. (2004) Contrast-enhanced ultrasound assessment of tissue response to high-intensity focused ultrasound, *Ultrasound Med Biol* **30**(6), 851–854.

64. Kinoshita M., and Hynynen K. (2005) A novel method for the intracellular delivery of siRNA using microbubble-enhanced focused ultrasound, *Biochem Biophys Res Commun* **335**(2), 393–399.

65. Kinoshita M., McDannold N., Jolesz F. A., and Hynynen K. (2006a) Noninvasive localized delivery of Herceptin to the mouse brain by MRI-guided focused ultrasound-induced blood-brain barrier disruption, *Proc Natl Acad Sci U S A* **103**(31), 11719–11723.

66. Kinoshita M., McDannold N., Jolesz F. A., and Hynynen K. (2006b) Targeted delivery of antibodies through the blood-brain barrier by MRI-guided focused ultrasound, *Biochem Biophys Res Commun* **340**(4), 1085–1090.

67. Klibanov A. L., Hughes M. S., Marsh J. N., Hall C. S., Miller J. G., and Wible J. H., et al. (1997) Targeting of ultrasound contrast material. An *in vitro* feasibility study, *Acta Radiol Suppl* **412**, 113–120.

68. Klibanov A. L., Hughes M. S., Wojdyla J. K., Marsh J. N., Hall C. S., and Miller J. G., et al. (1998) Targeting of ultrasound contrast material: selective imaging of microbubbles in vitro, *Acad Radiol* **5**(Suppl 1), S243–S246.

69. Klibanov A. L. (2005) Ligand-carrying gas-filled microbubbles: ultrasound contrast

agents for targeted molecular imaging, *Bioconjug Chem* **16**(1), 9–17.

70. Kolesnick R. (2002) The therapeutic potential of modulating the ceramide/sphingomyelin pathway, *J Clin Invest* **110**(1), 3–8.
71. Kolesnick R., and Fuks Z. (2003) Radiation and ceramide-induced apoptosis, *Oncogene* **22**(37), 5897–5906.
72. Kondo I., Ohmori K., Oshita A., Takeuchi H., Yoshida J., and Shinomiya K., et al. (2004) Leukocyte-targeted myocardial contrast echocardiography can assess the degree of acute allograft rejection in a rat cardiac transplantation model, *Circulation* **109**(8), 1056–1061.
73. Kong G., Braun R. D., and Dewhirst M. W. (2000) Hyperthermia enables tumor-specific nanoparticle delivery: effect of particle size, *Cancer Res* **60**(16), 4440–4445.
74. Kremkau F. W., Gramiak R., Carstensen E. L., Shah P. M., and Kramer D. H. (1970) Ultrasonic detection of cavitation at catheter tips, *Am J Roentgenol Radium Ther Nucl Med* **110**(1), 177–183.
75. Krizanac-Bengez L., Mayberg M. R., and Janigro D. (2004) The cerebral vasculature as a therapeutic target for neurological disorders and the role of shear stress in vascular homeostatis and pathophysiology, *Neurol Res* **26**(8), 846–853.
76. Kudo S. (1989)Thrombolysis with ultrasound effect, *Tokyo Jikeikai Med J* **104**, 1005–1012.
77. Lanza G. M., and Wickline S. A. (2001) Targeted ultrasonic contrast agents for molecular imaging and therapy, *Prog Cardiovasc Dis* **44**(1), 13–31.
78. Leen E., Ceccotti P., Kalogeropoulou C., Angerson W. J., Moug S. J., and Horgan P. G. (2006) Prospective multicenter trial evaluating a novel method of characterizing focal liver lesions using contrast-enhanced sonography, *AJR Am J Roentgenol* **186**(6), 1551–1559.
79. Leighton T. (1994) The acoustic bubble, Academic Press, San Diego.
80. Leighton T. G. (2007) What is ultrasound? *Prog Biophys Mol Biol* **93**, 3–83.
81. Leong-Poi H., Christiansen J., Heppner P., Lewis C. W., Klibanov A. L., and Kaul S., et al. (2005) Assessment of endogenous and therapeutic arteriogenesis by contrast ultrasound molecular imaging of integrin expression, *Circulation* **111**(24), 3248–3254.
82. Lim A. K., Patel N., Eckersley R. J., Taylor-Robinson S. D., Cosgrove D. O., and Blomley M. J. (2004) Evidence for spleen-specific uptake of a microbubble contrast agent: a quantitative study in healthy volunteers, *Radiology* **231**(3), 785–788.
83. Lindner J. R., Dent J. M., Moos S. P., Jayaweera A. R., and Kaul S. (1997) Enhancement of left ventricular cavity opacification by harmonic imaging after venous injection of Albunex, *Am J Cardiol* **79**(12), 1657–1662.
84. Lindner J. R., Wei K., and Kaul S. (1999) Imaging of myocardial perfusion with SonoVue (TM) in patients with a prior myocardial infarction, *Echocardiography* **16**, 753–760.
85. Lindner J. R., and Kaul S. (2001) Delivery of drugs with ultrasound, *Echocardiography* **18**(4), 329–337.
86. Lindner J. R., Song J., Christiansen J., Klibanov A. L., Xu F., and Ley K. (2001) Ultrasound assessment of inflammation and renal tissue injury with microbubbles targeted to P-selectin, *Circulation* **104**(17), 2107–2112.

87. Linker R. A., Reinhardt M., Bendszus M., Ladewig G., Briel A., and Schirner M., et al. (2005) *In vivo* molecular imaging of adhesion molecules in experimental autoimmune encephalomyelitis (EAE), *J Autoimmun* **25**(3), 199–205.

88. Lum A. F., Borden M. A., Dayton P. A., Kruse D. E., Simon S. I., and Ferrara K. W. (2006) Ultrasound radiation force enables targeted deposition of model drug carriers loaded on microbubbles, *J Control Release* **111**(1–2), 128–134.

89. Lynn J. G., Zwemer R. L., Chick A. J.,and Miller A. E. (1942) A new method for the generation and use of focused ultrasound in experimental biology, *J Gen Physiol* **26**, 179–193.

90. McDannold N., Vykhodtseva N., and Hynynen K. (2006) Targeted disruption of the blood-brain barrier with focused ultrasound: association with cavitation activity, *Phys Med Biol* **51**(4), 793–807.

91. McDannold N., Vykhodtseva N., and Hynynen K. (2008) Effects of acoustic parameters and ultrasound contrast agent dose on focused-ultrasound induced blood-brain barrier disruption, *Ultrasound Med Biol* **34**(6), 930–937.

92. Meairs S. (2008) Contrast-enhanced ultrasound perfusion imaging in acute stroke patients, *Eur Neurol* **59**, 17–26.

93. Medel R., Crowley R. W., McKisic M. S., Dumont A. S., and Kassell N. F. (2009) Sonothrombolysis: an emerging modality for the management of stroke, *Neurosurg* **65**(5), 979–993.

94. Miller D. L. (1987) A review of the ultrasonic bioeffects of microsonation, gas-body activation, and related cavitation-like phenomena, *Ultrasound Med Biol* **13**(8), 443–470.

95. Miller D. L., Bao S., and Morris J. E. (1999) Sonoporation of cultured cells in the rotating tube exposure system, *Ultrasound Med Biol* **25**, 143–149.

96. Molina C. A., Ribo M., Rubiera M., Montaner J., Santamarina E., and Delgado-Mederos R., et al. (2006) Microbubble administration accelerates clot lysis during continuous 2-MHz ultrasound monitoring in stroke patients treated with intravenous tissue plasminogen activator, *Stroke* **37**(2), 425–429.

97. Nakashima Y., Raines E. W., Plump A. S., Breslow J. L., and Ross R. (1998) Upregulation of VCAM-1 and ICAM-1 at atherosclerosis-prone sites on the endothelium in the ApoE-deficient mouse, *Arterioscler Thromb Vasc Biol* **18**(5), 842–851.

98. Neuwelt E. A., Barnett P. A., McCormick C. I., Frenkel E. P., and Minna J. D. (1985) Osmotic blood-brain barrier modification: monoclonal antibody, albumin, and methotrexate delivery to cerebrospinal fluid and brain, *Neurosurgery* **17**(3), 419–423.

99. Nyborg W. L. (2001) Biological effects of ultrasound: development of safety guidelines. Part II: general review, *Ultrasound Med Biol* **27**(3), 301–333.

100. Nishiie N., Suzuki R., Oda Y., Taira Y., Utoguchi N., and Maruyama K., (2009) Enhancement of antitumor effect by the combination of therapeutic ultrasound and immunotherapy, Proceedings 2009 International Society of Therapeutic Ultrasound Meeting.

101. O'Brien W. D. (2007) Ultrasound-biophysics mechanisms, *Prog Biophys Mol Biol* **93**, 212–255.

102. Palmowski M., Huppert J., Hauff P., Reinhardt M., Schreiner K., and Socher M. A., et al.

(2008) Vessel fractions in tumor xenografts depicted by flow- or contrast-sensitive three-dimensional high-frequency Doppler ultrasound respond differently to antiangiogenic treatment, *Cancer Res* **68**(17), 7042–7049.

103. Pardridge W. M. (2005) The blood-brain barrier: bottleneck in brain drug development, *NeuroRx* **2**(1), 3–14.

104. Paris F., Fuks Z., Kang A., Capodieci P., Juan G., and Ehleiter D., et al. (2001) Endothelial apoptosis as the primary lesion initiating intestinal radiation damage in mice, *Science* **293**(5528), 293–297.

105. Partanen A., Negussie A., Yarmolenko P., Frenkel V., Wood B., and Dreher M. (2009) Real-time monitoring of drug delivery with MrgHIFU and image-able low temperature sensitive liposomes, Proceeedings 2009 International Society of Therapeutic Ultrasound Meeting.

106. Peddu P., Shah M., and Sidhu P. S. (2004) Splenic abnormalities: a comparative review of ultrasound, microbubble-enhanced ultrasound and computed tomography, *Clin Radiol* **59**(9), 777–792.

107. Pena L. A., Fuks Z., and Kolesnick R. N. (2000) Radiation-induced apoptosis of endothelial cells in the murine central nervous system: protection by fibroblast growth factor and sphingomyelinase deficiency, *Cancer Res* **60**(2), 321–327.

108. Phillips P., and Gardner E. (2004) Contrast-agent detection and quantification, *Eur Radiol* **14**(Suppl 8), 4–10.

109. Piedra M., Allroggen A., and Lindner J. R. (2009) Molecular imaging with targeted contrast ultrasound, *Cerebrovasc Dis* **27**(Suppl 2), 66–74.

110. Plathow C., Lohr F., Divkovic G., Rademaker G., Farhan N., and Peschke P., et al. (2005) Focal gene induction in the liver of rats by a heat-inducible promoter using focused ultrasound hyperthermia: preliminary results, *Invest Radiol* **40**(11), 729–735.

111. Postema M., van Wamel A., Lancée C. T., and de Jong N. (2004) Ultrasound-induced encapsulated microbubble phenomena, *Ultrasound Med Biol* **30**(6), 827–840.

112. Postema M., Bouakaz A., Versluis M., and de Jong N. (2005) Ultrasound-induced gas release from contrast agent microbubbles, *IEEE transactions on ultrasonics, ferroelectrics, and frequency control* **52**(6), 1035–1041.

113. Qin S., Caskey C. F., and Ferrara K. W. (2009) Ultrasound contrast microbubbles in imaging and therapy: physical principles and engineering, *Phys Med Biol* **54**(6), 27–57.

114. Riesz P., and Kondo T. (1992) Free-radical formation induced by ultrasound and its biological implications, *Free Radic Biol Med* **13**, 247–270.

115. Rosenschein U., Furman V., Kerner E., Fabian. I, Bernheim J., and Eshel Y. (2000) Ultra -sound imaging-guided noninvasive ultrasound thrombolysis: preclinical results, *Circulation* **102**, 238–245.

116. Rychak J. J., Klibanov A. L., Ley K. F., and Hossack J. A. (2007) Enhanced targeting of ultrasound contrast agents using acoustic radiation force, *Ultrasound Med Biol* **33**(7), 1132–1139.

117. Saito M., Mazda O., Takahashi K. A., Arai Y., and Kishida T. (2007) Sonoporation mediated transduction of pDNA/siRNA into joint synovium *in vivo*, *J Orthop Res* **25**(10), 1308–1316.

118. Sakakima Y., Hayashi S., Yagi Y., Hayakawa A., Tachibana K., and Nakao A. (2005) Gene therapy for hepatocellular carcinoma using sonoporation enhanced by contrast agents, *Cancer Gene Ther* **12**, 884–889.

119. Sassaroli E., and Hynynen K. (2007) Cavitation threshold of microbubbles in gel tunnels by focused ultrasound, *Ultrasound Med Biol* **33**(10), 1651–1660.

120. Sathishkumar S., Boyanovsky B., Karakashian A. A., Rozenova K., Giltiay N. V., and Kudrimoti M., et al. (2005) Elevated sphingomyelinase activity and ceramide concentration in serum of patients undergoing high dose spatially fractionated radiation treatment: implications for endothelial apoptosis, *Cancer Biol Ther* **4**(9), 979–986.

121. Schlief R., and Deichert U. (1991) Hysterosalpingo-contrast sonography of the uterus and fallopian tubes: results of a clinical trial of a new contrast medium in 120 patients, *Radiology* **178**(1), 213–215.

122. Schumann P. A., Christiansen J. P., Quigley R. M., McCreery T. P., Sweitzer R. H., and Unger E.C., et al. (2002) Targeted-microbubble binding selectively to GPIIb IIIa receptors of platelet thrombi. *Invest Radiol* **37**, 587- 593.

123. Seidel G., and Meairs S. (2009) Ultrasound contrast agents in ischemic stroke, *Cerebrovasc Dis* **27**(Suppl 2), 25–39.

124. Sheikov N., McDannold N., Sharma S., and Hynynen K. (2008) Effect of focused ultrasound applied with an ultrasound contrast agent on the tight junctional integrity of the brain microvascular endothelium, *Ultrasound Med Biol* **34**(7), 1093–1104.

125. Sidhu P. S., and Dawson P. (1999) Adverse reactions, cause, prophylaxis and management, in *Textbook of Contrast Media* (ed. Dawson P., Cosgrove D. O., Grainger R. G.), ISIS Medical, Oxford, pp. 99–119.

126. Solbiati L., Goldberg S. N., Ierace T., Dellanoce M., Livraghi T., and Gazelle G. S. (1999) Radio-frequency ablation of hepatic metastases: postprocedural assessment with a US microbubble contrast agent—early experience, *Radiology* **211**(3), 643–649.

127. Stieger S. M., Dayton P. A., Borden M. A., Caskey C. F., Griffey S. M., and Wisner E. R., et al. (2008) Imaging of angiogenesis using Cadence contrast pulse sequencing and targeted contrast agents, *Contrast Media Mol Imaging* **3**(1), 9–18.

128. Stone M. J., Frenkel V., Dromi S., Thomas P., Lewis R. P., and Li K. C., et al. (2007) Pulsed-high intensity focused ultrasound enhanced tPA mediated thrombolysis in a novel *in vivo* clot model: a pilot study, *Thromb Res* **121**, 193–202.

129. Strauss A. L., and Beller K. D. (1999) Persistent opacification of the left ventricle and myocardium with a new echo contrast agent, *Ultrasound Med Biol* **25**, 763–769.

130. Stride E. (2009) Physical principles of microbubbles for ultrasound imaging and therapy, *Cerebrovasc Dis* **27**(Suppl 2), 1–13.

131. Tachibana K., and Tachibana S. (1995) Albumin microbubble echo-contrast material as an enhancer for ultrasound accelerated thrombolysis, *Circulation* **92**, 1148–1150.

132. Tartis M. S., McCallan J., Lum A. F., LaBell R., Stieger S. M., and Matsunaga T. O., et al. (2006) Therapeutic effects of paclitaxel-containing ultrasound contrast agents, *Ultrasound Med Biol* **32**(11), 1771–1780.

133. Tinkov S., Bekeredjian R., Winter G., and Coester C. (2009) Microbubbles as ultrasound triggered drug carriers, *J Pharm Sci* **98**(6), 1935–1961.

134. Tozer G. M., Kanthou C., and Baguley B. C. (2005) Disrupting tumor blood vessels, *Nat Rev Cancer* **5**(6), 423–435.

135. Treat L. H., McDannold N., Vykhodtseva N., Zhang Y., Tam K., and Hynynen K. (2007) Targeted delivery of doxorubicin to the rat brain at therapeutic levels using MRI-guided focused ultrasound, *Int J Cancer* **121**(4), 901–907.

136. Trillaud H., Bruel J. M., Valette P. J., Vilgrain V., Schmutz G., and Oyen R., et al. (2009) Characterization of focal liver lesions with SonoVue-enhanced sonography: international multicenter-study in comparison to CT and MRI, *World J Gastroenterol* **15**(30), 3748–3756.

137. Trübestein G., Engel C., Etzel F., Sobbe A., Cremer H., and Stumpff U. (1976) Thrombolysis by ultrasound, *Clin Sci Mol Med Suppl* **3**, 697s–698s.

138. Unger E. C., McCreery T. P., Sweitzer R. H., Shen D., and Wu G. (1998) *In vitro* studies of a new thrombusspecific ultrasound contrast agent. *Am J Cardiol* **81**, 58G–61G.

139. Unger E. C., Hersh E., Vannan M., Matsunaga T. O., and McCreery T. (2001) Local drug and gene delivery through microbubbles, *Prog Cardiovasc Dis* **44**(1), 45–54.

140. Unger E. C., Hersh E., Vannan M., and McCreery T. (2001) Gene delivery using ultrasound contrast agents, *Echocardiography* **18**(4), 355–361.

141. Unger E. C., Porter T., Culp W., Labell R., Matsunaga T., and Zutshi R. (2004) Therapeutic applications of lipid-coated microbubbles, *Adv Drug Deliv Rev* **56**(9), 1291–1314.

142. Vykhodtseva N., McDannold N., and Hynynen K. (2008) Progress and problems in the application of focused ultrasound for blood-brain barrier disruption, *Ultrasonics* **48**(4), 279–296.

143. Wei W., Zhengzhong B., Yongjie W., Lafeng Y., and Yalin M. (2004) A novel approach to quantitative ultrasonic naked gene delivery and its noninvasive assessment, *Ultrasonics* **43**(2), 69–77.

144. Weller G. E., Lu E., Csikari M. M., Klibanov A. L., Fischer D., and Wagner W. R., et al. (2003) Ultrasound imaging of acute cardiac transplant rejection with microbubbles targeted to intercellular adhesion molecule-1, *Circulation* **108**(2), 218–224.

145. Weller G. E., Wong M. K., Modzelewski R. A., Lu E., Klibanov A. L., and Wagner W. R., et al. (2005) Ultrasonic imaging of tumor angiogenesis using contrast microbubbles targeted via the tumor-binding peptide arginine-arginine-leucine, *Cancer Res* **65**(2), 533–539.

146. Wu J. (2002) Theoretical study on shear stress generated by microstreaming surrounding contrast agents attached to living cells, *Ultrasound Med Biol* **28**(1), 125–129.

147. Xie F., Tsutsui J. M., Lof J., Unger E. C., Johanning J., and Culp W. C., et al. (2005) Effectiveness of lipid microbubbles and ultrasound in declotting thrombosis, *Ultrasound Med Biol* **31**(7), 979–985.

148. Yu T., Wang G., Hu K., Ma P., Ba, J., and Wang Z. (2004) A microbubble agent improves the therapeutic efficiency of high intensity focused ultrasound: a rabbit kidney study, *Urol Res* **32**(1), 14–19.

149. Yu T., Li S. L., Zhao J. Z., and Mason T. J., (2006) Ultrasound: a chemotherapy sensitizer, *Technol Cancer Res Treat* **5**, 51–60.

150. Zhigang W., Zhiyu L., Haitao R., Hong R., Qunxia Z., and Ailong H., et al. (2004) Ultrasound-mediated microbubble destruction enhances VEGF gene delivery to the infarcted myocardium in rats, *Clin Imaging* **28**(6), 395–398.

Index